Adaptive Collaborative Approaches in Natural Resource Governance

The purpose of this book is to showcase a range of approaches that consider learning and collaboration as central processes in natural resource governance, innovation and development. It calls for a fundamental reappraisal of many of the accepted ways of thinking, through the analyses of the experiences of pioneering practitioners and researchers who have attempted to advance adaptive collaborative approaches in practice. Despite these being generated in different institutional domains with somewhat diverse epistemological and policy orientations, the authors show that there are common themes among these approaches.

The book presents a review of various adaptive and collaborative approaches developed to cope with the social and biophysical complexity of natural resource systems, and highlights the challenges faced by practitioners. It includes case studies from Bangladesh, Ecuador, Nepal and Zimbabwe and contexts ranging from farmer field schools, to floodplain management and community forestry. The authors provide rich accounts of how adaptive collaborative approaches were applied to synergise different types of learning, foster collaboration among stakeholders, and nurture innovative development processes.

It concludes that perspectives that have served us well in the past – such as participation, decentralization and empowerment – need to be re-thought. Through its introduction and conclusion chapters, the book establishes a clear theoretical approach and identifies a set of practical methodologies for combining different systems of knowledge in a way that generates and maximizes innovation and the translation of research into practice.

Hemant R. Ojha is a founder of Southasia Institute of Advanced Studies (SIAS) and ForestAction Nepal. He is also a fellow at the University of Melbourne and Australian National University.

Andy Hall is Director of LINK Limited and Senior Researcher at UNU-MERIT and Visiting Professor at the Open University, UK.

Rasheed Sulaiman V is Director of the Centre for Research on Innovation and Science Policy, Hyderabad, India.

Adaptive collaborative approaches in natural resource governance

Rethinking participation, learning and innovation

Edited by
Hemant R. Ojha,
Andy Hall and
Rasheed Sulaiman V

Routledge
Taylor & Francis Group
LONDON AND NEW YORK

earthscan
from Routledge

First published 2013
by Routledge
2 Park Square, Milton Park, Abingdon, Oxon OX14 4RN

Simultaneously published in the USA and Canada
by Routledge
711 Third Avenue, New York, NY 10017

Routledge is an imprint of the Taylor & Francis Group, an informa business

British Library Cataloguing in Publication Data
A catalogue record for this book is available from the British Library

Library of Congress Cataloging in Publication Data
Adaptive collaborative approaches in natural resource governance: rethinking participation, learning and innovation/edited by Hemant R. Ojha, Andy Hall and Rasheed Sulaiman V.
p. cm.
"Simultaneously published in the USA and Canada by Routledge" —T.p. verso.
Includes bibliographical references and index.
1. Natural resources – Government policy. 2. Natural resources – Management. 3. Natural resources – Management – Citizen participation. 4. Agriculture and state. 5. Agriculture – Management. 6. Agriculture – Management – Citizen participation. 7. Cooperation. 8. Political participation. 9. Learning. 10. Technological innovations. I. Ojha, Hemant R. II. Hall, Andy. III. Rasheed Sulaiman V.
HC85.A33 2012
333.7 – dc23 2012000004

ISBN 978-0-415-69653-1 hbk
ISBN 978-0-415-69910-5 pbk
ISBN 978-0-203-13629-4 ebk

Typeset in Baskerville
by Florence Production Ltd, Stoodleigh, Devon

Contents

Figures

Tables

Contributors

Mani Ram Banjade is currently the Ph.D. candidate at Fenner School of Environment and Society at Australian National University. He was formerly the Coordinator and Team Leader of Nepal RIU and ACM related projects at ForestAction Nepal. His Ph.D. focuses on deliberative governance of forest in Nepal. Email: mrbanjade@gmail.com.

Dr Robert Fisher is a Senior Lecturer at School of Geosciences at the University of Sydney. He was formerly the Deputy Director of Regional Community Forestry Training Center (RECOFTC), Bangkok, Thailand. He has conducted pioneering work on action research in the context of natural resource management in Asia. Email: rjfisher@ozemail.com.au.

Professor Andy Hall is Director of LINK Limited, Senior Researcher at UNU-MERIT and Visiting Professor at Open University, UK. Andy's principal area of research has been on the nature of agricultural innovation processes, policies and capacities. He has a particular interest in new sources and modes of innovation in emerging economies that are delivering services and improving the livelihoods of low-income households. He established the LINK initiative in an attempt to build bridges between the academic field of innovation studies, innovation practice and the policies of national governments and international agencies dealing with agricultural development. Email: andy.hall@innovationstudies.org.

Professor Cees Leeuwis is the Head of Communication and Innovation Studies Group at Wageningen. His research focuses on cross-disciplinary approaches, social learning and conflict management in networks, and the reflexive monitoring and evaluation of innovation support strategies and trajectories. Email: cees.leeuwis@wur.nl.

Dr Frank Matose is an academic staff member in the Department of Sociology at the University of Cape Town. He was formerly the team leader of CIFOR ACM team in Zimbabwe. His interests are in natural resource commons in southern Africa with particular focus on the intersection of local people, the State, conservation and protected areas. His interests in these areas are informed by intellectual projects around environmental governance, social justice, knowledge and power. He recently co-edited the book *Coping amidst chaos: studies on adaptive collaborative management from Zimbabwe* with Alois Mandondo and Ravi Prabhu. Email: frank.matose@uct.ac.za.

Dr Tendayi Mutimukuru-Maravanyika is a Focal Point Coordinator for the Evidence and Lessons from Latin America (ELLA) Project being implemented by Practical Action Consulting Southern Africa. She was formerly a member of CIFOR's Adaptive Collaborative Management (ACM) programme team in Zimbabwe. Tendayi has worked extensively in the Eastern and Southern Africa region and has vast experiences in using and training different stakeholders in adaptive management approaches. Tendayi's research interests include community-based natural resource management, adaptive management, participatory action research and climate change adaption. Email: tendayim07@yahoo.co.uk.

Dr Hemant R. Ojha is the founding Research Director of Southasia Institute of Advanced Studies (SIAS), and a founder of ForestAction Nepal, both based in Kathmandu, and a Senior Fellow at the School of Land and Environment, University of Melbourne. He blends ideas from social and natural sciences to undertake critical and participatory action research in natural resource management. He has pioneered new Critical Action Research approaches in Nepal to blend action research with civil society activism to strengthen policy processes and transform resource governance discourses and practices. His current research focuses on understanding when and how science, civic engagement and policy processes facilitate innovation, adaptive capacity building and accountable governance. Email: hojha@southasiainstitute.org.

Dr Carol J. Pierce Colfer is currently a Visiting Fellow at Cornel University and a Senior Associate with the Centre for International Forestry Research (CIFOR), Indonesia. She was formerly the Team Leader of Adaptive Collaborative Management programme at CIFOR. Email: c.colfer@cgiar.org.

Marc Schut is a Ph.D. candidate at Wageningen University, the Netherlands. Between 2005 and 2006, he conducted M.Sc. research on the diverse expressions of FFS in northern and central Ecuador. Marc's ongoing Ph.D. research in Mozambique centres on the country's biofuel sustainability framework. Email: marc.schut@wur.nl.

Dr Stephen Sherwood is Lecturer, Wageningen University, the Netherlands. He is also a specialist in Rural Innovation based at EkoRural in Ecuador. He is a co-founder of Groundswell International (www.groundswellinternational.org), a partnership that supports grassroots-level social change. He has over 25 years' experience with rural peoples' movements in Central and South America, including supporting the introduction of FFS methodology in the Americas. Email: ssherwood@ekorural.org.

Dr Rasheed Sulaiman V is Director of the Centre for Research on Innovation and Science Policy (CRISP) based in Hyderabad, India. He has extensive experience on research on agricultural extension and application of innovation systems in agriculture. He has also coordinated studies in South Asia as part of several consulting assignments. Previously he worked with the Indian Council of Agricultural Research (ICAR) as Senior Scientist at the National Centre for Agricultural Economics and Policy Research (NCAP), New Delhi. Email: rasheed.sulaiman@gmail.com.

Dr Parvin Sultana is Senior Research Fellow at Flood Hazard Research Centre, Middlesex University, based in Dhaka, Bangladesh. Her recent work has included action research on adaptive learning through networks of community organizations of poor floodplain users to promote and advocate improvements in practice and policy. She has also undertaken a wide range of consultancies in Bangladesh and Asia. Email: parvin@agni.com.

Dr Paul Thompson is Senior Research Fellow at Flood Hazard Research Centre, Middlesex University, based in Dhaka. Current research interests include integrated floodplain management, community-based fisheries management, wetland conservation and co-management, and institutional aspects of local management of floodplain resources. He is currently undertaking research into capacity building in Bangladesh for sustainable management of coastal natural resources in a changing climate. Email: paul@agni.com.

Acronyms

AAE	Agroecosystem Analysis
ACA	Adaptive Collaborative Approach
ACM	Adaptive Collaborative Management
ADB	Asian Development Bank
AGRITEX	Department of Agricultural Research and Extension Services
AIDAB	Australian Development Assistance Bureau (now AusAID)
ANR	Agriculture and Natural Resources
ANU	Australian National University
ANUTECH Pty Ltd	Australian National University's Business Arm
APRODIC	Asociación de Promotores de Desarrollo Integrado de la Comunidad (Association of Promoters of Integrated Development of Carchi)
AqDP	Aquaculture Development Project
AR	Action Research
AusAID	The Australian Agency for International Development
BELA	Bangladesh Environmental Lawyers Association
C&I	Criteria and Indicators
CA	Change Agent
CAMPFIRE	Communal Area Management Programme for Indigenous Resources, Zimbabwe
CAPRi	The CGIAR Systemwide Program on Collective Action and Property Rights
CBD	Convention on Biological Diversity
CBFM	Community Based Fisheries Management
CBNRM	Community-Based Natural Resource Management
CBOs	Community-Based Organizations
CEPIP-W	CERES Programme for Innovative Ph.D. Research at Wageningen University
CF	Community Forest/Forestry
CFUG	Community Forest User Group

CGIAR	Consultative Group on International Agricultural Research (CG centers in short)
CIAA	Commission for Investigation for the Abuse of Authority, Nepal
CIAL	Comites de Investigación Agrícola Local, a participatory research training offered by CIAT, Columbia
CIAP	Cambodia-IRRI-Australia Project
CIDA	Canadian International Development Agency
CIFOR	Center for International Forestry Research
CIP	International Potato Center
CM	Collaborative Monitoring
CNRS	Center for Natural Resource Studies, Bangladesh
COSUDE	Agencia Suiza Para el Desarrollo y la Cooperación
CRISP	Centre for Research in Innovation Policy, Hyderabad
CRT	Central Research Team, RIU
CTA	Technical Centre for Agriculture, DFID, UK
DA	District Administrator of Zimbabwe
DFC	FAO's Community Forestry Programme (in Spanish)
DFID	Department for International Development of the UK
DFO	District Forest Office(r), Nepal
DG	Director General
DITTE	National Directorate of Investigation and Technology Transfer
DNR	Department of Natural Resources
DOF	Department of Fisheries
EC	Executive Committee
ERAs	Agrarian Revolution Schools (in Spanish)
EU	European Union
FAO	Food and Agriculture Organization
FAP	Flood Action Plan
FC	Forestry Commission of Zimbabwe
FECOFUN	Federation of Community Forest Users, Nepal
FFP	Fourth Fisheries Project
FFS	Farmer Field Schools
FFS-FPR	Farmer Field Schools–Farmer Participatory Research
FHRC	Flood Hazard Research Center
FLORES	Forest Land Oriented Resource Envisioning System
FPU	Forestry Protection Unit, Zimbabwe

FRO	Forest and Range Organization
FTLRP	Fast Track Land Reform Programme (Zimbabwe)
GACF	Global Alliance of Community Forestry
GDP	Gross Domestic Product
GIF	Global IPM Facility
GTZ	The German Organisation for Technical Cooperation
IAASTD	International Assessment of Agricultural Knowledge, Science and Technology for Development
ICIMOD	International Centre for Integrated Mountain Development
ICM	Integrated Crop Management
ICRAF	International Center for Research in Agro-Forestry
IDRC	International Development Research Centre
IFAD	International Fund for Agricultural Development
IFM	Integrated Floodplain Management
INGO	International Non-Governmental Organization
INIAP	Instituto Nacional Autónoma de Investigación Agropecuaria (Ecuador)
IPM	Integrated Pest Management
IPM-CRSP	Integrated Pest Management–Collaborative Research Support Program, Ecuador
IRRI	International Rice Research Institute
ITDG	Intermediate Technology Group
IUCN	The World Conservation Union
IUFRO	International Union of Forestry Research Organizations
LGED	Local Government Engineering Department, Bangladesh
LLS	Livelihoods and Landscape Strategy of IUCN
LM	Landscape Mosaics
MACH	Management of Aquatic Ecosystems through Community Husbandry
MACRENA	Red de Manejo Comunitario de Recursos Naturales (Network of Community-Based Natural Resource Management)
MAG	Ministry of Agriculture, Ecuador
MAGAP	Ministry of Agriculture, Aquaculture and Fisheries of Equador
MDC	Movement for Democratic Change, Zimbabwe
MOEF	Ministry for Environment and Forests
MOFL	The Ministry of Fisheries and Livestock
MOL	Ministry of Land

MOU	Memorandum of Understanding
MRAG	Marine Resources Fisheries and Consultants
MWR	Ministry of Water Resources
MYGEC	Ministry of Youth, Gender and Employment Creation
NAFP	Nepal–Australia Forestry Project
NFMP	New Fisheries Management Policy
NFP	National Fishery Policy
NGOs	Non-Governmental Organizations
NPLG	National Level Policy Learning Group
NR	Natural Regions
NRC	National Research Council, US
NRM	Natural Resource Management
NRSP	Natural Resources Systems Programme
NTFP	Non-Timber Forest Products
ODI	Overseas Development Institute, London
OLPII	Oxbow Lakes Project phase II
ONPI	The Organization for Nomadic People
OP	Operational Plan
PAL	Participatory Action Learning
PAR	Participatory Action Research
PBAEP	Patuakhali-Barguna Aquaculture Extension Project
PBO	Planning and Budget Organization
PESACRE	Acre Research and Extension Group in Agroforestry Systems
PIC	Project Implementation Committee
PLA	Participatory Learning and Action
PNG	Papua New Guinea
PRA	Participatory Rural Appraisals
PROINPA	Fundación Promoción e Investigación de Productos Andinos
RDC	Rural District Council of Zimbabwe
RDDC	Rural District Development Committee
RECOFTC	Regional Community Forestry Training Center, Bangkok
RIU	Research Into Use programme of DFID
RLR	Rainfed Lowland Rice
RMC	Resource Management Committee, Zimbabwe
RNRRS	Renewable Natural Resources Research Strategy of DFID
RRA	Rural Rapid Appraisals
RSP	Resource Sharing Project, Zimbabwe
SDC	Swiss Agency for Development and Cooperation
SEMP	Sustainable Environmental Management Programme

SISA	South Asia Institute of Advanced Studies, Kathmandu
SMAFSP	Smallholders Market Access and Food Supply Project, Papua New Guinea
SSWRDSP	Small-Scale Water Resources Development Sector Project (1st phase)
SWRM	Society for Water Resources Management
T-for-T	Training for Transformation
ToT	Training of Trainers
UNDP	United Nations Development Programme
USAID	United States Agency for International Development
UWS-H	University of Western Sydney – Hawkesbury
VIDCOs	Village Development Committees, Zimbabwe
WADCOs	Ward Development Committees, Zimbabwe
WCS	World Conservation Strategy
WWF	World Wildlife Fund
ZANU-PF	Zimbabwe African National Union Patriotic Front

Acknowledgements

This book emerges through the contributions of many people in different organizations in different parts of the world. It is indeed an outcome of research conducted by the editors and authors over a period of 10–30 years during their professional lives, with numerous organizations. The final production of the book was made possible by Research Into Use (RIU) programme of DFID (UK) that enabled us to work together to produce this volume by offering financial assistance and also providing stimulating environment for the research. The first editor is also thankful to ForestAction Nepal, Center for International Forestry Research (CIFOR), and the Canadian International Development Research Centre (IDRC), which earlier supported the research into various aspects of adaptive and collaborative management, key insights of which helped to shape this book.

The editors of the book have benefited from the collaboration and shared learning with a number people in the course of compiling and editing this volume. We sincerely thank Carol J. Pierce Colfer for her excellent help while navigating through the community of researchers cum practitioners pursuing various strands of adaptive collaborative approaches (ACA). We also thank her for excellent comments on various pieces related to this research. Likewise, we highly acknowledge Cynthia McDougall not only for her comments on various chapters of the book, but for her engaging thoughts and inspiring conversations the first editor had with her on research into adaptive collaborative approach over the past several years.

We are thankful to Andy Frost (and his team at RIU) for being supportive and flexible to enable us to complete this work. We are equally thankful to Jesse Ribot of the University of Illinois for his useful comments on a draft version of the introduction chapter. We acknowledge the support and intellectual company of Arun Agrawal of University of Michigan, John Dryzek and Peter Kanowski of Australian National University, Rodney John Keenan of University of Melbourne, and Adrian Fisher of Victoria University while we conducted various aspects of the research that led to this publication. We are also thankful to members of Central Research Team (CRT) at RIU whose comments and feedback were helpful in refining our work. We also acknowledge the intellectual company with Ravi Prabhu, conversations with whom have shaped the idea of the book.

We thank Dr Hari Dhungana, Basundhara Bhattarai, Mani Ram Banjade and Tim Hardwick for reviewing the draft papers. We are very much thankful to the anonymous reviewers of Routledge/Earthscan for providing substantive comments to improve the outline and content of the book. The editorial assistance of Sangita Thapa, Kamal Devkota and Sneha Bhattarai is fully acknowledged. We are also indebted to Southasia Institute of Advanced Studies (SIAS), Kathmandu, and Center for Research in Innovation Policy (CRISP), Hyderabad, for the conducive working environment while doing research and editing that led to this book. We appreciate intellectual company and practical support provided by Dr Naya Sharma Paudel of ForestAction Nepal and Kumuda Dorai of LINK and CRISP.

We also thank the Routledge/Earthscan technical editorial team for their excellent work.

While we acknowledge contributions of various organizations and individuals, we take full responsibility for the content, and in no way reflect the opinions of any organizations or individuals other than the editors and authors themselves.

Hemant R. Ojha, Andy Hall and Rasheed Sulaiman V

1 Adaptive collaborative approaches in natural resource governance

An introduction

Hemant R. Ojha, Andy Hall and Rasheed Sulaiman V

A single conversation with a wise man is better than ten years of study.
Chinese Proverb

All of science is nothing more than the refinement of everyday thinking.
Albert Einstein (© Princeton University Press)

Why this book

This book is not about making one more call for participation, learning or collaboration – much cherished ideals over the past few decades in the field of development and natural resource management. It is instead about exploring how we can foster more effective learning and cooperative actions – two important elements of adaptive collaborative approaches (ACA). Our point of departure is that despite massive expansion of participatory, decentralized approaches to natural resource governance and management over the past three decades, success has remained limited. These approaches tend to shift the role or burden from one part of the system to another, or fail to forge genuine learning and cooperative actions among the actors residing at different parts of the socio-environmental system. It is now time to examine how we can link back the local with the national (and beyond) through learning and collaborative approaches. It is also time to go beyond circulating ideals from expert, donor or policy 'houses' and start exploring underlying limits and pitfalls of these popular strategies, so that we can advance our *learning* about learning and innovation in natural resource systems. In this book, we therefore refer to the world of practice to understand the possibilities of change. We drew inspiration from a Chinese proverb that we can learn important insights by listening to the people who have chosen the difficult way – at least to experiment or understand the world. This book is thus a compilation of stories told by practitioners or at least collected by the action-oriented researchers, from pre-existing practices and research, and not just research done by us, the editors. Although stories captured in the book are complex, the core message is simple: challenges to adaptive learning and collaborative governance are numerous and fundamental, and yet there is hardly an alternative route to effective natural resource governance other than to explore ways through which actors can engage with each other and take an adaptive approach to learning. This does not mean, however, that such approaches emerge automatically in practice, and but actors taking the lead have to face a number of constraints. This book advances the understanding of confronting diverse challenges to adaptive learning and innovation in the context of natural resource governance.

Clearly, this book is not an attempt to translate ideals from the world of policy and theory into practice, but to engage in particular real world experiments of learning and innovations to develop theoretical and policy understanding of what it takes to facilitate the learning process. Beyond proving or disproving specific hypotheses or providing specific prescriptions, we aim to present reflective insights of practitioners and researchers who have experimented with, or actively participated in, different ways of learning and innovation, and on various aspects of natural resource management, in different parts of the world, under diverse conditions. This is more like 'reflexive sociology' (Wacquant 1989) than hypothetic-deductive approach to research, in which data are collected to test pre-defined hypotheses.

While we (editors and contributors of this volume) have been associated with one or another aspects of learning and collaborative processes for years, the idea for this research emerged when we (editors) were undertaking research with the Research Into Use (RIU) programme of the UK Department for International Development (DFID). RIU was struggling to explore ways to support putting research into use, and we were tasked with exploring how that could happen. The programme as such was also trying to explore new frontiers of putting research into use, beyond the conventional approach of research and extension. We used this RIU interest as an opportunity to expand our learning through pooling lessons from wider world of practice. At the core of our interests was developing an understanding of how learning based and collaborative approaches are being applied in different contexts. We saw a clear opportunity to pool diverse experiences together from a wider 'community of practice' (Wenger 2000) – building on the experience of people and institutions already active in exploring innovative processes of learning, collaboration and change in governance and management of natural resources. This book is thus an attempt to learn from such innovative practices, but not necessarily 'success stories'.

The book advances the frontiers of adaptive learning and collaborative governance on two key fronts – understanding challenges, and exploring the possibilities. On the first front, it offers in-depth explanation of why adaptive collaborative approaches are slow to emerge and expand, and how different types of constraints and challenges affect the process. As the contributors highlight, major challenges experienced are related to dealing with traditional reductionist science, balancing research and action in the process, dealing with institutional environments, managing sponsorship to enable the works on the ground, and organizing collaborative actions. The experiments reported here remain at forefront in terms of the seriousness and commitment as well as professional competence deployed, and hence offer nuanced reflections, if not explanations, of how these constraints play out in the actual practice of advancing adaptive and collaborative approaches. Second, it documents and highlights on-the-ground struggles of promoters and facilitators of adaptive collaborative approaches, and describes how ACA actors used diverse strategies, tactics, tools and techniques, in different contexts, to address various issues impeding learning and innovation. Such reflective accounts of the actors provide concrete lessons and ideas for how we can go better in the future than in the past.

Building on some recent in-depth action-oriented research into learning and innovation in natural resource management (Colfer 2005; McDougall 2007), we see adaptive collaborative approaches as those in which there is a conscious and explicit attempt to embed learning while managing, and also an active predisposition to foster social cooperation across all legitimate claimants of material or symbolic aspects of a socio-ecological system. This takes us to an approach very different from the conventional practice of 'unmonitored experience' (see Lee 1999) or models of worldviews and strategies of learning that tend to ignore the systemic reality of socio-ecological system (Capra 1996). An in-depth review of conceptual foundations of adaptive collaborative approaches is presented in Chapter 2.

The book's contributors capture experiences of applying an adaptive collaborative approach in multiple domains of socio-ecological system. Empirical stories start with an analysis of the ups and downs of institutional learning in international research institutions with a biophysical focus (Chapter 3). Another contribution makes an analysis of how well-developed participatory learning methods in agriculture become distorted when they are put into the 'social wild' – comprising strategic actors trying to maximize their own interests and goals by modifying the methodologies (Chapter 4). Likewise, networking aspect of learning is captured in a chapter about Bangladesh floodplains, in which the prospect of 'learning through networking' among floodplain communities is assessed (Chapter 5). Similarly, we have included an analysis of what it takes to promote 'learning in the contested landscapes' in which the authors recollect experiences of applying 'adaptive collaborative management' in the most volatile and contested times in Zimbabwean history (Chapter 6). Still another dimension we have included is about 'learning to improve livelihoods' (Chapter 7) in which the author reviews the challenges of applying adaptive collaborative approaches to improve livelihoods of the poor through effective and equitable management of forest resources. In Chapter 8 an action research professional reflects upon the long and diverse experiences around 'learning through action' in diverse natural resource management initiatives in Asia.

Plethora of innovations, continuing challenges

Paradigms of natural resource governance have been shifting towards participatory, inclusive, decentralized and collaborative approaches. Despite three decades of reforms in policies and practices in the developing world, the achievement is limited. Still more than 1.3 billion people who base their livelihoods on fisheries, forest and agriculture (FAO 2004) are deprived of basic necessities and 'freedoms' (in Amaratya Sen's terms) (Sen 1999) that they have reason to choose. Moreover, the socio-ecological systems that generate these natural capitals are still fragile and in many cases in process of degradation, as evident in the Millennium Ecosystem Assessment (2005). The global climate change phenomenon is further adding stress and vulnerabilities to these socio-ecological systems and the poor residents. As a result, additional challenges have also surfaced: rising food prices, increased livelihood vulnerabilities, declining fish catches, erosion of agro-ecological systems, displacement of the poor from natural resource systems, and the like. As the Millennium Ecosystem Assessment (2005: 1) concludes:

> These problems, unless addressed, will substantially diminish the benefits that future generations obtain from ecosystems. The degradation of ecosystem services could grow significantly worse during the first half of this century and is a barrier to achieving the Millennium Development Goals.

This at least reminds us that our policy solutions to natural resources management have not worked well, and point to the need for revisiting the foundations and assumptions of current policy approaches and development strategies. The report further concludes that the challenge of 'reversing the degradation of ecosystems while meeting increasing demands for their services can be partially met. . . ., but these involve significant changes in policies, institutions, and practices that are not currently under way'. This conclusion is particularly noteworthy as the team of predominantly biophysical scientists who carried out the assessment came up with the conclusion that major problems lie in policy and institutions. In view of this, we must pause and ask: despite landmark natural resource policy reforms of the past three decades, with radical shifts involving decentralization and participatory management, why do we fail to achieve large scale impacts?

Undoubtedly, natural resource management (NRM) policy reforms with their participatory turn have been informed by quite nuanced understanding of the problems and the vision for change. Indeed, the policy change is informed by a series of 'reversals' (Chambers *et al.* 1993) – from professional to people first, from state-centric management to community-based management of resources, from techno-scientific to social and participatory approaches, from centralized planning to bottom up planning, from subsistence-oriented to market-oriented management, and so on and so forth. Recognizing the deepening socio-ecological crisis and growing contestations around resource control and management, more and more policy actors now agree that without involving the poor living in and around natural resource systems, neither poverty reduction nor environmental sustainability can be achieved. From this understanding, numerous attempts have been made worldwide, and as a result quite a few innovations and transformations have taken place in different parts of the world, empowering local communities, and facilitating fairer distribution of benefits from agriculture and natural resource management (Spielman and Pandya-Lorch 2011).

These successes have led some to argue that we have already arrived at paradigm capable of addressing human well-being and natural resources sustainability. We argue that innovations and working examples of natural resource governance are too small on a global scale to counteract the effects of human beings on ecosystems (which the Millennium Ecosystem Assessment revealed), and to demonstrate a way for inclusive resource management and fair distribution of benefits. In worst cases, these innovations are likely to revert back once active support systems are withdrawn (as the experiences of contributors in this volume show). Indeed, there is already a surge of trenchant critiques on the limits of participatory innovations in development and natural resource management. These solutions have overlooked the complex and multi-scale problems of natural resource governance (Cook and Kothari 2001; Mosse 2002; Colfer 2005). Many innovations at the local level have not emerged from the well-nourished breeding ground of a piloting stage (Hall 2007), and have at times extended the subtle instruments of state control rather than genuinely creating political space for local people (Li 1999). Such innovations, though widespread globally and found across diverse NRM contexts, often fail to find nurturing policy and institutional space for continued development and expansion (see Chapters 3 and 4 of this volume and Colfer *et al.* 2011).

From our own research into, and experience of moving through the journey of, participatory development and natural resource management, we strongly believe that we have not reached the 'end of history'[1] of NRM – such that a final policy approach has already been discovered and all that we need to do is to 'implement' it in practice. We are convinced that the problem lies at a more fundamental aspect of the policy approach – its faulty assumptions and misplaced expectations, where there is a huge potential to improve. While much of the practical methodologies and toolkits[2] can be saved for further use and innovations, we need a whole different way of thinking and acting about change.

We argue that there are three problems with the current approach that dominates the scene. First, the dominant actors following these approaches (be they policy makers or donors or conservation agencies) blame the local communities and institutions for the complex and cross-scale problems, as if the larger policy and institutional regimes are working just fine. Some have warned such an approach is riddled with 'dangers of localism', which tend to underplay local politics and the national and wider political forces (Mohan and Stokke 2000). The solution then becomes providing some incentives to local communities or decentralizing some power to local bodies, which hides the question of more fundamental political reform of the state institutions (Sundar 2001). Recognizing local people's rights to manage local natural resources is certainly a positive step, but in so doing the national policy actors cannot ignore the need to change their own planning and monitoring systems, and strategies for collaboration, and even the need for changes in institutional structures. Indeed, we argue that there is a need to rethink participation, learning and innovation as purely local phenomena, and conceive of more holistic and cross-scale phenomena of change, in which national institutions and policy systems are also seen as part of the problem of local level natural resource management.

Second, the current approach to change is still guided by a technocratic worldview that privileges experts and policy makers to make decisions for others, disregarding the agency and capability of the poor and affected local communities. Even the participatory approach has legitimized expert-led, Euro-centric, modernist visions and strategies of change – forcing everyone to think through West-centric lenses (Shiva 1988; Scott 1998) – and in effect creating 'participatory exclusions' (Agarwal 2001). These approaches privilege formal over informal, documented over tacit, project-based over evolutionary, time-bound over flexible, evidence over emotion, sectoral over systemic, disciplinary over holistic, uni-scale over multi-scale, pre-defined outputs over process, and so on and so forth. Advocates of indigenous knowledge have sought to challenge such hegemonic and technocratic conceptions and practices, but what we really lack here is an approach that engages with, and helps negotiate among, multiple worldviews and learning systems operating at multiple scales in this globalized world. This book builds and expands the ideas and strategies that emphasize deliberation (Ojha *et al.* 2010), communication and negotiation, transdisciplinarity (Rist *et al.* 2004) and social learning (Leeuwis and Pyburn 2002) between different sets of actors and their knowledge systems on the one hand, and interrogates how these normative and agency-led processes are pushed back by the rigid techno-scientific mindsets and strategies of the planners, public policy decision makers and even the donors of development.

Third, most reforms occur only on paper, not in practice. Governments, donors, international agencies, and even service providers – who are the key players in natural resource governance – all seem to be strong in getting things right on paper, such as through new policy documents, strategy papers, proposals, monitoring reports, management frameworks, project logframes, and so on. And they produce all of these for their clients and sponsors, not for themselves: governments prepare strategies for donors, donors do the same for recipient governments, and so on and so forth. There is little attention to reviewing how one's own institutions should change. Moreover, there is an important difference between what is said and what is done. For instance, a government declares publicly that it will provide rights to local people in wildlife management. In practice, it does just the opposite – chasing people away from the protected area. An underlying reason for all of these is that the dominant actors in the system benefit from either the status quo or even ensuring that a participatory policy never gets into practice, but is upheld for the positive image and symbolic power it lends. Given these complexities, it is not enough to 'talk' or 'write' about improving governance. Instead, interrogating and debating why nice talks and nice research outputs do not find their way into practice is more critical.

In view of these continued failures, recent attempts have begun to look at the process dimensions more seriously – looking at how policy and institutions emerge, function, change and improve to address the challenges (Hall and Clark 1995; Colfer 2005; Fisher *et al.* 2007). This book take sides with these attempts to advance an innovative approach to natural resource governance by focusing the analysis to understand the process of adaptiveness and collaboration dimensions. Whilst we refrain from prescribing a specific program of change, we aim for broad directions that various actors should consider in transforming the governance of natural landscapes. An important policy message coming out of this analysis is that policy systems should themselves be seen as learning systems, and we believe that there is indeed a great benefit in policy actors' considering the learning dimension seriously (Hall 1993).

Adaptive collaborative approaches to natural resource governance

Setting aside the busy policy makers and locally engaged resource managers, as members of the research community, we admit that we are facing a critical challenge to demonstrate solutions that help tackle poverty and human well-being (Serageldin 2002) as well as ensuing environmental change – two of the greatest challenges in the new millennium. But we do not believe that experts can fix the problems from the position of expert. At best social science experts can contribute to learning processes of decision makers. This book seeks to advance the learning route to natural resource governance reform, rather than prescribe grand solutions.

But the problem is that the *learning way* is easier said than done, as contributors of this volume say. In natural resources governance sectors, charting out the learning way is not an easy task, especially in the context of pre-existing deeply held mechanistic conceptions, fractured institutions and political asymmetry among the diverse and expanding stakeholders. As the nexus between poverty–environment is becoming even more entrenched with climate change (Casillas and Kammen 2010), a paradox emerges: learning and collaboration among NRM actors become more indispensable, and at the same time more difficult to achieve. In a globalizing world, natural resources are increasingly contested by an increasing number of actors for ever wider varieties of stakes (Leach *et al.* 1999). In such contexts, as post-structural social sciences have revealed (Rosenau 1991), simple categories of state, market and civil society do not help in understanding how things change (or why they do not change). We have a much more complex array of actors and interest groups within each of these meta institutions (Rose and Miller 1992; Arts 2006; Dryzek 2006). More than ever before, it has become important to look at relations and interactions more than substance and attributes, as innovation and change rest largely on how and to what extent these actors interact, and how actors and institutions co-evolve through the processes of 'transaction'.[3] The transactional view suggests that researchers recognize the dialogical relations among diverse actors of socio-ecological systems. As the techno-scientific approaches to understanding socio-ecological processes are facing trenchant critique, science is under pressure to help stimulate constructive dialogues (Fischer 1998) or facilitate agreement in conflict, rather than look for some ultimate truth (Rorty 2009). In this spirit, the collection of papers in this volume seek to shade lights on the problems, issues, dissonance around multiple forms of collaboration and transaction as NRM actors seek to find new ways of governing the resources.

Parallel to these science–democracy debates, more operationalizable concepts of learning and innovation have emerged around social and organizational learning related fields of knowledge and practices (Argyris 1993; Schon 2010), as well as around works that emphasize integrated analysis of society and natural systems, usually referred to as socio-ecological systems (Lee 1993; Holling 2001). Together these approaches have sought to consider learning and innovation aspects of not just 'resource management' or a particular organization, but the entire socio-ecological system (Berkes and Turner 2006) or the 'community of practice' (Wenger 1996) or 'public policy as social learning systems' (Hall 1993), involving multiple scales of time and space. This research blends ideas from such diverse fields of learning and innovation, and emphasizes the need to pay more attention to the process through which actors can negotiate and learn their way, instead of the prescriptive approach advocated by many techno-scientific groups.

In this research, we define these learning based approaches as adaptive collaborative approaches (ACA), which denote a family of concepts that seek to combine research and various other ways of learning as well as seek out collaborative actions among diverse stakeholders operating at multiple scales of decision-making and action. We see ACA as a suite of strategies, rather than an absolute approach, which are employed to learn and muddle through (Lindblom 1959) complex systems to generate and facilitate innovations on various aspects of resource governance and management. This seems unavoidable as environmental science itself is being seen as a politicized activity, and any innovation must pass through negotiation among competing and interlocking claims of power and truth (Forsyth 2003).

While doing this research, we started with a specific idea of adaptive collaborative management (ACM) developed within the Center for International Forestry Research (CIFOR) (reported in Chapter 3 by Colfer), and built further on the adaptive collaborative approach, which was coined by McDougall *et al.* (2009) within CIFOR and its diverse partners. The phrase 'adaptive collaborative approach' combines both normative (what needs to be done) and analytical elements (what needs to be understood) as NRM actors attempt to improve developmental outcomes and sustainability. The approach reflects the convergence of various traditions of learning, innovation and social cooperation that emerged in different practical and intellectual contexts. We also build on our own previous work in relation to innovation systems and deliberative strategies of science–practice communications (Hall *et al.* 2001; Ojha *et al.* 2010).

By adaptive we mean the human quality to learn in the face of the complexity and dynamism of socio-ecological systems, and this is inevitable in any human action given the inherent uncertainty involved in planned management actions. Likewise, the notion of collaboration stems from the fact that a) every resource management situation involves multiple actors, and b) collaboration among the actors has a greater potential to generate better (sometimes win–win) outcomes than without collaboration. This resonates with what is being popularized as 'social capital' in the development literature (Woolcock 1998; Luong and Daniere 2011). This does not however mean that actors are necessarily oriented to collaborate; indeed entrenched forms of conflict are a rule rather than an exception. In any case, resource management innovation is difficult without addressing resource conflicts and harnessing a certain level of cooperation. Between social capital and social conflict, we need to find an approach to engage actors in the process of learning, collaboration and innovation in the management of natural resources, if we are to address both poverty and ecological sustainability.

Adaptive collaborative approaches can be conceived in both active and somewhat passive senses (Ruitenbeek and Cartier 2001). In the active sense, it is facilitated by some actors external to the immediate system being worked on, and as such, these ACA facilitators are oriented to, and have the opportunity to, explicitly reflect and organize the process. In the more passive sense, actors internal to the system adopt, to varying degrees, elements and principles of ACA for organizing their actions and learning. They do this with or without conscious reflections of practice, and without necessarily being self-conscious of using this approach. In either case, the issues of learning and collaboration remain critical as NRM systems host a wide range of actors – resource owners, managers, conflicting claimants, extensionists, researchers, policy makers, value chain actors, technology suppliers and the like. The domain of learning and interaction is not narrowly confined to the members of a clearly defined group using or managing a natural resource system, but encompasses actors with both strong and weak interests, exercising direct and indirect influence, staying immediate or downstream in the value chain.

Resource governance systems are multi-scalar and interlinked economic, ecological and political systems (Holling 2001). This means that local level governance of natural resources, however sovereign or autonomous legally, has had to respond to and hence engage with forces, drivers and actors operating at higher levels, including the national policy regime. For instance, how much a local community group benefits from the management of forest is determined by the prices of forest products in the urban market, and national policies define who has access to local natural resources. These external forces not only shape constraints and opportunities but also mediate local level power relations. So conservation and management of natural resources is not entirely a free choice of local actors or communities. And this requires moving from an absolute local, community-centred approach to more multi-level and collaborative approaches. In such situations, issues related to cross-scale linkages as well as feedback systems from operational systems to policy systems become part of an adaptive collaborative approach to governance and innovation (Colfer 2005; Armitage *et al.* 2008).

A particular challenge to learning and collaboration in natural resource management is its unusually uneven playing field, with widespread processes of exclusion and marginalization (Blaikie and Brookfield 1987; Agarwal and Narain 1991; Peet and Watts 1996). The history of nation-state building and the expansion of the colonial regime are inextricably linked to this. Despite a history of exclusion and political asymmetry, we see that there is hardly an alternative than to reflect, review, discuss, negotiate, experiment and then learn from practice and collaborative inquiry if we are to ensure sustainability and fair systems of benefit sharing.

Overview of chapters

We have six contributions written by practitioners and scholars with long experience in the field of natural resources management. These chapters also cover a broad range of experiences in applying adaptive collaborative approaches in natural resource management. They encompass diverse resource sectors, country contexts (Asia, Africa, Latin America), and capture a diversity of strategies used to deal with varied concerns and issues of natural resource management and innovation. Almost all chapters share a conclusion that an adaptive and collaborative approach is not only inevitable in the contested world of resource governance, but it can also lead to better outcomes in terms of livelihoods, policy and resource sustainability, if applied appropriately. In particular, the authors reflect upon four categories of challenges to applying ACA – personal/attitudinal, institutional, cultural, and policy and underlying regimes.

What is common to all chapters is that all authors take a reflective approach to writing and analysis: making explicit their own assumptions, surprises, successes, failures and learning while applying the ACA in various situations. Here, we took an inspiration from Einstein that great science is the refinement of thinking in practice. The contributors bring unique strengths of various forms of writing agency and different vantage points to see, experience and reflect – from local practitioners, international programme leaders, academic researchers, action researchers, trainers to policy advisors. The authors also share a concern that moving away from traditional linear models of research and technology transfer to ACA involves even more challenges, in terms of managing complex social relations, framing learning and incorporating learning into action. This means that authors are not just describing cases out there, but also bring their own reflections to applying ACA approaches.

A brief description of each of these chapters is given below.

Chapter 2 presents a review of various adaptive and collaborative approaches developed to cope with the social and biophysical complexity of natural resource systems. It synthesizes key insights from different traditions to provide more holistic, pragmatic and unified theory of learning and change. As such, this chapter provides a conceptual reinforcement to the adaptive and collaborative approach, by forging dialogues among thinkers, practitioners, activists and researchers sharing similar concerns for embedding learning and collaboration in governing socio-ecological systems. Another objective of this chapter is to develop common threads running through various adaptive and collaborative approaches and contribute to synthesis of lessons in Chapter 9.

In Chapter 3, Carol J. Pierce Colfer, formerly a programme leader of the Adaptive Collaborative Management (ACM) programme at the Center for International Forestry Research (CIFOR), discusses how the idea of ACM was conceived in an institution with a biophysical research mandate, demonstrating the various ways in which conflicts and collaboration can occur among social and biophysical scientists. She reflects upon how ACM evolved within CIFOR describing the various struggles her team had to make with the CIFOR management. She outlines what worked and what did not in the course of applying ACM in 11 countries and over 30 sites over a period of about 10 years, and what challenges the ACM team had to face – within the team, in the organization, in the field and with the donors. The chapter is particularly strong in capturing the personal reflections of the author on her own engagement with the ACM project, and also in demonstrating how an ACM process emerges and is influenced by the underlying institutional, policy and knowledge systems environment within which ACM is applied.

While Colfer focuses on forested landscapes, Sherwood *et al.* in Chapter 4 document how the concept of Farmer Field Schools (FFS) has evolved into a variety of contextually grounded strategies of adaptive learning and innovation in the agricultural landscape of the Andes regions of Ecuador, South America. Unlike the CIFOR ACM case, which was coordinated by a single international organization, the analysis of FFS in Ecuador covers the ways in which it was promoted as a solution to agricultural development problems by several government and development organizations. They demonstrate how a successful innovation in one context (and time) has to go through the process of regeneration or modification when moved to or put into use in contexts different from the original contexts in which the innovation was developed (FFS initially emerged in East Asia and was brought to Ecuador). They also demonstrate how expert-led and people-led knowledge systems compete and converge in agriculture and natural resource management contexts, and how adaptive learning approaches can help reframe diverse knowledge interfaces and hence improve resource management practices. With better participation of the farmers, as Sherwood *et al.* show, the technology or innovation will undergo significant changes and modifications, beyond the technical conceptualization of the experts. These authors critically examine whether such methodology based interventions have the power to survive repressive institutional regimes, and in the process, to what extent the technology retains its integral character, and to what extent the regime of actors are also willing to embrace the learning focus of the innovation.

In Chapter 5, Sultana and Thompson explore the processes of networking local resource user groups as a key aspect of ACA. They describe action research processes with over 250 existing community based organizations (CBOs) managing floodplains resources in Bangladesh. These groups have been encouraged to adopt improvements in their practices that take a more system-wide view of the productivity of floodplains, which they call 'Integrated Floodplain Management'. Using the process of an adaptive learning network, they report how many CBOs have been able to improve their plans and practices, not just based on their own individual group learning, but also through the exchange of lessons and experiences with one another. They describe the process followed and how it evolved, and also provide an assessment of some of the outcomes from this adaptive learning network and lessons for potential wider adoption.

Chapter 6 is devoted to the experiential account of Tendayi Mutimukuru-Maravanyika and Frank Matose, who provide an assessment of applying ACM in the Zimbabwean forestry sector, based on a CIFOR-led project implemented in collaboration with the Forestry Commission of Zimbabwe. The ACM process reported here focused on enhancing collaboration among stakeholders at a range of scales, starting with local resource user groups, resource management committees, and forest officers and researchers. They were included in processes to develop visions and implement action plans.

The project sought to facilitate widespread use of self-improving and equitable forest resource management systems that build on local capacity and 'vertical' and 'horizontal' stakeholder interactions, and respond positively to external pressures. The authors report reflections of the revisit to the project site several years after the completion of the project, and critically reflect upon the initial assumptions and conceptualizations of ACM in light of the follow up study findings. They identify additional challenges to applying ACM related to political complexities and institutional conditions through a synthesis of almost a decade long experience in the region where ACM was applied.

In Chapter 7, Mani Ram Banjade reflects on the experience of having managed multiple ACA oriented initiatives in Nepal's community forestry as part of a search for solutions to enhance the effectiveness of forest management institutions in improving local livelihoods. By drawing on the experience of an action-research-focused non-governmental organization (NGO) which over the last decade applied a wide variety of tools to enhance adaptive learning and collaboration, he presents the challenges involved and how these were addressed. His experience covers working with various international institutions as well as local communities and government organizations from the community level to the national level in the context of forestry. Various ACA tools, he reports, have been instrumental in empowering the marginalized groups, forging multi-stakeholder collaboration to enhance the effectiveness of forest management systems, and to strengthen the links between operational processes and resource management and policy learning cycles. His experience also demonstrates how difficult it is to try a collaborative and learning based approach to natural resource management in the context of a highly stratified society such as Nepal where some of the world's poorest groups directly draw their livelihoods from forest and other natural resources.

In Chapter 8, Fisher presents a personal reflection of his over-30 years of experience on a variety of action research and adaptive collaborative management initiatives. These involve some of the pioneering work he did in setting up Nepal's community forestry back in the 1980s. As a keen promoter and engaged practitioner of action research, he reflects on a variety of roles he took as a promoter of this approach – undertaking community level research and extension through working as a trainer in international organizations to supervising action research projects in academic institutions. As an anthropologist with strong interests in human ecology and development, Fisher's experience has been somewhat unusual in that he has been engaged in action research and adaptive collaborative management both as an academic and as an active development practitioner. This has enabled him to experience the opportunities and challenges of these approaches in both academic and applied contexts, which is rare to find.

Chapter 9 summarizes diverse challenges reported by contributors and distils specific insights into confronting them while applying adaptive collaborative approaches. ACA actors have been found to muddle through multiple domains of adaptive learning and innovation, mainly: establishing epistemic credibility in the face of dominant traditional reductionist science, striking an intricate balance between research and action in the adaptive learning process, dealing with institutional orthodoxies, managing sponsorship to enable the works on the ground, and organizing collaborative actions among multiple and often conflicting actors.

Finally, Chapter 10 builds on the lessons of Chapter 9 and presents a bold new future direction.

Key message

At the risk of over-simplification, we now highlight the key messages coming out of this book of 'reflective essays' (key conclusions are elaborated more fully in Chapters 9 and 10). A simple conclusion is that, in the rapidly changing socio-environmental contexts and in view of the far too limited successes of the recent NRM policy reforms in the developing world, we need a wholly different way of approaching NRM and human wellbeing questions. The new thinking should not just be about how we implement policy, but more centrally about how the underlying policy and institutional regimes can and should be changed. We need changes in all the five key domains of NRM regimes – policy processes, management, administration, technology, and financing. This new approach should take learning and interactions among the NRM actors more seriously than ever before, and should always look for better ways to link learning and decision-making processes across scales. While we admit that these conclusions are not radically new, we certainly offer new insights as to what it takes to root these learning based approaches in practice. At least six points are noteworthy here:

1 We need to nurture and support new types of researchers who bridge, broker and facilitate change around specific poverty-NRM issues and at the same time connect the processes across the wider policy and institutional systems. The business-as-usual of developing technical solutions and then using extension systems to put the research into use does not work. We do not deny the role of carefully planned technical and more sophisticated research as integral to the process of innovation (such as a new variety or a new silvicultural technique). Yet our findings clearly suggest that innovation-centric research can still be organized within the context of application and with better communicative control by the stakeholders so that every research investment is best utilized. We have evidence in this compendium that embedded research strategies have worked in different contexts and are waiting for wider policy uptake.

2 Locally based NRM groups and entities should develop workable frameworks of learning, monitoring, review and reflections more explicitly, in order to find new and innovative ways to address the problems of poverty and NRM. Contributors report varying types of attempts to design and practise learning experiments at project level, institutional level, in collaborative undertakings, and even in university–community partnerships. These are also awaiting more enabling policy and financing environments.

3 Development agencies – both national and international – are still preoccupied with the strategy of developing technical models, with the belief that they can subsequently be scaled up and out widely. This strategy should be replaced by the one involving multi-scale learning approaches with simultaneous learning processes employed at different scales. There should also be efforts to facilitate linking these processes vertically and horizontally, through interactive institutional platforms, and two-way communications. Innovation and positive changes in NRM and livelihoods should be seen as outcomes of dynamic learning systems, and not as something resulting from the application or replication of technological packages. Strategies for change solely based on methodology, new tools or technology are unlikely to work unless there is a concurrent process of transforming the underlying institutional regime.

4 Not surprisingly, in most situations, practice has remained more innovative than policy. But surprisingly, at times, the policy environment has repressed such local innovations (and the actors driving such innovations), simply because the innovations did not comply with or at times even challenged the existing policy paradigm. There is a need to have a strategy to review national policy and institutional regimes from the perspectives of emerging local innovations and actors, and then link the findings with the local level innovations. This should be seen as a reversal of the current practice of evaluation being solely sponsored and undertaken from the perspective of policy makers and donors.

5 The policy system should itself be seen as a learning system, and should develop ways to organize action-reflection-review in relation to its own domains of planning and decision making, and also proactively seek out ways to enable multiple cycles of lower scale learning processes, across diverse threads of innovation. The learning emphasis should be reflected in treating policies as experiments – and hence having a strong element of monitoring, review and reflections. The accountability of the policy system to the larger citizenry should include an audit of how the system is improving over time, including the processes of learning. This is essential to ensure that public investment is put to effective use and saved from the current continuous series of failed attempts.

6 International development donors should not confine themselves to the delivery and scaling out of a standardized service package or generating consultant recommendations for public policy change. They should

support a process whereby each project, region or country learns to develop innovative solutions to their problems. Donors have the opportunity to help establish adaptive and collaborative learning systems in policy related agencies – so as to enable policy officials to reach out and talk to people struggling through innovation processes. Moreover, accountability of international funding should be seen not in terms of immediate, measurable values of tangible outcomes in the short run, but on the basis of promising plausible connections found between action, learning and possible outcomes. Evaluation of impact should not be left entirely to the consultants hired by the donors or appointed by the governments, but moved significantly to the reflective and self-critical assessment of the actors participating in the innovation process. As initially noted, this implies a need for a whole new approach to administration, management and financing in funding agencies.

In the chapters to follow, you will read how these conclusions emerged from a variety of adaptive collaborative initiatives in agriculture and natural resource management under diverse conditions. Our hope is that the rich and experiential accounts reported in the chapters will provide a strong basis to reframe innovation policies and practices for more equitable and sustainable management of natural resources and socio-ecological landscapes.

Notes

1 We draw a metaphor from Francis Fukuyama's (Fukuyama 1992) view that the world has finally discovered a system of governance based on market and liberal political system. We do not see such end of history at least in the context of natural resource governance.

2 A report lists tools on participatory methodologies for agriculture and natural resource management (de Zeeuw and Wilbers 2004).

3 Pragmatist philosopher John Dewey (Dewey and Bentley 1949) differentiated three types of interactions – 'self-action', 'inter-action' and 'transaction'. While the first is about individual entity becoming active, the difference between the last two is subtler. In 'interaction', two entities pre-exist and then interact, whereas in situations of 'transactions', two entities are in the constant process of creation and recreation through mutual interface. Natural resource governance studies have been dominated by the first two conceptions of reality, and less with the notion of transaction.

References

Agarwal, A. and S. Narain (1991) *Global warming in an unequal world: a case of environmental colonialism.* Earth Island Journal Spring 1991(39–40).

Agarwal, B. (2001) *Participatory exclusions, community forestry, and gender: an analysis for South Asia and a conceptual framework.* World Development 29(10): 1623–1648.

Argyris, C. (1993) *On Organizational Learning.* Cambridge, MA: Blackwell.

Armitage, D., M. Marschke and R. Plummer (2008) *Adaptive co-management and the paradox of learning.* Global Environmental Change 18(1): 86–98.

Arts, B. (2006) *Non-state actors in global environmental governance: new arrangements beyond the state.* In *New Modes of Governance in the Global System: Exploring Publicness, Delegation and Inclusiveness.* M. Koenig-Arcibugi and M. Zurn. Palgrave: Basingstoke, 177–200.

Berkes, F. and N. J. Turner (2006) *Knowledge, learning and the evolution of conservation practice for social-ecological system resilience.* Human Ecology 34(4): 479–494.

Blaikie, P. and H. Brookfield (1987) *Land Degradation and Society.* London: Methen and Co. Ltd.

Capra, F. (1996) *The Web of Life.* New York: Anchor Books.

Casillas, C. E. and D. M. Kammen (2010) *The energy–poverty–climate nexus.* Science 330(6008): 1181–1182.

Chambers, R., A. Pacey and L. A. Thrupp, Eds (1993) *Farmer First: Farmer Innovation and Agricultural Research.* London: Intermediate Technology Press.

Colfer, C. J. P. (2005) *The Complex Forest: Communities, Uncertainty, and Adaptive Collaborative Management.* Washington, DC: Resources for the Future; Center for International Forestry Research.

Colfer, C. J. P., E. Andriamampandry, S. Asaha, E. Lyimo, E. Martini, J. L. Pfund and J. Watts (2011) *Participatory action research for catalyzing adaptive management: analysis of a 'fits and starts' process.* Journal of Environmental Science & Engineering 5(1): 28–43.

Cook, B. and U. Kothari (2001) *The case for participation as tyranny.* In *Participation: The New Tyranny.* B. Cook and U. Kothari. London and New York: Zed, 1–15.

de Zeeuw, H. and J. Wilbers (2004) *PRA Tools for Studying Urban Agriculture and Gender.* Leusden, NL: Resource Centre on Urban Agriculture and Forestry (RCUAF).

Dewey, J. and A. F. Bentley (1949) *Knowing and the Known.* Westport, CT: Greenwood Press.

Dryzek, J. S. (2006) *Transnational democracy in an insecure world.* International Political Science Review/Revue internationale de science politique 27(2): 101–119.

FAO (2004) *The State of Food and Agriculture: Agricultural Biotechnology Meeting the needs of the poor?* Rome: Food Agricultural Organization of the United Nations.

Fischer, F. (1998) *Beyond empiricism: policy inquiry and post posivist perspective.* Policy Studies 26(1): 129–146.

Fisher, R., R. Prabhu and C. McDougall, Eds (2007) *Adaptive Collaborative Management of Community Forests in Asia: Experiences From Nepal, Indonesia and the Philippines.* Bogor, Indonesia: Center for International Forestry Research (CIFOR).

Forsyth, T. (2003) *Critical Political Ecology: The Politics of Environmental Science.* London and New York: Routledge.

Fukuyama, F. (1992) *The End of History and the Last Man.* London: Penguin.

Hall, A. (2007) *Challenges to Strengthening Agricultural Innovation Systems: Where Do We Go From Here?* Maastricht: United Nations University, Maastricht Economic and Social Research and Training Centre on Innovation and Technology.

Hall, A., G. Bockett, S. Taylor, M. V. K. Sivamohan and N. Clark (2001) *Why research partnerships really matter: innovation theory, institutional arrangements and implications for developing new technology for the poor.* World Development 29(5): 783–797.

Hall, A. and N. Clark (1995) *Coping with change, complexity and diversity in agriculture: the case of rhizobium inoculants in Thailand.* World Development 23(9): 1601–1614.

Hall, P. A. (1993) *Policy paradigms, social learning, and the state: the case of economic policy making in Britain.* Comparative Politics 25(3): 275–296.

Holling, C. S. (2001) *Understanding the complexity of economic, ecological, and social systems.* Ecosystems 4(5): 390–405.

Leach, M., R. Mearns and I. Scoones (1999) *Environmental entitlements: dynamics and institutions in community-based natural resource management.* World Development 27(2): 225–247.

Lee, K. N. (1993) *Compass and Gyroscope: Integrating Science and Politics for the Environment.* Washington, DC: Island Press.

Lee, K. N. (1999) *Appraising adaptive management.* Ecology and Society 3(2). http://www.consecol.org/vol3/iss2/art3/.

Leeuwis, C. and R. Pyburn (2002) *Social learning for rural resource management.* In *Wheelbarrows full of frogs: social learning in rural resource management: international research and reflections.* Assen: Van Gorcum Ltd.

Li, T. M. (1999) *Compromising power: development, culture, and rule in Indonesia.* Cultural Anthropology 14(3): 295–322.

Lindblom, C. E. (1959) *The science of 'muddling through'.* Public Administration Review 19(2): 79–88.

Luong, H. V. and A. Daniere, Eds (2011) *The Dynamics of Social Capital and Civic Engagement in Asia: Vibrant Societies.* London: Routledge.

McDougal, C., H. Ojha, R. K. Pandey, M. R. Banjade and B. H. Pandit (2007) *Enhancing adaptiveness and collaboration in community forestry in Nepal: reflections from participatory action research.* In R. J. Fisher, *ACM in Asia, CIFOR and RFF, Adaptive Collaborative Management of Forests: Experiences from Asia.* Bogor, Indonesia: CIFOR and RFF.

McDougall, C., B. H. Pandit, M. R. Banjade, K. P. Paudel, H. Ojha, M. Maharjan, S. Rana, T. Bhattarai and S. Dangol (2009) *Facilitating Forests of Learning: Enabling an Adaptive Collaborative Approach in Community Forest User Groups, A Guidebook.* Bogor, Indonesia: Center for International Forestry Research.

Millennium Ecosystem Assessment (2005) *Ecosystems and Human Well-being: Synthesis.* Washington, DC: Island Press.

Mohan, G. and K. Stokke (2000) *Participatory development and empowerment: the dangers of localism.* Third World Quarterly 21(2): 247–268.

Mosse, D. (2002) *People's knowledge, participation and patronage: operations and representations in rural development.* In *Participation: The New Tyranny?* B. Cooke and U. Kothari. London: Zed Books, 16–35.

Ojha, H. R., N. S. Paudel, M. R. Banjade, C. McDougall and J. Cameron (2010) *The deliberative scientist: integrating science and politics in forest resource governance in Nepal.* In *Beyond the Biophysical: Knowledge, Culture, and Politics in Agriculture and Natural Resource Management.* L. German, J. J. Ramisch and R. Verma. Dordrecht, Hiedelberg, London and New York: Springer, 167–191.

Peet, R. and M. Watts (1996) *Liberation ecology: development, sustainability, and environment in an age of market triumphalism.* Liberation Ecologies Environment Development Social Movements: 1–45.

Rist, S., A. Zimmermann and U. Wiesmann (2004) *From epistemic monoculture to cooperation between epistemic communities: development research and sustainability.* In Conference Proceeding of Bridging Scales and Epistemologies Millennium Assessment, Alexandria, Egypt.

Rorty, R. (2009) *Philosophy and the Mirror of Nature.* Princeton: Princeton University Press.

Rose, N. and P. Miller (1992) *Political power beyond the state: problematics of government.* The British Journal of Sociology 43(2): 173–205.

Rosenau, P. M. (1991) *Post-modernism and the Social Sciences: Insights, Inroads, and Intrusions.* Princeton, NJ: Princeton University Press.

Ruitenbeek, J. and C. Cartier (2001) *The invisible wand: adaptive co-management as an emergent strategy in complex bio-economic systems.* CIFOR Occasional Paper No. 34.

Schon, D. (2010) *Government as learning system.* In *Social Learning Systems and Communities of Practice.* C. Blackmore. Milton Keynes: The Open University Press, 5–16.

Scott, J. (1998) *Seeing Like a State: How Certain Schemes to Improve Human Condition Have Failed.* New Haven and London: Yale University Press.

Sen, A. (1999) *Development as Freedom.* Oxford: Oxford University Press.

Serageldin, I. (2002) *World poverty and hunger: the challenge for science.* Science 296(5565): 54.

Shiva, V. (1988) *Staying Alive: Women, Ecology, and Development.* London: Zed Books.

Spielman, D. J. and R. Pandya-Lorch (2011) *Proven Successes in Agricultural Development: A Technical Compendium to Millions Fed 2010.* Washington, DC: International Food Policy Research Institute.

Sundar, N. (2001) *Is devolution democratization?* World Development 29(12): 2007.

Wacquant, L. (1989) *Towards a reflexive sociology: a workshop with Piere Bourdieu.* Sociological Theory 7(1): 26–63.

Wenger, E. (1996) *How we learn: communities of practice: the social fabric of a learning organization.* The Healthcare Forum Journal 39(4): 20–26.

Wenger, E. (2000) *Communities of practice and social learning systems.* Organization 7(2): 225–246.

Woolcock, M. (1998) *Social capital and economic development: toward a theoretical synthesis and policy framework.* Theory and Society 27(2): 151–208.

2 Adaptive collaborative approaches

Traditions, foundations and frontiers

Hemant R. Ojha and Andy Hall

Introduction

How do diverse contributions in this volume speak with each other and how do we draw generalizable insights for improving natural resource governance? Why did we choose 'adaptive collaborative approach' as the main title of the book and why do we suggest the need for 'rethinking participation, learning and innovation' in the title? This chapter maps out the complex terrain of adaptive collaborative approaches, and then identifies some important conceptual issues running through the case study chapters.

We found that reviewing conceptual foundations of adaptive collaborative approaches is crucial in view of the fact that research, practice, policy and innovation processes have remained notoriously disjointed in the practice of natural resource management. This attempt is also oriented to synthesize key insights from different conceptual traditions to provide more holistic, pragmatic and unified theory of learning and change. As such, this chapter provides a conceptual reinforcement to the adaptive and collaborative approach, by forging dialogues, and to some extent 'epistemological alliance', among the thinkers, practitioners, activists and researchers, who at times are disjointed because of different use of language, despite sharing similar concerns for having learning and collaboration in governing socio-ecological systems.

Forging dialogue among different traditions has become particularly critical if we are to generate some generalizable insights. This book brings together multiple terms and concepts that the editors and authors have used to interpret their practice. And all of these are essentially concerned with enhancing learning and collaboration in natural resource governance. Some of these include: innovation systems (Hall *et al.* 2003), deliberative science (Banjade and Ojha 2005; Ojha *et al.* 2010), adaptive collaborative management (Colfer 2005; McDougal *et al.* 2007), monitoring systems (Malla *et al.* 2002; Paudel and Ojha 2007), action research (Edmunds and Wollenberg 2002), and adaptive learning (Agrawal and Ribot 1999). Building on these works, and aiming to further develop and clarify our approach, we did a review of related concepts and approaches in social sciences, and empirical innovations in agriculture and natural resource management fields. Four key traditions of learning and innovation in agriculture and natural resource management (although neither mutually exclusive nor collectively exhaustive) have been identified as relevant – social learning, adaptive management, participatory action research and innovation systems approaches.

In this book, we not only are interested in how systems adapt under natural conditions, but also seek to explore conditions, facilitative processes and policy frameworks that support or enable learning, collaboration and innovation in natural resource governance. This leads us to recognize two systems linked to each other – enabling system (such as policy) and operational system (local resource management system), much in the same way as Agrawal and Ostrom have conceptualized as 'constitutional choice' and 'collective/operational choice' in natural resource based collective action (Agrawal and Ostrom 2001). And as Bradbury and Reason argue, action researchers are there not only to 'implement' policy, but also to 'influence' the policies (Bradbury and Reason 2003). In an ideal situation, it may be expected that the higher order enabling system itself could be following its own adaptive and collaborative cycle of change and innovation. Our intention has been to survey the challenges of applying ACA at multiple scales and functional domains, albeit starting with local level innovations and then moving up to explore how these are related to multi-scale systems of change and innovation.

We have consciously used the term of *approach*, and not strategy, management, co-management or just adaptive management. This is because we would like to explore throughout the book possible ways of thinking about and acting on the natural resource systems as the coupling of social and ecological system, involving power, politics, conflicts as they play out in the process of learning and innovation. We also want to avoid being too methodological and tool-oriented, as this would detach the analysis from wider contextual settings. In the sections to follow, we provide conceptual foundations of adaptive collaborative approach and identify important issues that surround the body of knowledge around adaptive collaborative approaches.

Historical roots

From the historical point of view, we see that a major shift towards adaptive and collaborative approach (ACA) begins with the realization of the limits of science in facilitating social development. Since the European Enlightenment in the sixteenth century, science has emerged as a dominant way of understanding social and physical world. The triumph of science, which was triggered by experimental methods usually employed in the physical world, has lead to 'overscientization' of social and political life (Habermas 1970).

The emulation of natural science methods – which are oriented to prediction and control – by social scientists, along with the colonial expansion of nation-state, have led to technocratic governance, undermining the prospect of collaborative learning and deliberative innovations (Fischer 2000). Due to the inaccessibility and limited relevance of centralized, technocratic regimes of natural resource governance, local level agriculture and natural resource (ANR) actors have shown concerns for the recognition of local and traditional knowledge, and on that condition, developed willingness and commitment to social learning and policy innovation. There is a long tradition of scholarship to explore and promote how local knowledge and science can move together, while also forging collaboration and deliberative processes. Over the past several years, this search has widened and the initial idea of individual and small group focused learning to much wider processes of public policy making and societal innovations (Hall 1993). In such traditions, we see ACA as a potential approach for enabling researchers and practitioners to speak to each other, or at least provide a framework to bridge the communication gap, by allowing both formal and 'objectivist' on the one hand, and practical and 'constructivist' approaches on the other, to learning and institutional change. The promise of ACAs is further animated by growing concerns for learning while acting, and acting while learning.

While we believe that a unified theory of learning and action could be of some help to move forward, it is also important to understand how learning processes have been historically fragmented. In more philosophical discourse on knowledge, three paradigms of knowledge that nurture a society are differentiable (Habermas 1971) – a) technical knowledge (understanding nature), b) practical knowledge (knowledge for social reproduction and socialization) and c) emancipatory knowledge (knowledge for social change). The problem of state-centric, technocratic governance occurred when the technical knowledge – advanced by the development of positivist knowledge and technocratic research institutions – dominated the other two modes of knowing, undermining the prospect of emancipatory knowledge. Indeed, the dominant paradigm of the techno-scientific regime of knowledge and power has emerged and is sustained as 'ideological hegemony' in the Gramscian sense (Gramsci 1990), or doxa in Bourdieu's language (Bourdieu 1977), and 'governmentality' in the language of Michael Foucault (Foucault 2000).

In an adaptive collaborative approach, we are concerned with recovering multiple domains of learning – learning about the natural systems, social systems, socio-ecological systems – as well as learning beyond existing limits of socio-cultural codes. We are also concerned with harnessing the potential of interaction and collaboration among social actors with contesting claims over natural resources or having diverse visions about how things should change. In other words, the exploration of an adaptive collaborative approach to governance is a search for ways to explore ways to fully establish the communicative learning process in the context of the dominant techno-scientific world views in natural resource management (Fischer 1990;

Bäckstrand 2003; Ojha 2006), and also in the context of traditional forms of authority and hierarchy (such as hierarchical social institutions and unaccountable politics in natural resource governance) that inhibit free and open communicative engagement among actors. Our approach is to develop an integrated view of 'complex adaptive systems' in which research is seen as one function among many required to foster governance innovation (Hall *et al.* 2001; Hall and Clark 2010) – movement, resistance, brokering, negotiation and the like. We consider the primacy of the learning and interactive process in social change, rather than end goals determined through expert knowledge, much in the same way that John Dewey emphasized 'cooperative learning' as a strategy for improving democracy (Dewey 1916/1966). Considering these as starting points, we reviewed and analyzed various approaches to learning and collaboration, which are presented in the next section.

Adaptive collaborative approaches: key conceptual traditions

We trace the roots of ACA in four related conceptual traditions that stipulate various modes of learning and cooperation in diverse socio-ecological situations (see Table 2.1 for a summary). We briefly outline specific trajectories through which these approaches emerged, and explore shifting the contours of knowledge in relation to learning, social cooperation and innovation.

Social learning

Some of the early attempts to look into the questions of human learning were in the field of psychology, and then the research gradually moved to other fields of knowledge and practices such as organization science and community studies, network analysis, and to the context of natural resource management, research, practice and policy. Until the 1960s, psychologists emphasized human learning as the property of individual cognition and behaviour. But after the seminal work of Bandura (Bandura 1969), the social interactional dimension of human learning became well accepted. He coined social learning to emphasize how human agents learn through social engagement, claiming that human learning is not confined to the property of individual cognition and behaviour. This idea had a profound influence as it moved learning focus from individual research to social domains.

The social learning paradigm was paralleled with works that considered organizations and communities as units of learning, problem-solving and innovation, extending the critique of the assumption that held individuals at the centre. In the field of management, organizations began to be seen as the key blocks of social learning, and the ideas of a learning organization, organizational learning emerged (Argyris and Schon 1978; Senge 1990). These works discovered that both opportunities and constraints to learning

Table 2.1 Overview of various adaptive and collaborative approaches

ACA→ aspects	*Social learning*	*Adaptive management*	*Participatory action research*	*Innovation systems*
Emergence – time	1970s	1970s	1940s/1970s	1980s
Triggering factors for the emergence of the approach	Critique of individual focused, cognitive and behavioural theories of learning (Bandura).	Lack of monitoring resource management and problems of uncertainty, and need for experimental knowledge.	Lack of action orientation of science (Lewin), and development research and intervention (Fals-Borda).	Technology-driven agricultural development practices (Biggs and Clay 1981).
Major shifts	Organizational learning, action science, communicative action.	Co-management, collaborative management, collaborative inquiry, cross-scale dynamics, marriage with social learning, adaptive governance.	PRA to PLA, users to citizens, research and social movement, expert and democracy.	Innovation capacity, learning processes, institutional embedding and cross-system interactions.
Role of research	Researchers as co-learners in the social systems and networks.	Researchers set up monitoring plans, reflection and sharing.	Research agenda defined locally and put into use.	Technical research as part of wider innovation system, studies into innovation systems, not technology generation and adoption, relocate research in social and policy domain.
Key issues and gaps	Still limited consideration of politics, focus on normative and less pedagogic; silent on structural constraints to learning.	Confusing notion of 'management', little exploration of adaptation dynamics; limited treatment of conflicts.	Suffers from localism, limited replicability, limited policy link.	Focus on economic change, links with politics poor, limited conceptualization of the agency of change and innovation.
ANR sectors applied	Forest, biodiversity, water, NRM.	Forest, biodiversity, water, fisheries.	Agriculture, forest.	Agriculture.

and change lie at the level of organizations and communities rather than at the domain of individual life. The idea has inspired a wide range of organizational practices, including public learning processes (Schon 2010).

Extending the learning dimension beyond social organizations, the concept of 'communities of practice' (Wenger 2000) has also been widely appreciated. The argument is that the success of organizations depends more on 'communities of practice' connecting peoples and groups across a wide social domain, beyond the formal boundaries of the organizations. Such informal communities are seen not as peripheral elements but as the key crucial building blocks of social learning systems. This idea is more in favour of informal networks as the domain of social learning than the formal practices and processes taking place within organizations. Wenger makes a convincing case for this extrapolating the evolutionary history of humans – human beings have formed communities to share cultural practices and reflect their collective learning as a strategy for survival and development (Wenger 2000: 229). According to this view, communities have at least three shared elements (Wenger 2000): joint enterprise, mutuality and shared repertoire of communal resources – language, routines, sensibilities, artifacts, tools, stories, and style, all of which shape, influence and determine the nature and type of learning at individual, group and societal levels.

As the debate moves towards the social primacy of learning, a view challenging individual-social dichotomy is also on the rise inspired by Dewean (Dewey 1916/1966) synthesis of the social and individual. This approach proposes a third way wherein individual and the group are seen as being in the constant process of creation through experience and reflection (Elkjaer 2004), implying the need for focusing attention on the relations between individual and the group processes of learning.

There is also an argument to differentiate social learning as per the depth and level of change involved. Transformative learning theory emphasizes learning at the level of underlying assumptions and worldviews (Mezirow 1991, 1997). This conception resonates with 'double loop' and 'triple loop' learning concepts espoused by Argyris (Argyris and Schon, 1978) – which implies learning beyond the operational about the guiding assumptions and regime. At the core of transformative learning theory is the process of 'perspective transformation', with three dimensions: psychological (changes in understanding of the self), convictional (revision of belief systems), and behavioural (changes in lifestyle) (Clark and Wilson 1991).

In planning and organizational management fields, there is a preference for explicit, codified knowledge. But research shows that knowledge in group practice and as part of the interaction of the social world cannot always be reduced to knowledge that is generated as an outcome (Cook and Brown 1999). Existence of such knowledge was famously articulated by Polanyi (Polanyi 1983) as 'tacit knowledge', and as 'doxa' by Bourdieu (Bourdieu 1989). From this perspective, scholars have argued that the process of knowing is an important aspect of organizational development, which can be lost if there is an emphasis on knowledge as an outcome (Cook and Brown, 1999). Going beyond the rational choice models (which consider learning through bargaining and negotiation among the autonomous individuals), social learning perspective emerged as an approach to harness the communicative frontier among human agents (Rist *et al.* 2006).

In natural resource governance, it is seen as the process of framing issues, analyzing alternatives and debating choices, which enables constituencies to reflect on their own and others' values, orientations and priorities in the context of inclusive deliberation (Daniels and Walker 1996). Scientific advice itself is seen as a part of the social learning process (Gottschick 2008) with a reorientation of the role of expert to stimulate and clarify the debate rather than provide normative prescriptions (Fischer 2004).

While social learning perspectives have informed how learning and collaboration emerge in social systems, and also offered concrete conceptual elements for analysis and action such as open communication, diverse participation, unrestricted thinking, conflict bounding, multiple sources of knowledge, extended engagement and facilitation; an emphasis on learning sometimes misses the other modes of change (Schusler *et al.* 2003). In many situations, changes and innovations occur not through learning but through drastic political and economic transformation or crisis, and often these chaotic perverse situations do not appreciate and encourage learning based approaches. Bourdieu's (Bourdieu 1989) conception of social agent – as culturally inscribed and operating in structural harmony with the social systems (conceptualized as 'social fields') – also prompts us to see how crisis and dissonance parallel with learning and collaborative processes. In such contexts, we hold that frameworks of adaptive learning and collaboration should be analytically linked with the theory of lack of learning and non-cooperation or conflicts.

In the context of natural resource management, Lee (1993) linked the idea of social learning, combining the strategies of an experimental approach (adaptive management) with appreciation of the need to understand politics among actors. The work of Röling (Röling 2002) also challenged individually based cognitive models of learning and negotiation. More recently, the idea of social learning has been applied in the context of natural resource management (Colfer 2005; Fisher *et al.* 2007). We refer back to these while elaborating adaptive management traditions in the next section.

Adaptive management and adaptive collaborative management

Adaptive management approach emerged in response to the recognition of complexity in natural resource systems and the appreciation of the need to take an experimental approach in the management of natural resource systems (Holling *et al.* 1978; Walters 1986; Lee 1993; Gunderson 1999). It was initially applied in the context of river basin and natural ecosystem management in the United States and Canada, but then was extended to other parts of the world. In its original version, it represents the scientists' efforts to manage ecosystems in an adaptive way, by testing management assumptions through monitoring plans. They criticize contemporary resource and environmental management and associated policies, including economic instruments and incentives, as being preoccupied with the conception of steady-state views and assumptions, which reduces the capacity to deal with the change (Holling 1973; Folke 2006). At the fundamental level, adaptiveness is seen to emerge from the inherent properties of resilience in ecosystem and flexibility in social system (Gunderson 1999). However, the later approach marked a significant switch towards adaptive and collaborative dimensions – addressing how multiple interests and stakes can be reconciled in resource management, not just those of the scientists (Brunner *et al.* 2005). By now, the approach has gained sufficient maturity, and is often presented as a tool to frame the philosophical, methodological and practical challenges associated with the management of natural resources (Berkes *et al.* 2000; Armitage *et al.* 2007; Armitage *et al.* 2009; Berkes 2009).

In a broad sense, adaptive management embraces active learning, accepting surprises, embracing uncertainty, and continually modifying the management in the light of learning to enhance management effectiveness.

With the work of Lee (1993), adaptive management was more directly linked to the process of social learning around resource management. He used the metaphor of compass and gyroscope to emphasize the interactive process between scientific analysis and civic participation in adaptive management. Lee (1993: 9) has famously described adaptive management as 'An approach to natural resource policy that embodies a simple imperative: policies are experiments; learn from them'. He stipulated that being adaptive in management means: taking uncertainty seriously, treating human interventions in natural ecosystems as experimental probes, taking special care in collecting and analyzing information so that expectations can be compared with actuality, and finally, using the learning to correct errors, and change action and plans. He sees adaptive management as 'a compass for us to use in searching for a sustainable future'.

More recent works on adaptive management have further explicated the dimension of *collaboration* among actors and the links between social and ecological systems. Proponents of adaptive co-management argue that the approach has no closer alternative, given the fact that social and ecological systems are deeply interconnected and co-evolving across spatial and temporal scales (Folke *et al.* 2007). They hold that levels of organization and decisions in one place inevitably affect people elsewhere (Gunderson and Holing 2002). This is being conceptualized as cross-scale dynamics (Levin 1992), wherein the patterns at the macroscopic level are shaped by, and in turn affect, microscopic factors. New insights on cross-scale dynamics are coming from other research as well (Brown and Laland 2003) in which adaptive co-management is being seen more as problem-solving process, involving extensive deliberation, negotiation and joint learning (Carlsson and Berkes 2005).

As the idea of adaptive management began to spread through the developing country context, the dimensions of co-management and collaboration have become even more crucial (Colfer 2005). A particular strand that emerged in this context is the adaptive collaborative management (ACM) developed by the Center for International Forestry Research (CIFOR) in the context of forest management (Colfer 2005; Fisher *et al.* 2007) (applied and tested over 10 countries in Asia, Africa and Latin America (see Chapters 3, 6 and 7, this volume)). CIFOR ACM researchers concluded that ACM has three anchors: communication and creation of a vision, social learning and joint action (Prabhu *et al.* 2007). They further elaborate the three anchors into several operational requirements of an ACM approach – including the involvement of stakeholders in decision-making and implementing actions, communicating and sharing learning and skills, conflicting management, planning based on human-ecological systems analysis, and an element of visioning and forward-looking beyond the current state.

Having worked for about a decade on ACM research, and putting it into more complex field situations, the CIFOR research resonated with the Habermasian concept of communicative rationality as being vital in the ACM process (Prabhu *et al.* 2007: 18):

> Communication is critical in ACM, not only for enabling the shared vision to emerge, but also because effective communication is the foundation for creating a whole that is greater than the sum of its parts. Effective communication enables diverse actors to share – and ultimately negotiate and create synergies from – their world-views, goals, values and knowledge.

In a slightly different version, Colfer (Colfer 2008) outlines an adaptive and collaborative management that has three key elements (see Chapter 3 for details on how these aspects worked in her own ACM work with CIFOR): a) horizontal theme in which stakeholders interact and negotiate goals around a resource management context; b) a vertical theme in which local communities and actors at other scales develop mechanisms for two way communication, cooperation and conflict resolution; and c) an 'iterative' or progressive theme wherein stakeholders learn, over time, about the management of their resources and their communities, in the course of actions evolving out of that growing understanding.

As the experience with ACM grows, especially in the complex forested landscapes in the tropical regions, there is added sensibility among those advancing ACM strategies. Wollenberg *et al.* (2007), through an experience of facilitating ACM in forest management in Malinau, East Kalimantan of Indonesia, argue that it is not always easy to apply ACM, because of the weak, uncertain institutional setting and complex and shifting political landscape, which made formal cooperation among the local actors problematic. They report the need for more flexible and spontaneous strategies of cooperation and engagement. They further argue that, 'ironically, ACM is intended to cope with complex external contexts, but has not addressed how to cope with complexity internal to decision-making and management'. They report that their capacity to be flexible in their facilitation strategy was the key to working with spontaneous cooperation. They made adjustments by 'muddling through'. This resonates with the reflections of ACM researchers who worked in the turbulent period of Zimbabwe during the years before and after the onset of the new millennium: 'the journey is far more important than the destiny' to find the way into the complex, chaotic, uncertain socio-ecological world. As Mandondo and Matose (Mandondo and Matose 2008) remark at the end of the book on Zimbabwean ACM:

> Pre-determined technical solutions are unlikely to succeed if they are mere prescriptions, without room for adaptation . . . it is not just the solution developed outside the place where it is applied that undermines its potential for success; rather . . . prescriptive approach precludes two-way communication. In many ways, ACM is simply about focusing more sharply and effectively on fundamental human processes of communication, observation, learning and collaboration, and accepting at the same time that failures and surprises are the bumps that will be found along any road.

Participatory action research

The third strand of the adaptive collaborative approach we selected is participatory action research (PAR). A range of initiatives aimed at reforming traditional research and extension model of natural resource management is what is now variously known as action research, participatory research and participation action research. While there is a great variation in the institutional contexts in which they emerged, knowledge imposed from outside is much less relevant for change in the local domains. All of them share a key concern that at times, this leads to an extension of external control. So, from this perspective, what is required is the creation of knowledge through the dialogue and cooperation between researchers, facilitators and the local communities or beneficiaries. Of different strands within this family of approach, participatory action research is more encompassing in scope, as it originated from worldwide practices and diverse fields of knowledge – adult education, agrarian reform and agricultural technology (McIntyre 2008).

In social science, the conception of 'action orientation' goes back to the work of Lewin (Lewin 1946) who stipulated that researchers should try to change the system as well as generate knowledge. The value of action-orientation in research has been appreciated not only in terms of its contribution to action, but also in enhancing the epistemological quality of knowledge itself (Whyte 1989). Social science has now a well-accepted tradition of action-oriented research (Small 1995; Small and Uttal 2005). In the field of development, Robert Chambers (Chambers 1994) popularized the approach to participatory research, initially through rural rapid appraisals (RRA) and participatory rural appraisals (PRA), which was later improvised in the form of participatory learning and action (PLA).

A more critically engaged approach is participatory action research (Fals-Borda and Rahman 1991). PAR embraces various key tenets (McIntyre 2008: 1): a) a collective commitment to investigate an issue or problem, b) a desire to engage in self- and collective reflection to gain clarity about the issue under investigation, c) a joint decision to engage in individual and/or collective action that leads to a useful solution that benefits the people involved, and d) the building of alliances between researchers and participants in the planning, implementation and dissemination of the research process. These are achieved through a cyclical process of exploration, knowledge construction and action at different moments through the research process (McIntyre 2008). As Bradbury and Reason define, action research is a participatory, democratic process concerned with developing practical knowing in the pursuit of worthwhile human purposes, grounded in a participatory worldview. It seeks to reconnect action and reflection, theory and practice, in participation with others, in the pursuit of practical solutions to issues of pressing concern to people. More generally it grows out of a concern for the flourishing of individual persons and their communities (Reason and Bradbury 2001: 1).

Fals-Borda (Fals-Borda 1987: 329) elaborates the tenets of PAR thus: 'Interest in PAR has grown worldwide duc to its pertinence to the initiation and promotion of radical changes at the grassroots level where unsolved economic, political and social problems have been accumulating a dangerous potential'. He argues that PAR combines scientific research, adult education and potential action.

Learning to interact and organize is a crucial aspect in PAR process. Fals-Borda elaborates this aspect thus (1987: 332):

> through actual experience of something we intuitively apprehend its essence, we feel, enjoy and understand it as reality, and we thereby place our own being in a wider, more fulfilling context. In PAR such an experience is complemented by another one – that of authentic commitment.

The political stance taken by the practitioners of PAR is confirmed through wider review of literature as well. While reviewing thc commonalities of diverse participatory research, Cornwall and Jewkes (1995: 1671) found that 'practitioners take explicitly political stance, focusing on empowering disenfranchised and marginalized groups to take action to transform their lives'. Recent work has drawn on feminist research and critical theory to further develop the theoretical basis of PAR (Maguire 2001; Cornwall 2003).

PAR distinguishes two kinds of change agents: those from within exploited groups and researchers from outside them. Recognizing the power relations within which the activity of research is located, practitioners of PAR work towards a process whereby the conventional subject/object relation is confronted. Different actors, each with their own knowledge, techniques and experiences, work together in dialectical process, through which new forms of knowledge are produced. An emphasis is placed on people's history, within which people can locate themselves and their experiences, and on indigenous conceptualizations. Local people are involved in a process through which they are empowered to take charge of the research process and to organize to implement potential solutions or to take action on concerns. Ideally, through this process the initial agents of change 'become redundant . . . that is, the transformation process continues without the physical presence of external agents, animators and cadres' (Cornwall and Jewkes 1995: 1617).

The rich experience and repertoire of insights that exist around PAR bring together a set of assumptions underlying 'new paradigm' science and in contrast to those of traditional or 'old paradigm' science (Wadsworth 1998). The PAR approach comes in sharp contrast to the notion of positivism that arose to bring certainty and verifiability to a world with enormous complexity (Wadsworth 1998). If conventional science wanted to give a group of people the power to determine 'truth' for and on behalf of others, the new science arose from a world of multiple and competing versions of truth and reality as a way of assisting people both come to the truth of their own reality, and also to embrace that of others (Wadsworth 1998). Wadsworth (1998) further argues:

> The old adage about science being 5% inspiration and 95% perspiration may hold true enough! The hard sweat and toil comprises the long hours of talking and thinking and sharing the results of our 'fieldwork' with one another. The moment of inspired thinking is when collective values are expressed in a new way of connecting ideas or a new way of 'naming' the world, that advances the collective situation of participants.

Despite these elegant claims, PAR is not free from criticisms. Critiques find participatory research as being too concerned with the here and now, and less political and theoretical (Kapoor 2002), suffering from the problem of 'localism' (Hickey and Mohan 2005). While the dimension of research has paid significant attention in PAR, the issue of participation is less linked. While multiple levels, such as contractual, consultative, collaborative or collegiate participation are identified in PAR (Cornwall and Jewkes, 1995) there is now an emerging debate towards relocating participation from the paradigm of 'users and choosers' to effective citizenship through which people can exercise their agency to define visions and strategies of change, without necessarily being constrained by the logic of market or the prescriptive boundaries espoused by the experts (Cornwall and Gaventa 2000).

Innovation systems frameworks

The innovation systems concept emerged in the field of industrial development in the 1970s and 1980s, primarily around the national policy context in the West (Sharif 2006). It was a response to the debate over the last 40 years about whether scientific research should drive innovation or there has to be more interactive process between science and many other actors and processes in the economy. Besides, several other theoretical developments contributed to the popularization of innovation systems perspectives (Hall *et al.* 2006: 15) such as new growth theory stressing the importance of increasing returns to knowledge accumulation; increasing consensus that knowledge generation is an evolutionary process involving the interaction between research and non-research actors; and social capital theories revealing the possibilities of reducing transaction costs in the economy.

Recent empirical work has extended the application of innovation systems perspectives. It is now being used to study not only national innovation systems in industrialized country manufacturing but also developing country agriculture.

At the most basic level, the concept of innovation draws attention to (Hall *et al.* 2006): 1) the totality of actors needed for innovation and growth, 2) consolidating the role of the private sector and the importance of interactions within a sector, and 3) emphasis on the outcomes of technology and knowledge generation and adoption rather than the strengthening of research system and their outputs. As Spielman *et al.* (2009: 399) argue, 'an innovation system framework shifts the analytical emphasis from a conventional linear model of knowledge and technology transfers (from researcher to extension agent to farmer) to a more complex, process-based systems approach'.

The popularity of innovation systems perspectives on agricultural research and technological change in the developing countries is growing. This is because of both application of concepts in understanding and facilitating agricultural development, as well as serious attention and endeavours of scholars to promote the approach as a corrective measure to the failure of conventional research and extension driven approach to agricultural development (Biggs and Clay 1981). The perspective also challenges earlier claims that technological change drives social and economic development, suggesting instead that technological change is embedded in institutional contexts of learning, power, trust and interactions among various subsystems (Hall *et al.* 1998; Hall *et al.* 2003).

According to Spielman *et al.* (2009), it was Biggs and Clay (1981) and Biggs (1990) who pioneered the approach by introducing several key concepts of institutional learning and change, and the relationship between innovation and the institutional milieu in which innovation occurs. This has become central to later innovation systems studies on developing country agriculture. Hall *et al.* (2002) provide an in-depth study of the institutional and organizational learning processes and hence offer empirical foundation of how innovation systems perspectives can help understand socio-economic change. In all these studies and reflections, proponents claim that the shift in the emphasis from a unidirectional technology transfer approach to a more complex, process based systems approach is appropriate for the study of developing country agricultural development and natural resources management because it can help policymakers, researchers, research managers, donors, entrepreneurs and others identify and analyze new ways to encourage innovation. These approaches can work even with the data limitations in developing countries, and can go a long way in fostering development and reducing poverty in developing country agriculture. Hall and Clark (2010) argue that innovation systems thinking is a major epistemological departure from the traditional, neoclassical studies of technological change that are often used in research driven by current national agricultural research systems in the developing world.

Since agricultural innovation involves organisations beyond the public sector research, and innovation emerges from the interface between research and practice, partnership really matters in innovation development (Hall *et al.* 2001). This approach has inspired and reframed the whole way of doing development from defining problem to doing research to facilitating change. This is exemplified in the conclusion of Hall *et al.* (2007) that fodder scarcity in India and many other regions is indeed a problem of innovation capacity scarcity, and not the problem of lack of technology or individual production capacity of a farmer. This conclusion implies a radical shift in development research and capacity building interventions.

While there is a resurgence of local knowledge movement in the development, proponents of innovation systems approach claim that an effective strategy should accord primacy neither to local knowledge nor to technology, but to the system itself. As Hall and Clark (2010: 320) assert: 'Our account does not give primacy to technology as the driver of change, nor does it give primacy to the knowledge of farmers; rather, while recognizing the importance of these it gives primacy to the capacity to respond to changing circumstance through adaptation and innovation.'

Innovation systems thinking has opened up new ways of thinking about public policy in development. As Hall and Clark (2010: 322) argue:

> The challenge for public policy is therefore two-fold. Firstly, to find ways to be alert to emerging innovation practices that, by definition, are invisible to most mainstream thinking and sources of information that

policy draws upon. Secondly, to find ways to provide the nurturing environment that can move these new products and services and new innovation capacities from the margins to the mainstream and in doing so accelerate the learning process through which innovation capacity is enhanced. An implication of this is that public policy needs to shift from an orchestration role in which it sets the conditions from which innovation will emerge, to a more pro-active role where it supports new patterns of innovation behavior.

Pulling diverse threads together

Looking at the rich traditions of the four approaches to learning and innovation we reviewed above, the following important ground-setting conclusions emerge. First, traditional social science is facing ethical, epistemological and political dilemmas in the context of its failure to address some of the critical natural resource governance challenges. If social science is to be brought into the service of society, it has to take some kind of action orientation. This is important not only from the *relevance* point of view, but also from the *epistemic quality* point of view – for better understanding the reality beyond the positivist conception of the truth. What we need is not still another theory of truth, but a theory of learning that helps to improve the sustainability and performance of the socio-ecological system. Second, social actors have a huge potential to engage in a more collaborative process than the standard practice. Not all natural resource governance situations are zero-sum games, and actors in natural resource governance are likely to change their positions and preferences through collaborative learning processes.

Third, looking at the ways socio-ecological systems are experiencing change, some of which are irreversible, we cannot simply wait for a system to change itself in an evolutionary way in response to selection pressures that built up naturally. Interventions could take a wide variety of forms, including research itself, policy, entrepreneurship, actor interfacing coalitions and the like. Fourth, learning and collaboration processes are inherently inter-contextual. Learning from some contexts will offer insights into others though not necessarily in the form of prescription. Translating lessons of one context into reflective insights in another requires a different paradigm of action and learning – for which there is already such a huge repertoire of insights from around the four approaches reviewed, and beyond.

Fifth, the cross-scale nature of natural resource governance means that neither localism nor globalism alone can address the problems of natural resource governance. We need to ground action and learning processes at the local context, but also engage with global discourses that produce or frame local knowledge systems and power relations. Sixth, power matters in many diverse ways, including in the very process of transformation and empowerment. Processes of empowerment should be linked with constructing social realities. Seventh, the ecological dimension needs to be better integrated with the social; we need to explore ways to understand and facilitate how socio-ecological interactions take place at multiple scales.

In view of these broad conclusions, what exactly does an ACA comprise and what should we consider while examining an ACA? We now turn to elaborating key components of an ACA which would help pool useful conceptual resources from the broad review done in the previous section. More importantly, this would also allow us to pose continuing questions and delve into outstanding puzzles relating to employing learning and collaborative strategies in practice. In particular we synthesize the above mentioned approaches to distil key insights in relation to: a) underlying assumptions upon which an adaptive collaborative approach is construed and applied, b) the nature and diversity of immediate innovation outcomes anticipated by an ACA, c) the general programme of actions and interventions, and d) general strategies for production and use of knowledge in the process of natural resource management and governance. An approach to management and governance cannot be formulated without assumptions, and it should also indicate a general programme of action. Since one of the central problems has been about linking research and practice, we specifically distil how an ACA addresses this. Likewise, based on the assumptions of how change is triggered, as we also identify key outcomes targeted by an ACA.

Assumptions of ACA

The review of the four families of ACA approaches reveals that ACA researchers have certain commonly held assumptions about the world, processes and the prospect of change. These are summarized as follows:

- ANR management and change are complex system phenomenon, where linear conceptions of input and output do not work.
- Social reality is historically constructed (following critical theory traditions), and the learning process should be oriented more towards renegotiating the existing truth than looking for a new, ultimate truth.
- Social and ecological systems are not separate but are coupled, and hence understanding the links between the two is critical for any management interventions to succeed.
- Any process of change has to confront multi-level dynamics, and power relations and conflicts are inevitably linked to the processes of learning and collaboration in any natural resource governance situation.
- Local communities around natural resources have strong agency but may be unaware of the constraints to freedom – especially drawing on PAR (e.g. Ozanne and Saatcioglu 2008). Critical pedagogic tools and critical dialogues have the potential to discover useful knowledge and transform perceptions and attitudes (Freire 1993; Fals-Borda 1987) which are at times considered barriers to innovation (Hall *et al.* 2006: ix).

- An emphasis on experiential roots of knowledge, and knowledge actually derived from trying to change the system, has greater likelihood of contributing to change.
- Learning is related to interactions between people, and between people and the natural world, not just when scientists measure the natural world through scientific equipment and techniques.

Outcomes of ACA

The review of various ACA strategies presented above suggests that immediate outcomes are diverse and varied, including:

- improved innovation capacity of the resource organisations and actors, including a greater ability to identify emerging innovation practices and enable their adaptation;
- behavioural changes of actors in relation to learning processes, including increased appreciation of conscious learning, an attitude that seeks surprises and learns from failures and an approach to learning as an intrinsic process of human life (lifelong learning);
- enhancement of reflexivity among actors and organizations;
- institutional change and economic efficiency. These improvements could be technical, managerial, institutional or of a policy nature;
- empowerment of local actors and disadvantaged groups and improved participation in the process of decision-making and collaborative actions;
- changes in technological aspects around particular economic opportunities and resource management practices.

ACA based actions

General principles of ACA-oriented action and interventions include:

- Innovation can comprise radical improvements, but usually consists of many small improvements and a continuous process of upgrading (Hall *et al.* 2006: 15).
- The strategy of muddling through is inevitable in complex and uncertain situations (Diaw *et al.* 2008: 468).
- There should be a variety of possible interventions: initiating, experimental, interventions to build on or nurture emerging opportunities, remedial, maintenance, etc.
- There are important roles for a broad spectrum of actors outside government, and the actors' relative importance changes during the innovation process.
- Some approaches such as PAR prefer to start with a local problem or issue and at the community level, and then build up the process of change through higher levels.

- There should be an emphasis on the importance of facilitating iterative feedback loops between research, local experience and policy adjustments as a means of strengthening positive interactions between cooperation and adaptation (Diaw *et al.* 2008: 470).
- Policy itself should be considered as an experiment to be framed under the learning and action approach.
- The concept of 'bounded conflicts' that recognizes conflict as an essential part of learning if properly bounded and put within the limit should be embraced.
- While an ACA recognizes that the interventionist has to be clear about goals, it is PAR that explicitly advocates for PAR practitioners to align themselves with poor and disadvantaged groups while producing knowledge and action.

Production and application of new knowledge through ACA

An ACA also has specific strategies for production and application of knowledge in the process of resource governance and management:

- applying an experimental approach to learning and management;
- emphasizing communities of practice, networking and communication as major strategies of producing knowledge;
- using the questions of local people and those to be empowered to guide the research process;
- focusing not on specific knowledge or technology but the systems of interaction among actors who share a common system of resource management and production value chain;
- failure is also recognized as a source of learning while experimenting;
- formal research is only one among many functions of innovation, and it is often other functions rather than research that kicks off change;
- ACA scholars follow 'engaged scholarship' or action-oriented social science;
- increasing focus on tacit knowledge as a worthwhile resource for learning and innovation.

Despite their origins in different institutional contexts, the four traditions of learning and collaboration demonstrate a number of similarities. First, all these approaches appear to be challenging some orthodox conventional counterparts. Innovation systems approaches aim to challenge linear technology transfer models that are dominant in the agricultural development sector. Adaptive management approaches challenge linear, science-driven models of resource management. Social learning approaches challenge cognitive, individually focused, organizationally bounded approaches to learning. Participatory action research approaches aim to challenge extractive science and lack of empowerment elements. Second, all four approaches go beyond epistemological positivism, and seek communicative engagement, but vary with respect to defining the levels of actions/ interventions in knowledge making.

In all the approaches reviewed above, there is a broad distinction between insider and outsider interactions in the process of learning and action. PAR also believes in local control of the process, without any continuity of support from outside, at the later stages of the cycle. In situations of conflict, PAR advocates that the external facilitator/researcher take the side of the exploited could generate controversy over the actions of outside facilitators and could lose the benefit of political neutrality. Other approaches remain neutral at the level of action, and focus on the system or policy level.

In every human institution, knowledge and power are inextricably linked. Thus, the treatment of knowledge in isolation gives an incomplete view of learning and of the innovation system (Ojha *at al.* 2008). Power is grounded in diverse dimensions of social class – such as economic assets, symbolic capital (such as social status), gender and ethnicity – to varying degrees. The knowledge–power nexus is enacted, contested and resisted in day-to-day governance practices. Several of the reported tensions in innovation processes – such as scientific versus indigenous/lay knowledge, theoretical versus practical knowledge – are actually a result of underlying power relations among the social agents.

The approaches also leave some remarkable conceptual gaps. They are not explicit in providing a possible typology or taxonomy of actors, and do not consider the possibility of conflicts and collaboration among them in various action situations. Adaptive management presupposes experts or scientists doing the management, while ACM engages all stakeholders to be engaged. But this could also dilute the political stakes of right holders – the actors are structurally and politically differentiated in the field of power, and the question arises as to how the approach should actually enable these differentiated actors to participate in the process such that equity is realized. Overall, adaptive collaborative approaches are confronting the following critical remarks in the wider literature:

- There is also a danger of localism when acting locally (Hickey and Mohan 2005) – especially in PAR.
- Local micropolitics may affect knowledge generation – the elite may benefit disproportionately.
- Not all behavioural change results from social learning; some results from crisis and structural constraints.
- Consensus and shared learning may not be feasible and even if they are, they may not lead to progress, as we need competition and conflict for change (Muro and Jeffrey 2008). Penalties or incentives might prove better than social learning (ibid).

- There can be an overt focus on systems at the expense of the agency of change.
- There can be little consideration of issues related to rights, power and inequality.
- There is a tendency to emphasise on interventionist and research functions and less exploration of the theory of spontaneous change.

A reflection on case study chapters

Given the complex terrain of adaptive collaborative approaches, what new contributions can the case study chapters make? While we aim to synthesize actual lessons in the conclusion chapter, here we briefly reflect on various conceptual threads running through the chapters. The contributors together address several important issues in relation to conceptualizing and applying adaptive collaborative approaches in the context of natural resource governance. In particular, they identify and map out how collaborative learning engages the professional community as well as organizations, sponsors, beneficiaries and the accountability, national policy and bureaucratic environment. The chapters will throw new light on the following key aspects of ACA in relation to natural resource governance:

- How do the facilitating agency and local agents of change interact? (Chapters 5, 6, 7)
- What are the emerging frontiers of research and action linkages? (Chapters 3, 8)
- How to deal with conventional decision-makers in the agency promoting ACA? (Chapters 3, 4, 6, 7)
- What institutional and entrepreneurial innovations emerge through ACA – nature, types and pathways? (Chapters 4, 5, 6, 7)
- What should be the key functions of a facilitative agent – brokering, evidencing, services, empowerment? (Chapter 4)
- How can ACA practitioners become accountable from the point of view of both contributing to practice and generating evidence and knowledge for wider sharing? (Chapters 4, 6, 7)
- What are the strategies and lessons for self- and collaborative monitoring, new ways of monitoring and testing assumptions? (Chapters 3)
- How do ACA collaborators deal with power at different levels? (Chapters 3, 4, 6)
- What pushes the shift from a learning package to a technological package in the course of expanding ACA? (Chapter 4)

Reporting the experiences from such diverse contexts as floodplains management (Chapter 5), community forestry and forest governance (Chapters 3, 6 and 7), agricultural technology development through farmer participation (Chapter 4), and watershed management (Chapter 8), the contributions review experiences of applying ACA towards different types of anticipated outcomes – improving equitable access to resources, enhancing productivity of the resource system, and contributing to resource sustainability. More immediate outcomes targeted include: improved monitoring systems, capacity building, networking, changes in management practices, and facilitating reflective and deliberative processes.

The chapters present diverse domains of engagement and contestations as part of ACA engagement. This includes engagement with multiple epistemic communities (such as contestations between ACA researchers and biophysical scientists described in Chapter 3), donors and sponsors of projects (Chapters 3, 4, 5 and 6), local communities (Chapters 4, 6 and 7), national governments (Chapters 3 and 6), private sector (Chapters 4 and 7), academic institutes (Chapters 3 and 7), and training and facilitating organizations (Chapters 3 and 7).

Chapters also span a range of evolutionary pathways through which ACA is developed, regenerated or expanded over time and space. The case of scaling up Farmer Field Schools (Chapter 4) has demonstrated how a successful innovation in one context (and time) has to go through the process of regeneration or modification when moved to or put into use in other contexts. With better participation of farmers, as the authors show, the technology or innovation will undergo significant changes and modifications, beyond the technical conceptualization of the experts. But it is not a straightforward and linear process; it involves complex trajectories of conflicts, collaboration and learning. Enabling adaptive learning networks among floodplain communities through multi-year efforts shows that community empowerment is critical for multi-actor engagement especially with the government and policymakers (Chapter 5). Political complexities have also had a major influence on the trajectories and pathways of ACA in natural resource governance (Chapters 6 and 7). The contributions also include observations and reflections from almost three decades of action learning practice, pooling surprising parallels and continuities of forces that tend to undermine the use of the ACM approach across time and space, while contributing some insights for those who wish to promote the approach (Chapter 8).

Contributors tackle a broad range of common issues while facilitating an ACA in ANR contexts – ranging from expert domination and sponsorship of learning to empowerment. A particular issue has been in relation to how expert-led and people-led knowledge compete and converge in an ANR context. Linked to this is the key question of exploring the conditions for the emergence of adaptive learning approaches, and then their potential to reconcile contesting knowledge and political claims around resource governance practices. Temporal dynamics of learning and collaboration have been particularly revealing (Chapters 7 and 8).

Contributions have brought new insights into the severity of the challenges that are inevitable in moving an ACA from the local to the national level especially when trying to link policy system with the local action and learning processes (Chapters 6 and 7). Contributors also discuss the process and challenges faced while attempting to facilitate local and community-based actors to incorporate more system-wide considerations of ecological systems (such as in the case of integrated floodplains management in Bangladesh, Chapter 5).

Interestingly, contributors have made contradictory claims about the prospects of methodological toolkits in facilitating adaptive collaborative approaches. Some report that methodological strategies and tools have been instrumental in empowering marginalized groups, forging multi-stakeholder collaborations to enhance the effectiveness of forest management systems, and to strengthen the links between operational processes and resource management and policy learning cycles (Chapters 6 and 7). These attempts sought to facilitate the widespread use of self-improving and equitable forest resource management systems that build on local capacity and 'vertical' and 'horizontal' stakeholder interactions, while responding positively to external pressures. But the case of FFS in Ecuador does not support the efficacy of methodology-based interventions (Chapter 4). Clearly, there is a need to further explore the appropriate mix of ACA actions at different scales, and also in different contexts.

Concluding remarks

This chapter has mapped out the wider conceptual terrain of adaptive and collaborative approaches to natural resource governance. We identified and pooled four key approaches and strategies related to learning, collaboration and innovation that have emerged either in social sciences or in the actual practice of natural resource governance. Despite using different language and having diverse contexts of origin, these approaches imply a clear need for radical reframing of ongoing research practices, policy processes and advocacy coalitions around ANR governance, if we are to bring more effective changes in improving human well-being and environmental sustainability.

A simple assumption on which ACA rests is that there is no predefined final solution to fix the complex problems of natural resource governance. There is a need to frame governance as a learning system, and to look for strategies to address challenges and constraints that keep actors from being collaborative and adaptive while governing natural resource systems. There is also a need for an innovative combination of research, action, learning, movement and policy beyond the traditional disciplinary focus. This means going beyond methodologies of participation and science-informed innovation, to look for more substantive and multi-scale engagement among actors with diverse roots of knowledge and power. This will also imply a need for innovative bridging language and mechanisms through which practitioners and researchers can better communicate with each other as they engage in shared learning and more active innovation processes. Likewise, promoters of ACA have to confront preexisting forms of knowledge as well as power differences that prevent meaningful collaborative learning. In such contexts, they have to confront complex social systems and realities of power relations, and hence should emphasize both a normative framework of change as well as an analytical framework that provides critical reflections and empowerment opportunities to those occupying disadvantaged positions in the innovation system.

Yet, the current body of knowledge around ACAs is ambiguous on several fronts. First, dilemmas persist on how to integrate knowledge and action at different scales. Some rely more on the endogenous political pressure for change (as in the case of PAR), while others on generating hard evidence for policy change (as in the case of adaptive management). The actual mix of workable strategies should certainly vary across contexts, and finding the right suite of power-centric and knowledge-centric strategies should itself be a part of the adaptive collaborative approach. Second, despite rich conceptual explorations, there are still relatively few empirical studies demonstrating the causal impacts of these diverse adaptive learning and collaboration strategies. There is still a shortfall of thinking and methodology that can allow such approaches to work in more difficult situations. Third, despite prolific innovations on methodological fronts, there is still a dominance of linear thinking and input–output conceptual models in policy and strategic decision-making, where there is a lack of appreciation of ACA based strategies for management of natural resources.

While ACAs recognize the need to identify and harness the opportunities for minimizing conflicts and enhancing collaboration among actors at different scales and domains of the innovation system, there is still a long way to go in terms of making this practically effective and successful. More importantly, there is still limited dialogue among diverse threads of ACAs and there is an opportunity to forge interactive learning among their promoters. In the rest of the book, case study chapters will provide rich insights into how adaptive learning and collaborative actions are made possible in different contexts. In particular the chapters will focus on how ACA promoters have struggled with different sets of constraints and challenges they faced, and what can be learnt from these experiences. Contributors will not provide a grand technological package, but a framework of learning, with suggestions on what questions to ask and how in different contexts, rather than creating a simplistic recipe of an ACA. Indeed, there is a whole suite of new experiences and evidence against 'methodologism', pointing to the need for tackling more fundamental governing factors hindering learning and collaboration.

References

Agrawal, A. and E. Ostrom (2001) *Collective action, property rights, and decentralization in resource use in India and Nepal.* Politics and Society 29(4): 485–514.

Agrawal, A. and J. Ribot (1999) *Accountability in decentralization: a framework with South Asian and West African cases.* Journal of Developing Areas 33(4): 473–502.

Argyris, C. and D. Schon (1978) *Organizational learning: a theory of action perspective.* Reading, Addison-Wesley.

Armitage, D. R., F. Berkes and N. Doubleday, Eds (2007) *Adaptive co-management: collaboration, learning, and multi-level governance.* Vancouver and Toronto, University of British Columbia Press.

Armitage, D. R., R. Plummer, F. Berkes, R. I. Arthur, A. T. Charles, I. J. Davidson-Hunt, A. P. Diduck, N. C. Doubleday, D. S. Johnson, M. Marschke, P. McConney, E. W. Pinkerton and E. K. Wollenberg (2009) *Adaptive co-management for social–ecological complexity.* Fontiers in Ecology and the Environment 7(2): 95–102.

Bäckstrand, K. (2003) *Civic science for sustainability: reframing the role of experts, policy-makers and citizens in environmental governance.* Global Environmental Politics 3(4): 24–41.

Bandura, A. (1969) *Social learning of moral judgments.* Journal of Personality and Social Psychology 11(3): 275–279.

Banjade, M. and H. Ojha (2005) *Facilitating deliberative governance: innovations from Nepal's community forestry: a case of Karmapunaya.* The Forestry Chronicle 81(3): 403–408.

Berkes, F. (2009) *Evolution of co-management: role of knowledge generation, bridging organizations and social learning.* Journal of Environmental Management 90(5): 1692–1702.

Berkes, F., J. Colding and C. Folke (2000) *Rediscovery of traditional ecological knowledge as adaptive management.* Ecological Applications 10(5): 1251–1262.

Biggs, S. D. (1990) *A multiple source of innovation model of agricultural research and technology promotion.* World Development 18(11): 1481–1499.

Biggs, S. D. and E. J. Clay (1981) *Sources of innovation in agricultural technology.* World Development 9(4): 321–336.

Bourdieu, P. (1977) *Outline of a theory of practice.* Cambridge, Cambridge University Press.

Bourdieu, P. (1989) *Social space and symbolic power.* Sociological Theory 7(1): 14–25.

Bourdieu, P. (1998) *Practical reason: on the theory of action.* Cambridge, Polity.

Bradbury, H. and P. Reason (2003) *Action research.* Qualitative Social Work 2(2): 155.

Brown, C. and K. N. Laland (2003) *Social learning in fishes: a review.* Fish and Fisheries 4(3): 280–288.

Brunner, R. D., T. A. Steelman, L. Coe-Juell, C. M. Cromley, C. M. Edwards and D. W. Tucker (2005) *Adaptive governance: integrating science, policy, and decision making.* New York, Chichester and West Sussex: Columbia University Press.

Carlsson, L. and F. Berkes (2005) *Co-management: concepts and methodological implications.* Journal of Environmental Management 75(1): 65.

Chambers, R. (1994) *The origins and practice of participatory rural appraisal.* World Development 22(7): 953–969.

Clark, M. C. and A. L. Wilson (1991) *Context and rationality in Mezirow's theory of transformational learning.* Adult Education Quarterly 41(2): 75.

Colfer, C. J. P. (2005) *The complex forest: communities, uncertainty and adaptive collaborative management.* Washington and Bogor, RFF and CIFOR.

Colfer, C. J. P. (2008) *From understanding to action: building on anthropological approaches to influence policy making.* Against the grain: the Vayda tradition in human ecology and ecological anthropology. B. B. Walters, B. J. McCay, P. West and S. Lees. Lanham, Altamira Press: 273–286.

Cook, S. D. N. and J. S. Brown (1999) *Bridging epistemologies: the generative dance between organizational knowledge and organizational knowing.* Organization Science: 381–400.

Cornwall, A. (2003) *Whose voices? Whose choices? Reflections on gender and participatory development.* World Development 31(8): 1325–1342.

Cornwall, A. and J. Gaventa (2000) *From users and choosers to makers and shapers: repositioning participation in social policy.* IDS Bulletin 31(4): 50–62.
Cornwall, A. and R. Jewkes (1995) *What is participatory research?* Social Science & Medicine 41(12): 1667–1676.
Daniels, S. E. and G. B. Walker (1996) *Collaborative learning: improving public deliberation in ecosystem-based management.* Environmental Impact Assessment Review 16(2): 71–102.
Dewey, J. (1916/1966) *Democracy and education: an introduction to the philosophy of education.* New York, The Free Press.
Diaw, M. C., R. Prabhu and T. Aseh, Eds (2008) *In search for common ground: adaptive collaborative management of forest in Cameroon.* Bogor, Indonesia, CIFOR.
Edmunds, D. and E. Wollenberg (2002) *Disadvantaged groups in multistakeholder negotiations.* Development and Change 32(2): 231–253.
Elkjaer, B. (2004) *Organizational learning: the 'third way'.* Management Learning 35(4): 419–434.
Fals-Borda, O. (1987) *The application of participatory action-research in Latin America.* International Sociology 2(4): 329–347.
Fals-Borda, O. and M. A. Rahman (1991) *Action and knowledge: breaking the monopoly with participatory action-research.* London, Intermediate Technology Publications, The Apex Press.
Fischer, F. (1990) *Technocracy and the politics of expertise.* Newbury Park, CA, Sage.
Fischer, F. (2000) *Citizens, experts and the environment: the politics of local knowledge.* Durham and London, Duke University Press.
Fischer, F. (2004) *Professional expertise in a deliberative democracy.* The Good Society 13(1): 21–27.
Fisher, R., R. Prabhu and C. McDougall, Eds (2007) *Adaptive collaborative management of community forests in Asia: experiences from Nepal, Indonesia and the Philippines.* Bogor, Indonesia, Center for International Forestry Research (CIFOR).
Folke, C. (2006) *Resilience: the emergence of a perspective for social-ecological systems analyses.* Global Environmental Change 16(3): 253–267.
Folke, C., L. Pritchard, F. Berkes, J. Colding and U. Svedin (2007) *The problem of fit between ecosystems and institutions: ten years later.* Ecology and Society 12(1): 30.
Foucault, M. (2000) *The subject and power.* Readings in contemporary political sociology. K. Nash. Massachusetts and Oxford, Blackwell: 8–26.
Freire, P. (1993) *Pedagogy of the oppressed. 1970.* New York, Continuum.
Gottschick, M. (2008) *Participatory sustainability impact assessment: scientific policy advice as a social learning process.* Systemic Practice and Action Research 21(6): 479–495.
Gramsci, A. (1990) *Culture and ideological hegemony.* Culture and society: contemporary debates. J. C. Alexander and S. Seidman. Cambridge, Cambridge University Press, 47–54.
Gunderson, L. (1999) *Resilience, flexibility and adaptive management: antidotes for spurious certitude?* Conservation Ecology 3(1).
Gunderson, L. H. and C. S. Holing, Eds (2002) *Panarchy: understanding transformation in human and natural systems.* Washington, DC, Island Press.
Habermas, J. (1970) *The scientization of politics and public opinion.* Toward a rational society: 62–80.
Habermas, J. (1971) *Knowledge and human interests.* Boston, Beacon Press.
Hall, A. (2002) *Innovation systems and capacity development: an agenda for North–South research collaboration?* International Journal of Technology Management and Sustainable Development 1(3): 146–152.

Hall, A. and N. Clark (2010) *What do complex adaptive systems look like and what are the implications for innovation policy?* Journal of International Development 22(3): 308–324.
Hall, A., M. V. K. Sivamohan, N. Clark, S. Taylor and G. Bockett (1998) *Institutional developments in Indian agricultural R and D systems: emerging patterns of public and private sector activities*. Science, Technology and Development 16(3): 51–76.
Hall, A., G. Bockett, S. Taylor, M. V. K. Sivamohan and N. Clark (2001) *Why research partnerships really matter: innovation theory, institutional arrangements and implications for developing new technology for the poor*. World Development 29(5): 783–797.
Hall, A., R. Sulaiman, N. Clark, M. V. K. Sivamohan and B. Yoganand (2002) *Public–private sector interaction in the Indian agricultural research system: an innovation systems perspective on institutional reform*. Agricultural research policy in an era of privatization. D. Byerlee and R. Echeverría, Eds. Oxon, UK, CABI: 55–176.
Hall, A., V. R. Sulaiman, B. Yoganand and N. Clark (2003) *Post-harvest innovation systems in South Asia: research as capacity development and its prospects for impact on the poor in developing countries*. Outlook on Agriculture 32(2): 97–104.
Hall, A., W. Janssen, E. Pehu and R. Rajalahti (2006) *Enhancing agricultural innovation: how to go beyond the strengthening of research systems*. World Bank, Washington.
Hall, A., V. Rasheed Sulaiman, M. Dhamankar, P. Bezkorowajnyj and L. Prasad (2008) *Reframing technical change: livestock fodder scarcity revisited as innovation capacity scarcity*. Maastricht, ILRI/UNU-M.
Hall, P. A. (1993) *Policy paradigms, social learning, and the state: the case of economic policy making in Britain.* Comparative Politics 25(3): 275–296.
Hickey, S. and G. Mohan (2005) *Relocating participation within a radical politics of development.* Development and Change 36(2): 237–262.
Holling, C. S. (1973) *Resilience and stability of ecological systems.* Annual Review of Ecology and Systematics 4: 1–23.
Kapoor, I. (2002) *Deliberative democracy or agonistic pluralism? The relevance of the Habermas-Mouffe debate for third world politics.* Alternatives 27(4): 459.
Lee, K. N. (1993) *Compass and gyroscope: integrating science and politics for the environment.* Washington DC, Island Press.
Levin, S. A. (1992) *The problem of pattern and scale in ecology: the Robert H. MacArthur award lecture*. Ecology 73(6): 1943–1967.
Lewin, K. (1946) *Action research and minority problems.* Journal of Social Issues 2(4): 34–46.
Maguire, P. (2001) *Uneven ground: feminisms and action research.* Handbook of Action Research: Participative Inquiry and Practice: 59–69.
Malla, Y. B., R. Barnes, K. Paudel, A. Lawrence, H. Ojha and K. Green (2002) *Common property forest resource management in Nepal: developing monitoring systems for use at the local level.* Kathmandu and Reading, ForestAction and The University of Reading.
Mandondo, A. and F. Matose (2008) *A comparative analysis of adaptive natural resources management in Zimbabwe*. Coping amidst chaos: studies on adaptive collaborative management from Zimbabwe. A. Mandondo, R. Prabhu and F. Matose. Bogor, Indonesia, CIFOR: 119–132.
McDougal, C., H. Ojha, R. K. Pandey, M. R. Banjade and B. H. Pandit (2007) *Enhancing adaptiveness and collaboration in community forestry in Nepal: reflections from participatory action research.* R. J. Fisher, *ACM in Asia, CIFOR and RFF. Adaptive Collaborative Management of Forests: Experiences from Asia.* Bogor, Indonesia, CIFOR and RFF.
McIntyre, A. (2008) *Participatory action research.* London, Thousand Oaks and New Delhi, Sage Publications, Inc.

Mezirow, J. (1991) *Transformative dimensions of adult learning.* San Franscisco, CA, Jossey-Bass.

Mezirow, J. (1997) *Transformative learning: theory to practice.* New Directions for Adult and Continuing Education 1997(74): 5–12.

Muro, M. and P. Jeffrey (2008) *A critical review of the theory and application of social learning in participatory natural resource management processes.* Journal of Environmental Planning and Management 51(3): 325–344.

Ojha, H. (2006) *Techno-bureaucratic doxa and the challenges of deliberative governance: the case of community forestry policy and practice in Nepal.* Policy and Society 25(2): 131–175.

Ojha, H., N. Timsina, R. B. Chhetri and K. Paudel, Eds. (2008) *Knowledge systems and natural resources: management, policy and institutions in Nepal.* New Delhi, Cambridge University Press and IDRC.

Ojha, H. R., N. S. Paudel, M. R. Banjade, C. McDougall and J. Cameron (2010) *The deliberative scientist: integrating science and politics in forest resource governance in Nepal.* Beyond the biophysical: knowledge, culture, and politics in agriculture and natural resource management. L. German, J. J. Ramisch and R. Verma. Dordrecht, Hiedelberg, London and New York, Springer: 167–191.

Ozanne, J. L. and B. Saatcioglu (2008) *Participatory action research.* Journal of Consumer Research 35(3): 423–439.

Paudel, K. and H. Ojha (2007) *Imposing Indicators or Co-Creating Meanings in Nepal.* Negotiated learning: collaborative monitoring in forest resource management. I. Guijt, Ed. Washington, DC, Resource for the Future: 49–57.

Polanyi, M. (1983) *The tacit dimension.* Gloucester, MA, Peter Smith.

Prabhu, R., C. McDougall and R. Fisher (2007) *Adaptive collaborative management: a conceptual model.* Adaptive Collaborative Management of Community Forests in Asia: 16–49.

Reason, P. and H. Bradbury (2001) *Handbook of action research: participative inquiry and practice.* London, Thousand Oaks and New Delhi, Sage Publications Ltd.

Rist, S., M. Chiddambaranathan, C. Escobar and U. Wiesmann (2006) *'It was hard to come to mutual understanding . . .' The multidimensionality of social learning processes concerned with sustainable natural resource use in India, Africa and Latin America.* Systemic Practice and Action Research 19(3): 219–237.

Röling, N. (2002) *Beyond the aggregation of individual preferences: moving from multiple to distributed cognition in resource dilemmas.* Wheelbarrows full of frogs: social learning in rural resource management. C. Leeuwis, and R. Pyburn. Assen, the Netherlands, Koninklijke Van Gorcum.

Schon, D. (2010) *Government as learning system.* Social learning systems and communities of practice. C. Blackmore. The Open University Press: 5–16.

Schusler, T. M., D. J. Decker and M. J. Pfeffer (2003) *Social learning for collaborative natural resource management.* Society and Natural Resources 16(4): 309–326.

Senge, P. M. (1990) *The fifth discipline: the art and practice of the learning organization.* London, Century Business.

Sharif, N. (2006) *Emergence and development of the National Innovation Systems concept.* Research Policy 35(5): 745–766.

Small, S. A. (1995) *Action-oriented research: models and methods.* Journal of Marriage and Family 57(4): 941–955.

Small, S. A. and L. Uttal (2005) *Action oriented research: strategies for engaged scholarship.* Journal of Marriage and Family 67(4): 936–948.

Spielman, D. J., J. Ekboir and K. Davis (2009) *The art and science of innovation systems inquiry: applications to Sub-Saharan African agriculture.* Technology in Society 31(4): 399–405.

Wadsworth, Y. (1998) *What is participatory action research?* Action Research International 2: 1–18. http://www.aral.com.au/ari/p-ywadsworth98.html, accessed 5 March 2012.

Walters, C. J. (1986) *Adaptive management of renewable resources.* New York, Macmillan, Collier Macmillan.

Wenger, E. (2000) *Communities of practice and social learning systems.* Organization 7(2): 225–246.

Whyte, W. F. (1989) *Advancing scientific knowledge through participatory action research.* Sociological Forum 4(3): 367–385.

Wollenberg, E., R. Iwan, G. Limberg, M. Moeliono, S. Rhee and M. Sudana (2007) *Facilitating cooperation during times of chaos: spontaneous orders and muddling through in Malinau district, Indonesia.* Ecology and Society 12(1): 3. http://www.ecologyandsociety.org/vol12/iss11/art13/, accessed 5 March 2012.

3 The ups and downs of institutional learning

Reflections on the emergence and conduct of adaptive collaborative management at the Center for International Forestry Research (CIFOR)[1]

Carol J. Pierce Colfer

Introduction

This chapter presents the challenges experienced in developing and applying the concept and methods of adaptive collaborative management (ACM) in forest governance and management in a number of countries. The analysis is primarily based on my reflections on the experience of working as the team leader of the ACM program at the Center for International Forestry Research (CIFOR). These reflections focus on how ACM evolved within CIFOR, what worked and what did not in the course of applying ACM, and what challenges the ACM team had to face – within the team, in the organization, in the field and with the donors. In so doing, I demonstrate how ACM processes emerge and are influenced by the underlying institutional, policy and knowledge systems environment within which ACM is applied.[2]

At CIFOR, our adaptive collaborative management effort (see Box 3.1) entailed a hope a) that our research would improve environmental and human conditions, initially where we worked, and b) that ultimately we would be able to contribute some processual, methodological insights that could be applied more widely.[3] Our observations of the divergent goals and management strategies at play in tropical forests suggested that better collaboration among stakeholders would be important. This essay reflects my agreement with Taylor, who writes of 'unruly complexity' in ecological and social situations, and urges us to become more self-conscious about our engagement with our own social contexts (Taylor 2005).

This chapter begins with a discussion of the philosophy of science that dominated at CIFOR in the late 1990s, when CIFOR's ACM work began. I then introduce the CIFOR ACM team's perceptions of the 'real world' problems we wanted to address, and the program of research we developed in response. The bulk of the chapter focuses on the actual experience of ACM

Box 3.1 Adaptive collaborative management – CIFOR's original definition, plus

Adaptive collaborative management (ACM) is a value-adding approach whereby people who have interests in a forest agree to act together to plan, observe and learn from the implementation of their plans while recognizing that plans often fail to achieve their stated objectives. ACM is characterized by conscious, facilitated efforts among such groups to communicate, collaborate, negotiate, and seek out opportunities to learn collectively about the impacts of their actions.

[Later, based on our experience, we added the following phrase:]

Work with a given group of people requires involving actors at multiple scales – usually at least one level down and one level up (e.g., user groups within a community and district officials above).

practitioners, including attention to institutional and interactional elements, as well as outcomes of our efforts. I conclude with an analysis of what we have learned and how we might proceed with similar efforts more effectively in the future.

Genesis within a dominant scientific paradigm

Paradigmatic differences

The Consultative Group on International Agricultural Research (CGIAR) centers (or CG centers for short) emerged in the 1980s to cater to the demand for scientific and technological research in agriculture and rural development. CIFOR is among the latest to address natural resource management issues (including production, conservation and human issues in sustainable development). Most of the 15 CG centers focus on specific commodities. Several – e.g., CIFOR, the International Water Management Institute, the World Fish Center – have been formed, more recently, to address broader, natural resource issues; but there has been a strong and sometimes successful history of reductionist science as the backbone, particularly of the commodity-based centers' approach to research. The problems integrating social sciences into these centers are clearly articulated in the collection by Cernea and Kassam (2006).[4] CIFOR has represented a far more hospitable environment for social sciences than have the other CG centers (Colfer *et al.* 2006; Rhee 2006); but it has not been able to escape the reductionist bias in research. In CIFOR's context, agricultural production and hypothesis-testing, experimental, biophysical research have been highly regarded. Indeed, despite the progress reported below, in recent years there seems to have been some backsliding, as the CG's coordinating Science Council has developed ever-more stringent, top-down and 'bean counting' approaches to center evaluation.

CIFOR's first Director General (DG) (a wildlife biologist) and Board of Trustees recognized the inadequacy of an exclusive, reductionist focus and worked hard in the mid-1990s to create an institution where social and biophysical scientists could work together and address institutional constraints to effective forest management. This effort was initially somewhat hampered by the dominance of senior, biophysical scientists[5] and CG assumptions and requirements about the nature of research have continued to present obstacles to effective interdisciplinarity. In the early years, there was a will, but not necessarily the knowledge among decision makers, to address human issues related to forestry.

At the risk of over-generalizing, I have created Table 3.1, which summarizes (in 'ideal type' form) some of the goals, beliefs, preferences and emphases of many CG and CIFOR scientists, on the one hand, contrasted with tendencies within the ACM team.

As with any social grouping, there was variation in perceptions, both within CIFOR's scientific and administrative staff, and within the ACM team itself; and opinions and understandings changed over time. But issues related to the differences listed in Table 3.1 emerged repeatedly, particularly during the ACM program's development process (1998–2000).

CIFOR's emphasis on, and the team's own commitment to, interdisciplinarity meant that minimizing disciplinary jargon was a must. Although

Table 3.1 Theoretical underpinnings and orientations (CG vs. ACM)

	CGIAR/CIFOR	*ACM Tendencies*
seeking	one truth	multiple truths/perceptions/ perspectives
belief in	pure objectivity	impossibility of pure objectivity in social research
belief in	conventional science	linking science with other local knowledge systems
belief in	apolitical nature of science/research	inevitability of political implications of social research
preference for	hypothesis testing, deductive approach	research questions; holistic, inductive approach, strengthened by testing many small hypotheses determined in the course of action
preference for	experimental approach	contextual, holistic, pluralistic approaches
preference for	quantified results	combination of qualitative and quantitative results
emphasis on	generalizability	understanding local processes (for ultimate extension and wider use)
emphasis on	components of systems	systems as a whole

I certainly agreed with what Rudel and Gerson (1999) summarized as the key elements in postmodernism,[6] even I (a social scientist) was put off by the complexity of the language used in many postmodern writings. We struggled to simplify important social science concepts; yet even our simplified jargon mystified many colleagues. 'Can't you say it more simply?' or 'Use plain English!' they'd complain.

Another contextual struggle pertained to the general acceptance within forestry of the institutional and political status quo. Many scientists saw their roles as technical, and decidedly *not* political (cf. Ojha 2008, on related doxa in Nepal's forestry bureaucracy). Yet within ACM there was broad recognition that a central goal was empowerment – in Foucault's positive, 'power to', sense (Foucault 1980) – of the women and men in forest communities, in their interactions with others. There was also a backdrop of recurring debates within CIFOR related to the institution's central mandate. These were phrased as contrasting science and advocacy; quantitative and qualitative approaches; research and development; biophysical and social. ACM was seen as 'pushing the envelope' toward the second, more controversial of these contrasting pairs; we saw ourselves as straddling the pairs.

CIFOR has fairly consistently expressed the desirability of working at multiple scales; but our efforts to begin at the village or landscape scale met with recurrent resistance. The Board of Trustees and CIFOR administrators repeatedly asked 'How will you scale up from single cases?' and said 'But your results won't be generalizable.' They worried that 'CIFOR scientists are too expensive to focus on single villages. National research institutions should be doing this work.' But most national forestry institutions had no social scientists, and foresters were typically working within bureaucratic, militaristic and technocratic frameworks ('doxa'), uncongenial to the empowerment of rural peoples.

Finally, our internal emphasis on empowerment elicited considerable discussion amongst ourselves. How, for instance, could we genuinely work on leveling the playing fields in tropical forests, without sparking resistance within the governments whose permissions we needed (and even whose enthusiasm we sought)? In the sections to follow, I discuss the challenges we experienced, and our attempts to cope with them as we tried to move forward, applying ACM in forest contexts.

Creating the rationale for adaptive collaborative management

In December 1994, Ravi Prabhu, the leader of a project on 'Criteria and Indicators (C&I) for Sustainable Forest Management' invited me to work with him. My previous work had primarily been long-term, ethnographic research (a logging community in the US State of Washington; a village and hospital in the central Sultanate of Oman; and a number of forest communities in East and West Kalimantan and West Sumatra). The CIFOR (C&I) project introduced me to comparative international work, with month-long field visits to communities in or adjacent to timber concessions in Brazil, Indonesia, Ivory Coast and Cameroon. Such visits made the ubiquity of the relative powerlessness of forest dwellers vis-à-vis timber companies increasingly clear to me. In that and subsequent research, I noted similarities in such people's problems with plantation companies, conservation projects, and the people's own governments. There was thus a limited prospect of collaborative management without some sort of active facilitation from outside to strengthen the roles of local people and create a more level playing field.

Our interests in the C&I project involved a dual, equivalent emphasis on sustainable forest management and enhanced human wellbeing. My job was to develop a conceptual framework for looking at the latter issue (Colfer and Byron 2001) – a task that rendered forest dwellers' comparative powerlessness ever clearer. We were not yet focused on making a contribution to the process of local empowerment. The C&I project's second phase (1996–1998) focused on the social component, and both Ravi and I began to struggle with how to make use of what we had learned in a way that would directly improve forest conditions and people's wellbeing. We were both increasingly skeptical about the most common model by which research was designed to reach policymakers and affect real-world conditions. We had seen the slow pace by which research planned, implemented and analyzed by scientists – such as our own – got implemented; and we began designing an approach that we felt would simultaneously contribute to knowledge, improve local conditions, and build on local people's knowledge, energy and interests. We began developing this concept, spurred by ideas from scientists such as Lee (1993) about adaptive management and others writing on participatory action research, farming systems, gender and diversity, etc., all of which called for linking diverse knowledge, forging collaboration and working in a participatory way with diverse stakeholders.[7]

Meanwhile, Lini (Eva) Wollenberg was working from CIFOR with colleagues at Cornell University. She'd coordinated research on social learning (Wollenberg *et al.* 2001a), devolution (Edmunds and Wollenberg 2003), and livelihood issues (Wollenberg and Ingles 1998) – in response to many of the same conditions Ravi and I had observed in the field. In the spring of 1998, we realized that both our projects, each unbeknownst to the other, had developed a similar concept (see Buck *et al.* 2001). CIFOR's Board suggested merging our respective efforts, under my leadership. Ravi and Lini both agreed.

In light of the dire situation of both forest peoples and their tropical forest landscapes, we all wanted to take a more pro-active approach that integrated participatory action research (PAR) – not something any other CIFOR projects were doing at the time. We wanted to see changes during the course of our project, rather than waiting until others could (or would) implement our findings (Colfer 2008). CIFOR's commitment to approaching issues at multiple scales was one we also considered important. We considered rural peoples to be a resource – their energy, motivation, goals, knowledge – that had been largely ignored. We wanted to work with them to catalyze this energy and creativity in pursuit of locally-appropriate and -determined ends that could improve lives and environments.

The C&I we had developed seemed a reasonable place to start in helping local communities to take more of a learning approach, to plan, monitor and improve their own management of local resources (in collaboration with other actors). The devolution project had shown that formal decentralization efforts often resulted in local peoples' being *more* under the thumb of their governments rather than less. We all felt the need for greater empowerment of local peoples. We concluded though that emphasizing 'empowerment', in our strategic planning documents, was probably too politically sensitive, that it might frighten the governments with whom we knew we would have to work: we used different words.

Both original projects (C&I and devolution) had been designed and implemented by researchers, though they had both involved collaborating partners. Neither had involved long-term, direct involvement in specific communities and landscapes. Both had focused on the local level (communities, forest management units, plantations), but also addressed multiple scales. Both had also included researchers from various disciplines. Although members of both projects tried to look holistically at their subjects of interest, this was difficult with short visits and multiple sites. There was no mechanism in either project for local level learning, only researcher learning; and no options for action on the ground, except in very peripheral, unsystematic ways. Indeed, such action would have been counter to the more conventional, reductionist conceptual frameworks within which the projects were designed. The fact that these two projects were designed in this way reflected not only research-designer wishes and expectations, but also donor and CIFOR institutional culture-issues that we would have to confront head-on when we began creating the ACM program.

The new program's eventual name, 'Devolution, Livelihoods and Adaptive Collaborative Management of Forests', reflected some early dissension, or at least malaise, between the two proto-projects (an early plan had been to call it only Adaptive Co-Management). I drafted 17 versions of a central concept note (Bogor ACM Team 1999) and engaged in multiple explanations, both within CIFOR itself and to the Board of Trustees, before our plan was formally accepted in the spring of 1999. Some of the circulation and resulting revisions came out of our team commitment to involve all members (including administrators), both to improve its quality and to strengthen team buy-in to our eventual action.

One of my goals was for all team members to value their own role, to be genuinely motivated by what we were trying to do. In the course of these revisions, we strengthened the role of extractive, cross-site research; moved toward harmonizing/synthesizing the ongoing activities and conceptual frameworks inherited from the original C&I and devolution projects; developed clearer steps in the processes we envisioned for ACM; and more clearly delineated responsibilities within, and the structure of, our program. One positive result was a shared vocabulary and conceptual framework within the Bogor-based team.

Whereas holistic and collaborative action research long ago won acceptance within many agricultural circles (e.g., Shaner *et al.* 1982), it has seemed strange, somewhat bizarre, to the forestry community. The historical forestry perspectives from Europe, wherein forests have been seen as the preserves of elites, with an assumed quasi-military approach to forest management, have been well documented (cf. collections by Leach and Mearns (1996) on Africa; Poffenberger (1990) on Southeast Asia). Local people have often been seen by foresters as 'encroachers' or 'squatters'. These notions have rendered people-oriented approaches to forest use alien to foresters (see also Ojha 2008 on related 'habitus' and 'doxa'). The landscape level emphasis within forestry may also have militated against concerns for people living there; many foresters have focused on the 'forest massif' rather than on smaller forested plots that local people may claim (but often not legally own).

Board members and many colleagues outside our project agreed that CIFOR research findings were not getting to users, that we needed to work at multiple scales and that local people needed to be involved. But it still made them very nervous. How would the scientific forestry community respond to this 'radical' kind of research? What if nothing came of it, when we were allocating millions of dollars to it? Where were the experimental designs, the hypotheses to be tested? What would constitute controls?[8] What if local people just wanted to cut down the forest?

By the time we began – taking into account the complexity and dynamism of local systems, both human and ecological – our conceptual framework had evolved to include three major 'legs':

1 We were interested in strengthening the vertical links between communities and wider scale actors (government, industry, conservation projects). This included the ideas of local level empowerment, the potential for local people to act as monitors of political actors at broader scales, and for better local access to outside resources (expertise, funds, links to others, support for local initiatives).
2 We also focused on horizontal links among communities and other local actors. This was deemed crucial because of the frequent lack of harmony among local management goals; we reasoned that enhanced collaboration would identify win–win situations, complementary desires, and in some cases negotiated solutions to difficult conflicts.
3 Finally we included an iterative element to integrate social learning (cf. Wollenberg *et al.* 2001a, 2001b) and the C&I experience (cf. CIFOR

1999). We believed that effective adaptation to change requires people to learn more systematically from their own experience.

We planned to use participatory action research, look at local situations holistically and interdisciplinarily, and build on indigenous knowledge, insofar as possible. Our version of PAR – learned most systematically in a 1999 training workshop coordinated by Bob Fisher (Fisher and Jackson 1998) – involved the identification of long-term goals through visioning processes (e.g., Wollenberg *et al.* 2000) and sustained facilitation of iterative processes of self-analysis, planning, implementation, monitoring and revision, among small, locally constituted PAR groups (see McDougall *et al.* 2009, or Mutimukuru-Maravanyika 2010, for thorough, country-level explications of its implementation).

Our differentiation between adaptive collaborative management (ACM, an umbrella framework or approach) and participatory action research (PAR, a method) needs clarification. PAR shares with ACM the emphasis on facilitated and shared social learning in iterative cycles and on human resource development. But, unlike ACM, it can take place within a single group and address any topic of importance to the participants. CIFOR's version of ACM, on the other hand, *requires* attention to improving human wellbeing, inter-group equity, health of forests (or other natural resources), policy concerns, and collaborative work with other stakeholders or actors (at other levels or in other communities). ACM is simply a recombination of existing methods and concerns, a value-adding approach. To some degree, we structured ACM to include these elements to address a certain level of oblivion within forestry concerning human elements. PAR was 'necessary but not sufficient', given the problems of communities and the professional world of forestry in the late 1990s. We felt that bringing these ideas together into one reasonably coherent, forestry-oriented approach – with a relevant *name* – would make its ultimate use by foresters more likely.

I believed that the goal of developing one 'technology' that could improve people's relationships to their forests and be applied in multiple locations was a chimera. The technology transfer model is based on this assumption, and it has sometimes worked. It had a particularly long and – to some – glorious history within the CGIAR. But 'development', 'conservation', 'sustainable forest management', these are concepts that require recognition of and linking with complex and dynamic systems. I argued that, given such system features, we would have to develop a *process* that could be replicated, one that allowed local people to solve their own problems (with outside facilitation and networking assistance), based on their own concerns, opportunities, realities and goals. So CIFOR's ACM effort involved both an informal experiment of sorts with this method (does it work, under what conditions, and what are the impacts?) and an effort to initiate real change in actual communities predominantly by forest users.[9]

We developed a framework with two prongs: an action and a research prong. All communities selected were involved in managing local forests, had low to medium levels of conflict, and more than one stakeholder. The research prong eventually involved the selection of 30 communities in 11 countries, with varying values for seven features (devolution status, conflict levels, population pressure, forest type, management goals, diversity and social capital). The team identified these variables (see Bogor ACM Team (1999) for full description) as likely to be important in comparing ACM success across sites. Our plan – one that proved unrealistic, due to time constraints – was to make yearly assessments of these various characteristics.

The second, 'action prong' – in its ideal form – involved one or more facilitators/researchers entering a forest community with the community's agreement.[10] Such facilitators first conducted a series of context studies of that setting (linked to the research prong above) and provided opportunities for reflection and deliberation about visions, strategies and practices. The context studies included identifying stakeholders, documenting historical trends, assessing collaborativeness and adaptiveness in the community, and biophysical and social assessments using C&I (e.g., Indriatmoko *et al.* 2006). In many cases, the facilitator would organize a future scenarios workshop with community members (and often other stakeholders). Such results served as a 'guiding star' for the community's subsequent actions. The next step would be the development of PAR groups (small groups of individuals interested in working on some problem they had identified in their community); and the initiation of a process of iterative social learning within that group – always keeping the long term goal (the 'guiding star') in mind. By an 'iterative social learning process', we referred to cycles of self-analysis, planning, implementation, monitoring, and revised self-analysis, planning, etc.

The ride was exciting but not easy (see Table 3.2).

Institutional evolution of our approach

Early battles within CIFOR

As already noted, qualitative approaches were not conventional within the CGIAR. Although our DG expressed his interest in the approach, he and I had long arguments about it. Our relationship, which began on a rocky footing, had improved by the late 1990s, and I felt free to express my opinion.[11] But he loved to play 'devil's advocate'. He would reiterate his skepticism about anthropology, descriptive methods, participatory approaches and holistic analyses. He reminded us of the Board's (to me quite strange) insistence that all research at CIFOR test hypotheses, a directive I had consistently ignored. One of CIFOR's 'claims to fame' was that it was an interdisciplinary institution; although I continued to ignore this directive, its existence necessitated periodic rear-guard action on my part (Box 3.2).

Table 3.2 An exciting but not easy ride for CIFOR participants

Excitements	*Related difficulties*
Learning about local human and environmental contexts	Researchers recognize their lack of needed skills
	Communities/partners/officials may be skeptical, antagonistic, or greedy and manipulative
	Living conditions may be difficult (bad roads, schools, medical facilities, housing)
Learning about other disciplines and spheres	Researchers recognize their lack of needed skills
	Miscommunications can easily occur among team members
	Genuine differences of opinion, procedure, conclusions require resolution
Interaction across sites can be stimulating and productive of insights	Distances complicate communication and collaboration
	Cross-cultural communication can lead to misunderstandings and conflict
	Differences in access to information can lead to inequities within teams
Involvement in local level and international teams can stimulate cross-team fertilization and excitement	Researchers needed to maintain (at least) two vocabularies, contexts, sets of interactional norms, etc.
	Researchers were often overwhelmed with demands from their communities combined with demands from Bogor
Seeing progress on site	Confronting difficult barriers and having to help figure out how to overcome them
	Explaining successes (particularly qualitative ones) and failures to managers, partners, donors and evaluators

Besides recurrent explanations and dialogue with our DG, I was grilled twice a year at Board meetings (unlike other programs), throughout the first four years of our effort.[12]

The same questions and concerns outlined above (on hypothesis testing, experimental designs, replicability, controls see Table 3.1) would surface every time – some of it stimulated by the DG himself, who remained anxious about our efforts.[13] Again, I would explain why we were arguing for a different approach. I stressed the complexity, dynamism and geographical variability of human and environmental systems; the human resources that

Box 3.2

27 July 2010

> a crucial distinction is between a *practical failure* to comply and *declared refusal* to comply . . . [P]ublic refusal . . . constitutes a . . . symbolic declaration of war.
>
> (Scott 1990: 203)

We – the scientists in CIFOR's governance program – were gathered together in a conference room in Bogor, Indonesia. The acting director of the program was explaining the importance of academic citation indices in determining where we should publish, despite the fact that two scientists had just presented evidence of the biased nature of such indices. I also felt strongly that such indices were inappropriate as primary guides for our action, given our mandate to have an impact on conservation and development globally. I had expressed this view diplomatically without apparent effect. He asked us to what extent we took these indices into account in deciding where to publish. Recognizing that I was being somewhat belligerent and undiplomatic, I responded rather vehemently, 'Not at all.' I admit to having felt the thrill of disobedience to authority, but I suppose I felt somewhat protected by my imminent departure and also by my age and related prestige. The director turned to me and said pointedly and disrespectfully, 'No one can expect to be a senior scientist at CIFOR without taking such indices into consideration in their publications.' I was actually the most senior and one of the most widely published scientists in the entire institution at that time. His response was clearly stated for the audience's sake as an attempt to alter our practice, and to 'put me in my place'. It was a clear insult and contrary to the prevailing Governance Program norms, a declaration of war on our 'institutional culture'. I had (perhaps unwisely) responded with my own declaration of war.

local people represented; the inequity of current approaches, which tended to deal with local male elites if they involved any local people at all; the lack of sustainability of short-term projects that conducted conventional research and left; the need to have some immediate results. I also explained about PAR, its long-term collaborative nature, its opportunistic use of relevant research methods, its iterative nature. Each time the DG and the Board would rule in our favor. But many remained worried. The presence of several qualitatively sophisticated Board members was helpful to us in this effort.

Still, when we were a year or two into our program, the DG surreptitiously and informally arranged for an additional, independent review of our work, urging the reviewer to be as critical as possible. I was on sick leave when the team discovered this, but it proved to be a demotivating event and reduced team members' confidence in our institutional support.

As ACM was approved and supported by donors, we began building our team. I was committed to making certain that we made full use of the various kinds of expertise our existing staff represented. We wanted to replicate the adaptive aspects of our field-based program within the team itself; we wanted to 'walk the talk' (specifics described below).

Response to CIFOR's reductionist bias

We all recognized that for continued viability within our institution, we needed to respond substantively to our colleagues' and the Board's concerns. Some team members also felt mildly uncomfortable with a purely PAR approach; as we were sometimes reminded, CIFOR's mandate was 'research', not 'development'. I personally saw the field-based collaborative effort as central, and imagined supplementary, more conventional research as ideally designed in response to field conditions and needs. Some team members wanted pre-planned conventional research because that was a central interest of theirs; and others agreed to design some cross-site studies, as a form of 'insurance', a means of allaying wider internal CIFOR fears.

We considered the idea of selecting 'experimental' and 'control' communities with which to work; but discarded the idea on careful inspection. Some of our team members felt it was unethical to 'experiment' with communities (cf. McDougall *et al.* 2007b; see also Fennella *et al.* 2008). My own perspective was that people are experimenting all the time (cf. the Chambers *et al.* 1993 collection). I argued that honesty and humility could absolve us of much responsibility, particularly if we made it clear that any final decisions rested with the communities.[14] But I found even more convincing the argument that people who were similar enough were inevitably geographically close to each other; contamination from experimental to control group seemed highly probable. If the changes were found to be beneficial, then it definitely *would* be unethical to deny the 'control' community access (even if possible).

In the end, rather than try to design such an experiment, we strengthened the 'research prong' mentioned above to include the matrix of seven features we considered, after much debate, likely to be important in analyzing differences among sites. We planned to place sites on these continua, and monitor the changes. We also concluded that the series of 'context studies' could serve as more conventional research outputs. Both of these efforts allayed the fears of CIFOR administrators and the Board, as well as providing an insurance function as we undertook this atypical action research.

We knew that our work would involve an element of risk; community action is impossible to fully predict.

The variety of conceptual and analytical elements (from ACM and PAR to context studies and cross-site analyses of pre-determined features) allowed the various site-based teams considerable flexibility in terms of their own emphases and analyses. We purposely designed the framework to maximize flexibility, seeing the framework as an umbrella within which many different kinds of useful activities, approaches and emphases could be accommodated, with valuable results. This proved to be true.

One of the important disadvantages of this approach, however, was the demands on field-based personnel. The process of facilitating PAR groups involves a great deal of time in the field and unpredictable responses to local demands. Field personnel had to monitor their own and the groups' actions, as well as gathering data (both the context studies and information on the seven variables). The context studies took considerably longer than anticipated – usually six months rather than the two to three we had planned. Initially, we expected to repeat the context studies yearly. This proved impossible; there was simply not time.

Our intended 'two-pronged approach' evolved a third prong. The field personnel had to divide their brains into thirds – one third facilitating a successful PAR process; another monitoring and analyzing what they were doing; and the final third providing materials for cross-site analyses taking place elsewhere. Most original ACM field teams (and the later, smaller CAPRI/ACM project on elite capture in Sumatra) managed to do this, with consistent reinforcement and encouragement from both national coordinators[15] and Bogor-based staff.

In the more recent Landscape Mosaics (LM) project (Landscape Mosaics Team 2008a, 2008b), with less external reinforcement and encouragement, the teams were unable to fulfill both PAR and the more demanding, parallel set of tasks (a complex cross-site database plus separate studies for three of the four project 'themes' – livelihoods, governance and the biophysical environment). PAR pretty much lost out.

Pragmatically, there has been a difficult balance among 1) professional (institutional, disciplinary and, in some cases, political) pressures for conventional, reductionist and extractive research; 2) donor requirements for immediately demonstrable plans and accountability; and 3) what I (and other ACM advocates) believe to be the potential and creativity resident in forest communities. Different actors opt for different balances, also at different junctures. For me, from the standpoint of enhancing people's livelihoods and quality of life, the third element is the most crucial. But in our context, we had to placate the first two in order to advance the third. From a research standpoint, I also saw the value of the cross-site work and the attempt at disinterested self-observation within the PAR process. Because we hoped to apply what we learned more widely, we needed to understand if there were specific conditions and approaches that made success more likely, and what kinds of impacts were possible.

Getting funded

We also had some, though comparatively few, problems with donors. Environmental issues were still popular in the late 1990s, which made accessing funding simpler than was the case later; and of course working in an international center can also enhance access to funds. In some cases, members of our team knew project officers in funding agencies; in Bolivia and Kyrgyzstan, we were able to piggyback on other existing forestry-related projects; in a few cases, we obtained partial funding because of end-of-fiscal-year excess funds available (both within and outside CIFOR). USAID, at one time a common source of such end-of-year funds, would have been unlikely to *plan* to contribute to ACM – the institution was very wedded to quantitative and reductionist approaches. Political changes in the US (George W. Bush was elected) and (perhaps related) changes in project officers within USAID led to their reduced interest in CIFOR's work generally. Much donor response depends on the personal biases of project officers.

Prabhu, Wollenberg and McDougall were excellent proposal writers, and they generated about 4.5 million dollars in 'soft money' over the first three years. We knew that three years was insufficient – we expected the processes we were trying to initiate to take 10 to 15 years. But no donors emerged, willing to fund our work for longer time periods. 'What are your impacts?' queried our donor from the Asian Development Bank; 'How many hectares of forest have you saved?' asked our USAID and CGIAR donors. Evidence of research impacts, convincing to donors, tends toward numbers of hectares or species saved from deforestation, dollars of increased income for x number of families. Although we were convinced that our impacts – in social learning, empowerment, increased equity, political and analytical savvy – were absolutely crucial to improving human wellbeing and environmental management, we did not have the units, the measurements that donors sought. Additionally, many of the impacts took longer periods of time to emerge than had transpired. These issues (time needed and measurement capabilities) remain problematic.

We hoped we would get further funding; and we reasoned that we could learn something significant even in a shorter time frame. Several sites were able to get subsequent funding (Indonesia, Nepal) – though in some cases, the second phase was in bits and pieces (Zimbabwe, Cameroon, Brazil) and/or for related approaches that spun off full scale ACM (Bolivia, Nepal).

Although most donors granted us considerable autonomy, the Asian Development Bank (ADB) required that we change one research site. We had selected truly tropical forests, globally – CIFOR's focus at the time. But ADB insisted we exchange a tropical Asian site for Nepal, arguing that Nepal's cutting-edge experience of collaborative management would be an important addition to our analysis. We exchanged the site, as requested; and certainly the Nepal work has proved fruitful.

The same man from ADB later came to CIFOR to work as the Research Director. He was not sympathetic to the ACM approach. We later learned that he had convinced ADB *not* to fund our work for a second phase, but rather to invest in other CIFOR programs. By this time, there was also a new DG, an economist who understood ACM, but was less interested in it.

Our primary donors included the ADB, the European Union (EU), the United Kingdom's Department for International Development (DFID), the Canadian International Development Research Centre (IDRC), Ford Foundation and the United States Agency for International Development (USAID); more recently the Swiss Agency for Development and Cooperation (SDC) and several German institutions have become involved (see Table 3.3). Our most intimate relations with German institutions were via CAPRi (the CGIAR's Systemwide Initiative on Collective Action and Property Rights), through which German Organisation for Technical Cooperation (GTZ) funds were channeled. We found IDRC, CAPRi and Ford particularly congenial; their program officers were knowledgeable about and supportive of participatory approaches, and their reporting requirements were flexible, their attitudes constructive. In the late 1990s, none of the program officers from the ADB, the EU, or USAID appeared to be familiar with participatory approaches; and the institutions had very cumbersome and bureaucratic reporting requirements.

DFID and, more recently, SDC have been inordinately attached to logframes (decidedly difficult to adhere to with PAR), though their program officers were more in tune with our approach than many. Still, a fundamental feature of ACM, as originally planned and implemented at CIFOR, was the freedom for local community PAR groups to determine their own priorities for action. This presents a formidable obstacle to writing fundable proposals (and sometimes to complying with contractual obligations). Donors typically want to know detailed plans before they allocate funds; an ACM approach makes such advance knowledge improbable. Within the original ACM project, our proposals were sufficiently broad (focused on sustainable forest management, for instance) that we could creatively link donor and community interests. The issue was more difficult in some of the recent LM

Table 3.3 ACM experience of donor qualities

	Overly bureaucratic	*Overly committed to logframes*	*Troublesome project officers*	*Particularly helpful project officers*	*Flexible about plans, successes and timelines*	*Open to participatory approaches*
ADB	x		x			
CAPRI				x	x	x
DFID		x		x		x
EU	x					
Ford				x	x	x
IDRC				x	x	x
SDC		x	x			x
USAID	x	x	x			

project sites. SDC, the donor, was concerned about biodiversity conservation; the local people in most sites were far less so. The project leader (actually seconded from SDC) struggled to comply with both needs; but the donor proved difficult to resist. Similar problems emerge with very stringent and pre-defined budgetary categories; budgetary flexibility is, predictably, very useful in an ACM project.

One successful strategy has been to tie the ACM approach to a 'sexy' development or conservation fad. We have drawn on such fads as C&I, community forestry, sustainable forest management, conservation and development, governance, multi-scale issues. At the moment, the most likely source of such funding is climate change research (see CIFOR 2008). I believe that effective approaches to address climate change are likely to be location-specific, given the variability in projected changes and of course the same dynamism and complexity that prompted our original interest in ACM.

In late 2002, as much of our initial funding was ending, CIFOR experienced its first real budgetary shortfall. The need to analyze our early results and CIFOR's money problems conspired to make a sabbatical for me at Cornell a win–win proposition. As I left and Ravi took over from his base in Zimbabwe, the new Governance Program was being formed in Bogor. A combination of factors conspired to reduce our presence and influence: vastly reduced funding, distant ACM leadership, and new CIFOR administrators unfamiliar with ACM and confronted with an increasingly interfering CG (with its biases toward reductionist science). In 2003, when I returned to Bogor, Ravi and I became ACM 'point persons' (an informal title designed to maintain some continuity of effort and allocate some limited responsibility); CIFOR's ACM program, per se, spun off into other centers of action.

Our approaches in the field

Important ACM actors can be distinguished, in 'ideal type' form, as team members (those directly hired by CIFOR), partners (researchers and institutions with whom we work), and community members, though the boundaries can be fuzzy. The individuals are sometimes the same, but there are different issues to address when a person is acting in these different roles. To some extent, team members are supposed to be 'on the same side'; partners and community members can be expected to have more serious differences. All categories do in fact periodically manifest different interests.

Teamwork

We used a variety of techniques to build and maintain teamwork. The Myers-Briggs Personality Index helped us identify our respective strengths and weaknesses; we developed a philosophy that recognized the usefulness, equality and rights to speak of all team members, from team leader to drivers; we discussed ways to ensure equality by discipline, ethnicity, gender, religion; we learned facilitation techniques to ensure that the quiet (and/or culturally respectful) team members had a chance to voice their opinions and encouraged all to do so; we hired an external group to evaluate our teamwork and organized a retreat to address a significant trust issue that emerged; we experimented with various methods of evaluating workshops we planned (both for iterative improvement and to share our findings (e.g., 'peer assist' (Hailu 2002)); we developed an authorship policy that explicitly stated our equity goals, including guidance on order; we published *ACM News*, a newsletter designed to share information among teams, to build our global teamwork, and to give young researchers experience with analysis and writing (http://www.cifor.cgiar.org/acm/pub/news.html) – recipients later evolved into a widely circulated, and still functioning, 'ACM listserve'; we brought together steering committees, composed of academics, policymakers and NGO activists, in each country and globally to routinely evaluate our efforts, give us feedback and provide guidance on relevance and utility.

Within our teams (both global and within-country), we tried to replicate some of the approaches we hoped to use with our partners and collaborating communities. My own leadership philosophy was 'democratic'; I emphasized the value and potential contribution of each team member, seeing my own role as one more of service to the team than of authority over it. I knew we would need both a broad, unifying framework and considerable freedom and flexibility, given the extreme diversity we anticipated in the field.[16] We recognized the importance of taking risks and learning from 'failure'. In the CAPRi project in Jambi, we couldn't find women applicants for our field researcher positions, though we had intended to hire at least one. We reluctantly hired two apparently qualified men, but it soon became clear that our fears about their ability to work with local women were well founded. They were simply not doing it. We sent in a gender consultant to train them and to work with them for some weeks, to ensure their improved competence. In some cases, difficult choices about project direction had to be made: In Indonesia, the original intention had been to focus on the nomadic, hunter gatherers (*Orang Rimba*); but once in the field, the logistical requirements and research needs convinced the team reluctantly to switch to working only with a settled population.[17]

Although I was aware that, as program leader, I might have to defend these 'failures', in fact that did not prove necessary. Program leaders at that time had sufficient autonomy to make such decisions (or allow field leaders to do so); however, later, in my lesser role as one of two 'point persons' for ACM, my decisionmaking authority significantly diminished (also related to my dramatically reduced level of funds to allocate). Certainly the CIFOR institutional context in the late 2000s was less conducive to risk-taking than had been the case earlier. Recent leadership has been more directive, less tolerant of experimentation and related 'failures', more volatile in conveying displeasure, with a predictable loss of willingness to take risks overall, in the institutional culture. These have probably all been factors in the lack of ACM success in the LM project.

Some of our partnerships did not pan out well, which meant ACM did not succeed. In Kyrgyzstan, the primary facilitator's doctoral committee changed, and he lost the prime proponent of the ACM approach. Lack of enthusiasm from the revised doctoral committee, the country's forestry establishment and collaborators conspired to convince the student and the donor to abandon the effort. In another, the qualified social scientist got another job after CIFOR had signed a contract with the Forestry Institute, which neglected to replace him. The foresters who remained did not have the expertise or interest to implement the approach (nor were we able adequately to supervise their work effectively from afar).[18] In two other countries, the links with the Forestry Departments resulted in personnel with less motivation than we hoped (and found elsewhere), though in both of these countries ACM was implemented; and it may be that sufficient changes in the institutional cultures occurred to justify the frustrations.

One partnership that worked exceptionally well evolved in response to a language problem we had. No one on our team or at CIFOR headquarters at that time spoke Portuguese, yet we had selected two sites in Brazil. Marianne Schmink and her graduate students at the University of Florida proved to be invaluable partners, consistently contributing to the program with minimal financial incentives. Their group took on aspects of ACM, and continued to work with us in a parallel and very constructive fashion, over the years. Some excellent ACM team members came from their group.

Our flexibility granted the teams important freedom and spawned creativity. In Nepal, for instance, the team was able to respond to the government's desire for a separate, complementary, multi-site study addressing some of the issues identified in the ACM sites (Ojha *et al.* 2002); in a second phase, the teams responded again to Nepali concerns, in experimenting with linking district administration with village and Forestry Department structures. Local ACM researchers also articulated and marketed the innovative ideas on learning, monitoring and collaborative processes in the local contexts (e.g. by ForestAction Nepal which conducted training on a cost-recovery basis). The African teams built on FLORES (Forest Land Oriented Resource Envisioning System) modeling work begun in Indonesia to examine forest policy implications in Cameroon and Zimbabwe (see Vanclay 2003 or Vanclay *et al.* 2006). In Indonesia, where the government and communities were struggling with new problems linked to decentralization, the Sumatra team was able to respond to the desires of both, by facilitating the development of community election procedures that more equitably clarified community wishes (Kusumanto 2007).[19] Such team creativity was helpful in Brazil, where an anthropologist and a forester had quite different ideas about how to proceed. Each coordinated a separate process, in Acre and Pará, respectively: The Acre team worked closely with an NGO (Acre Research and Extension Group in Agro-forestry Systems (PESACRE)) on the specific task of developing community timber management, including explicit use of C&I (Cunha dos Santos *et al.* 2007). Later the NGO, Technical Centre for Agriculture (CTA), adapted these methods as a standard operating procedure. The Pará team focused on communities' links with timber companies and experimentation with community monitoring, rejecting the C&I approach as inappropriate (Pokorny *et al.* 2007). Both teams included significant capacity building of graduate students. In another case where researchers disagreed about the best approach, we were free to switch one of these researchers to another country; both produced excellent (different) results.

We also had a wide variety of expertise in the different field teams – both in terms of discipline and level of formal qualifications. This too brought dilemmas: An early issue was the appropriate unit of analysis. Throughout the process of developing our conceptual framework, we assumed we understood each other.[20] It was only after we began doing the research that we realized that the biophysical scientists were thinking in terms of landscapes as the central unit of analysis, the social scientists were thinking of villages. There was initial dismay, as we realized our error, and some passionate defense of respective positions. However, ultimately we realized there was benefit to both approaches. Such an agreement would have been much more difficult had we not built coherent and cooperative teams. After considerable discussion, we opted to let each country team determine its own unit of analysis; we would simply make sure we explained the different orientations in our writings – a solution that has worked well.

In the later LM project, different valuation and interpretation of what PAR and ACM meant and how they could contribute, vis-à-vis the contribution of the reductionist and extractive science, was a bigger problem. I believe PAR was seen by many key actors in the LM project as a handmaiden, to be used or not at will. There was ongoing tension within the project, at least on my side, as I felt responsible and wished to conduct ACM in a manner parallel to our other efforts, to learn more about its utility at the landscape level and with more serious attention to broader scale actors. I felt we were missing an opportunity on most of the sites to make a real difference locally, and also to learn how better to implement the approach on a broader scale.

Scale presented another dilemma. Early on, most of us assumed we would be focusing primarily on a very local scale for the PAR work – though all recognized and agreed that our 'vertical' interests were also key. The Cameroon team decided that, for the impact they sought, they should begin at the 'meso-level'. Rather than starting with community work, they began working with government people to identify relevant policy arenas. They structured their community work to provide input to these policy questions (including forest minorities, agro-forest mosaics, biodiversity conservation, sustainable forest management, communal and community forests, and fiscal decentralization). I heard little discussion or analysis, for instance, of mechanisms for social learning or the use of C&I within communities, from that site (Diaw *et al.* 2009). Their focus was on a broader scale. The Nepal team worked up to a similar strategy of working at the meso-level, in its second phase ACM work funded by IDRC.

Both with the initial ACM program and in the subsequent, smaller CAPRi project in Indonesia (2004–2006), we learned the most when there was a recurrent back and forth between ideas from the center (Bogor, in this case) and from the field. In the CAPRi project, this was also true of ideas emanating from PAR groups in the village vis-à-vis those formed with District level officials (an innovation we tried to continue, less successfully, with the LM project).

As we progressed, some of the differences in skills became more obvious. As an anthropologist, I'm more aware of the shortcomings related to social skills (facilitation, participant observation, ability to set aside one's assumptions); but certainly each team member had his/her own strengths and weaknesses. We did effective, internal team building – described earlier – to help us recognize people's strengths and fill in the gaps in our teams' knowledge/abilities. In retrospect, after participating in an ACM project led by a (sympathetic) biophysical scientist, I think perhaps having a social scientist lead the group was an advantage for interdisciplinary communication, given the respective prestige of biophysical and social sciences. Also, if a social scientist sympathetic to PAR approaches is in charge, it may be easier to assure that methods like PAR are implemented (Colfer *et al.* 2011).

As the program progressed, the different emphases on different aspects of ACM by different team members became clearer. The flexibility we tried to build in worked. Taylor asks himself in *Unruly Complexity* (2005), 'What would it mean . . . to take seriously the creativity and capacity-building that seems to follow from well-facilitated participation . . .?' In his response to his own question, he talks of 'flexible engagement'. Our philosophy of devolved responsibility and encouragement of open and revolving leadership helped to inspire people to take the lead in a variety of relevant directions:

- One sub-group focused on formal modeling (participatory modeling, system dynamics, agent-based and more conventional modeling methods) of the environment and the problems confronted by local people (see the 13 journal articles in Vanclay 2003, by this group, or Vanclay *et al.* 2006).
- In Mafungautsi, Zimbabwe, for instance, this group worked with local women broom grass collectors to model their experience and help them plan future action.
- In Bulungan, Indonesia, a multi-stakeholder group came together to model the uses to which local resources were put. The ACM group hooked up with the FLORES group, which used modeling (à la *Sim-City*) to better understand how relevant natural resource and human systems worked in Indonesia, Cameroon and Zimbabwe.

- Another sub-group pursued the C&I track (Hartanto *et al.* 2002b; Hartanto *et al.* 2003; McDougall *et al.* 2009; McDougall *et al.* 2010), adapting and developing more suitable C&I with communities for monitoring and improving their management. The Cameroon team used the C&I successfully as a tool to enhance communication among stakeholders (e.g., Diaw and Kusumanto 2005; Tiani *et al.* 2001), including government officials, timber company representatives, and local people.
- In the Philippines, though the team's focus was the community, its interest in C&I fit very well with NGO and government interests in assessing sustainability of forest management on Palawan, broadening local commitment to their efforts. Initial successes eventually met a stumbling block in the lack of incentives for implementing the monitoring (Hartanto 2007). I suspect this could have been overcome had we been able to continue active facilitation longer.
- In Pará, Brazil, Pokorny and colleagues (2007) found the formal C&I, as developed by CIFOR, too complex for local people to use; these researchers recommended simplifying them for local use (a conclusion reached by several teams, and one, in fact, envisioned by the Bogor team; see also Cunha dos Santos *et al.* 2007, on Acre).
- One Nepal team had a similar experience, and converted their approach to a more qualitative and locally appropriate one (Paudel and Ojha 2007); see also Pandey (2002) for a more positive experience of C&I in Nepal.
- Another sub-group focused on social learning, closely related to monitoring, as the central concern of the project (Kusumanto *et al.* 2005; Wollenberg *et al.* 2001b; Guijt 2007), approaching adaptive management in a more qualitative manner.
- In Jambi, Indonesia, the team addressed political processes for fairness (Kusumanto 2007); Sithole (2001, 2002) examined micro-level power issues in Zimbabwe, both contributing to project concerns with equity.
- Yet another group saw collaboration as the key (Wollenberg *et al.* 2005; Yuliani *et al.* 2007), recognising the role of conflict (see Yasmi 2007, in Indonesia; de Koning 2009, in Cameroon) and of equity issues (Colfer 2005a; Indriatmoko *et al.* 2007; Wiliam-deVries 2006). The Nepal team found collaborative monitoring and PAR to be useful in encouraging equity in Nepal (Milne *et al.* 2001; McDougall *et al.* 2007a; Dangol 2005), as did the Bolivia team (Cronkleton 2005).

These sub-groups within the global ACM team were not at all coterminous with individual field teams; and their results reflected enormous amounts of individual creativity and adaptation of our initial framework – though none of the above examples *contradicts* the original ACM plan. Rather, the experiences of the field teams served to improve on and adapt the suggestions originally made. Indeed, these varying interests among the team members in a single country served the function hoped for (though rarely achieved) with 'matrix management' (wherein an institution is divided in two ways, sometimes by discipline and by topic).

Periodic facilitated meetings reminded us all of the links between our respective bits. The core team in Bogor had yearly meetings with our global Steering Committee, attended by a small number of varying field personnel. These meetings were intense, involving serious, critical assessment, advice, and some training, about our methods and our progress. In each country, the teams also convened a national Steering Committee, with similar intentions (and varying utility).

One particularly effective, internal technique was a half-day workshop in which team members created pictorial images of how the ACM components fit together.

The writing workshop we convened, led by Bob Fisher, in November 2001 in Bogor, was extraordinarily valuable (and something we repeated several times). This two week long, facilitated workshop granted 35 competitively selected team members (from 10 countries) the opportunity to take a breather from fieldwork, interact with others involved in parallel kinds of work in other parts of the world, enhance their analysis and writing skills, and take advantage of CIFOR's other personnel and resources (*CIFOR News* 2002). This kind of opportunity (expensive as it is) served as a powerful motivator, in terms of continuing their field work, facilitating cross-site collaboration and capacity building.

One problem to which no truly successful solution was found related to the distances involved among our sites. That 'the squeaky wheel gets greased' is a truism; and distance reduces the volume of the 'squeak'. In Bogor, we agreed intellectually to prioritize field teams. I posted team leaders' names in large letters on my wall, to remind me to respond to their needs first. Still, we relied on email, and many of the teams were (had to be) in remote areas without regular access. We tried a web-based collaboration tool that was inaccessible to most teams. We put such contents on CDs and sent them to the field, but also with marginal utility.

A related problem was more fundamental for researchers and their professional development. Sited in Bogor (with excellent internet connections, reliable electricity, air conditioning and equipment, a good library, stimulating colleagues, access to materials from all the sites), I could pull together materials far more easily than could any of our field team members. Although we all agreed that any team member could lead a collaborative research effort, edit a cross-site book, analyze his/her own materials, there is no denying that such efforts were far easier for me and the other Bogor-based team members than for those in the field. This inequity troubled me; we never found an adequate solution.

Later in the project, the different disciplinary traditions relating to publications emerged. Many biophysical science journals require a standard succinct format (introduction, methods and materials, results, discussion), with each section clearly differentiated and with expensive page charges. Many anthropologists, on the other hand, are more comfortable 'telling a story', and their journals reflect this freedom. Even simple formatting requirements differ from journal to journal – creating further hurdles for interdisciplinary and cross-site analyses.

One potential source of problems (which we managed largely to ignore) was CIFOR's institutional preference for refereed journal articles in English.[21] Our team agreed that to have impact (part of our mandate) one could not rely only on such documents. We needed an array of documents for different audiences. In the Philippines, where the population was largely literate, the team helped local people reflect their experiences and concerns to NGOs, donors and government officials. They also helped the PAR group (a local timber managing coop) publish a community newsletter, *Balangaw* (Hartanto *et al.* 2002a); as did Indonesian teams. In Zimbabwe, the Forestry Commission, inspired by the ACM work, decided to expand it to nine other districts; the ACM team provided a training workshop, including a 'how to' video on ACM (Zimbabwe ACM Team 2004).

Prabhu commissioned a wonderful series of PowerPoint slides, combining theoretical knowledge of how ACM works with beautiful artwork, for use as curricular materials (Ruitenbeek and Cartier 2001). The Indonesia team has produced a number of documentary videos: one documents the ACM process in Sumatra (Jambi ACM Team 2006); others are a series of interviews with local stakeholders, explaining their perceptions of local problems and issues in Borneo (Malinau ACM Team 2005; Danau Sentarum ACM Team 2005). The teams' published works range from short articles in *ACM News* (which I edited until 2003, helping junior staff with format, English, scientific convention, etc., and distributed, mostly via the web), to manuals on the ACM process in single countries to edited collections to formal journal articles and theses.

As will be clear from the next section, there was considerable interaction between our teams and local partners. These in turn, along with the National Steering Committee members, served as local conduits for our findings. In some countries this was quite effective, in others, less so. Some places, budgets were allocated for national level summary workshops. With the benefit of hindsight, I would devote more attention to country-level dissemination.

Partnership

ACM 'partners' included NGOs, academic and government institutions, and independent researchers. The core team members selected their partners during the early days of their efforts, depending on the availability of needed expertise, willingness/interest to collaborate, and knowledge about the area in question. Some teams opted for close links with the government forest establishment (Bolivia, Ghana, Kyrgyzstan, Malawi, Nepal, Zimbabwe); others kept their distance, at least in terms of formal affiliations, preferring NGO links (Brazil, Indonesia). Some teams made extensive opportunistic use of students (at least Brazil, Bolivia, Indonesia, Kyrgyzstan, Philippines, all the LM teams); others didn't. In Zimbabwe, links established during the ACM process in Mafungautsi resulted in methodological sharing with researchers on other sites (Chivi and Mutoko), which ultimately resulted in comparative analyses (Mandondo *et al.* 2008). A similar pattern emerged in Cameroon (Diaw *et al.* 2009).

In the early days, our Philippine ACM team learned of the work of Delbert Rice and the indigenous group with whom he worked. The team contracted Rice to write up his successful experiences, as background for use in developing their own activities. His input convinced the Philippine team, and then the global team to change the relevant part of our program's name from 'Adaptive Co-Management' to 'Adaptive Collaborative Management'. He argued that the use of the 'co' implied equivalence of rights among stakeholders, and that indigenous groups in fact had stronger rights, which we should be supporting. We agreed.

As noted above, we tried to coordinate our work through a democratic process. We wanted input from our partners and we wanted to involve them in planning, implementation, analysis, all phases of our work. In some countries, the scientific research establishment is organized hierarchically (see Ojha's 2008 analysis from Nepal), with supervisors having the right to publish the work of their subordinates. This was against our philosophy of 'sweat equity'. Although in most cases we were able to overcome such hierarchical preferences, in Nepal, in at least one instance, we failed. The lead researcher, not involved in actual fieldwork, insisted that his name be first on a paper that the fieldworker had written about her own experience. We discussed the problem in Bogor and gently reminded the lead researcher of our 'authorship policy', but he did not budge. We reluctantly capitulated, feeling that continued collaboration was more important than this one issue.[22]

Most funds came directly to CIFOR, and we were responsible for financial management and oversight, but were also free to dispense them as we chose. In one of the Indonesian sites where we got second phase funding, DFID proposed a different arrangement. DFID expected to enhance equity – a concern we shared – by contracting directly with both CIFOR and the NGO partners; we readily agreed. However, it soon became clear to DFID that our Indonesian partners were unfamiliar with standard, western-style bookkeeping norms. Western-defined corruption was the norm, and it is even highly probable that our partners, at least in some instances, did not even realize that the donor (and CIFOR) considered certain practices corrupt. DFID decided that CIFOR should monitor the partners' expenditures, thereby invoking both the inequity of the old arrangement and the transaction costs of the new. A general distrust among NGOs of CG centers was exacerbated by this arrangement, necessitating a huge expenditure of time and energy for soothing partner sensibilities. CIFOR's financial section did not help matters. They were comparatively uncooperative about speeding up reimbursements, even when the NGO's bankruptcy seemed imminent and the core Bogor team urged them daily to reimburse expended funds.

Again, I am ambivalent about the implications of this experience. Clearly giving funding to an external institution like CIFOR exacerbates international inequities and misses an opportunity for capacity building within-country; yet the transaction costs were extraordinary. CIFOR's site coordinator was frequently at her wits' end, and CIFOR personnel had to make a number of emergency trips to the province to soothe ruffled feathers[23] and provide needed financial training. Still, work continued on the site, much quite valuable. My conclusion was that it *was* worth it; the Indonesian site coordinator's was that it was not.

In my early imaginings of the project, I expected that the qualified people with whom we were partnering would live in the field, working closely with communities and other stakeholders. I had my own field experience in mind: I'd lived in villages and worked with the people, for one to three years at a time. My reasoning was that, given the three-pronged nature of our effort – facilitating a PAR process on the one hand, and conducting research on that, as well as conducting context studies and assessing cross-site variables – we should have a seriously qualified person. I imagined that ultimately a less qualified person (such as an extension agent), with focused training, could conduct ACM alone, *without the research*. Research would be less central, once we'd confirmed that the ACM approach was working as anticipated (see e.g., McDougall *et al.* 2009, a manual specifically designed for this kind of use in Nepal; or Kusumanto *et al.* 2005 for Indonesia; Hartanto *et al.* 2003 for the Philippines).

The scenario I'd envisioned was rare among our partners. The formally qualified person we hired simply hired others with fewer qualifications to do the fieldwork. This meant that the work I considered the 'real work' was conducted by the least qualified team members. Since we were uncertain about the actual level of expertise needed, I re-calibrated my expectations, and consoled myself that we could learn from the experience whether such expertise was critical or not. My conclusion was that there needs to be someone, actively involved, who has such capabilities and can help those with lower levels of training.[24] Some people with minimal training are 'naturals' at facilitation – but that's a matter of luck, in the real world.

Another problem arose during analysis and write-up. A consultant produced a booklet on micropolitics that was considered by the local ACM team to be too sensitive to publish (Sithole 2002). The team feared that the analysis would adversely affect their relationships with their Forest Commission partners. Several solutions were considered (removing the offending section, sharing it as written, or waiting until after the project ended). The solution, not entirely satisfactory to anyone, was to withhold the publication for some time, only distributing it years later. This stimulated anger in the author, anxiety among the team members, and loss for the scientific community as the paper was of value. Still, given the tense political situation on site, its publication might well have caused serious problems for the team. Such dilemmas, common in anthropological work, require careful consideration of ethics and likely practical outcomes for local people, researchers and the project itself.

The most successful partnerships were those in which both parties were able to pursue important goals they valued. Our work in Brazil and Bolivia, for instance, benefited greatly from the congruence of our interest in testing the method and the University of Florida's Latin American Studies Center's interest in conducting research on a combination of environmental and human concerns and in training their students (e.g., Noemi Porro, Peter Cronkleton, Samantha Stone) and NGO partners (Magna Cunha dos Santos). In Nepal, CIFOR occupied an intermediary role, between that of our committed NGO partners (one quite academic; three others more action-oriented)[25] and the government's forestry establishment.[26] Although this linking role sometimes involved a balancing act, it granted access to experienced and motivated NGO personnel, and the possibility of both expanding the work geographically and, in later years, incorporating it more meaningfully into the institutional setting in the country.

Another key feature that affected success was the availability in each country of ACM team members with strong commitments to and knowledge/skills related to the approach. This and the degree to which they were personally integrated into the global program affected motivation. The Zimbabwe team was led by Ravi Prabhu, one of the originators of ACM's central ideas, and a creative researcher personally committed to making it work; Cameroon (Chimère Diaw), Nepal (Cynthia McDougall), Indonesia (Lini Wollenberg, Linda Yuliani), and the Philippines (Herlina Hartanto) benefited from similar informed leadership and enthusiasm. All of these leaders had been involved from the beginning in developing the program. The lack of success in Ghana and Kyrgyzstan (and to a lesser extent, Malawi) is, in my view, directly related to those teams' lack of access to such integration, expertise and backup. We tried to coordinate activities in those countries from afar, with short training programs but little in the way of routine supervision. It didn't work.

There was a marked differentiation in the kinds of research results/products desired by different partners. Graduate students and collaborators from other research institutions were most interested in scientific papers (preferably, in many cases, published in refereed, English language journals,[27] and often in electronic formats). Many also assumed (mistakenly, for many of our partners) that electronic materials would be more widely available. Developing country partners, particularly those with rural forested research interests, wanted hard copies (reprints, books, manuals, policy briefs)[28] because of the expense or unavailability of internet access and, in some cases, frequent losses of electrical power. NGO practitioners leaned toward practical manuals, showing less compulsion to finalize scientific products in English. Government officials worked well in the shared production of policy briefs in some cases, particularly in local languages. Both NGO and government researchers appreciated workshops where results were shared orally and/or by means of videos.

Working with communities

We were cognizant of the discrepancies in status and rights within communities, whether related to gender, ethnic group or caste (see collections by Colfer 2005b; Indriatmoko *et al.* 2007). In Sumatra, as noted earlier, our 1999 plan to work with the hunter gatherer *Orang Rimba* failed because of the incompatibility of their mobility and the research demands on the team's time. Instead the team focused on the settled community of Baru Pelepat. In the subsequent CAPRi follow-up work in Indonesia (2004 and 2006), one team member, fearing dietary and other kinds of personal contamination, refused to work with the *Orang Rimba*; they were not Muslim (as was he). We discovered this prejudice so far into the research that hiring a new facilitator was impractical, and other aspects of his work were going well – a second failure with this particularly marginalized group.

Similar problems occurred in other countries. Educated team members tended to be from the country's dominant ethnic groups – not the majority groups usually residing in forests. This granted community members from those dominant ethnic groups better access to the team, usually better communication with them (through shared language, culture and experience), and the teams were better able to express the interests and viewpoints of such groups. This was not our intention; but we were never able fully to overcome it, despite serious efforts to do so. Like working with women, attending to such ethnic and other kinds of marginalization demands special effort.

In several countries the population's experience of outsiders was primarily as a source of funds or material goods; such contributions were welcome, given their minimal incomes. Our teams were seen to come 'bearing gifts'. A big part of both the ACM approach and rationale has to do with the capacity of the people to continue the work after the team leaves,[29] so it has been important to clarify (and insist upon) the team's inability and unwillingness to fund the various activities or construction projects local people have in mind. My conclusion from our experience is that, though difficult at first for people to believe and accept, the concept of financial sustainability is one that villagers readily understand.[30] On the other hand, small amounts of somewhat discretionary funds can be very helpful in moving community plans forward (e.g., a cross-visit to a site where people have accomplished something the group wants to do (Arda-Minas 2002); funds to visit the district capital to pursue a possible source of routine funding; a prize for a community-planned competition; etc.).

Another issue that affected our work in communities pertained to national level politics. In two countries, Zimbabwe and Nepal, there were serious conflicts marked by violence during our work. In Zimbabwe, the teams were unable to address many politically sensitive topics, because to have done so would have endangered their lives. In Nepal, during the Maoist insurgency, there were long periods during which our teams could not safely go to the field (though sometimes, team members simply braved the danger). Interestingly, these two countries had vibrant ACM processes, at least during the early phases of the violence. Still, although Mutimukuru-Maravanyika and colleagues (2008) conclude that 'Using the ACM approach in politically volatile situations will not yield lasting results', they go on to emphasize its

> potential for enhancing the lives of local people and the status of natural resources. Under [stable] conditions, making joint learning processes deliberate and conscious enhances adaptiveness in resource management.
>
> (Mutimukuru-Maravanyika *et al.* 2008: 57)

In my own 2003 cross-site assessment, I concluded that 'the degree to which conflicts in fact emerged as *opportunities* for bringing about collaborative behavior was considerably greater than expected' (Colfer 2005a, italics in original: 151). The difference between these two conclusions may well be in the intensity, spread and duration of the conflict.

Much has been written about the depth and meaning of the term, 'participation' (Arnstein 1969, Cooke and Kothari 2002; Greenwood and Levin 1998) – ideally a concept closely related to Habermas' 'deliberation'; and the issue remains crucial. A lot of community work that is underway around the world represents very minimal kinds of participation – doing a survey, calling a meeting, meeting with the headman once. ACM, as originally conceptualized, is founded on a deep and serious attempt to involve local people in meaningful ways; yet the recognition that one needs special skills to catalyze people's involvement and understand their systems remains under-recognized (cf. Colfer *et al.* 2011). We have had varying degrees of success at reaching these deeper, more fundamental levels of understanding, but I believe that the degree to which we have been able to tap into such knowledge has been positively correlated with our successes.

Outcomes of our practice

In the 2003 evaluation mentioned above (Colfer 2005a), I tried to ascertain our impacts at that point on each site. These impacts would basically reflect the success or failure of our research, since our central goal was to catalyze beneficial changes to communities and landscapes. Although a number of people have evaluated what we have done, and only one donor ever expressed dissatisfaction with our results,[31] we remain dissatisfied with our ability to reflect adequately either the real impacts we have seen or their importance for improving forest management. We satisfied CIFOR and our donors by producing abundant publications.

To some extent, of course, the process of participating in an ACM process involves ongoing self-evaluation – as part of the iterative cycles mandated by the approach. People checked to see if they were meeting their self-defined goals, routinely (did the women participate in the forest user group meetings actively in Nepal? Was broom grass growing better in the managed plot than in the control in Zimbabwe? Were the community elites responding to the community's collective efforts to reduce their corruption in Jambi? etc.).

But stepping back and looking at the approach as a whole, the most significant changes fell into four categories (livelihoods, environment, empowerment and social learning). The impacts were most impressive in the areas of learning (improved adaptive ability), empowerment and increased knowledge/awareness relating to livelihoods and the environment. Empowerment included the following topics: strengthened social capital/trust/cooperation within the community or between groups, increased skills at negotiation/expression/representation, greater equity, and improved internal governance (transparency, checks and balances). It was very clear at that stage that length of time in the community typically translated into more significant improvements. But actual, visible livelihood and environmental improvements were minimal. The teams were particularly frustrated by the difficulties of both bringing about and measuring environmental changes. Even tracking livelihoods, many of which included major subsistence elements, was difficult. Prabhu *et al.* (2009) found similar strengths to those outlined above, in a subsequent analysis of three sites over four years. Milne *et al.* (2001) conducted an early methodological comparison between extractive and participatory approaches in Sumatra and Nepal, noting the strengthened buy-in and learning that was taking place, even at that early date in the participatory efforts.

There have been several other formal assessments of outcomes on specific sites. Steve Rhee devoted several years, as a Yale graduate student, in the early 2000s to an ethnographic analysis of ACM in Malinau, CIFOR and the contexts in which they operated. Although Indonesian chaos at the time (related particularly to the central government's 2001 decentralization policy) was far less violent than Zimbabwe's or Nepal's, still local institutions were in a continual state of flux, with corresponding unpredictability for the team (see Rhee 2006 and/or Wollenberg *et al.* 2007). Rhee emphasized the excellent relationship the ACM team had with local communities, and the team's accepted and appreciated roles as brokers, conveners and sources of useful information; he also noted the community (and team) disappointment in their inability to develop a constructive relationship with district government or to catalyze the development of concrete improvements in livelihoods. He contrasted a) the comparative powerless of both the ACM team and CIFOR in interaction with local government,[32] with b) CIFOR's global stature and influence. Rhee also noted CIFOR's institutional culture at that time that focused on what CIFOR teams could learn from their grounded research, and pass on to the global scientific and development worlds. He saw CIFOR's 'being there', being in the field and thereby conducting firmly grounded research, as an important contributor to the institution's global authority and legitimacy. Certainly valuable analyses have come out of the Malinau experience he examined.

At the end of the formal ACM project in Jambi, Indonesia, Indriatmoko *et al.* (2006) used locally developed C&I to assess the overall impacts of our work there, comparing conditions in 2000 and 2006. As with my own earlier conclusions, improvements in a number of cognitive, social and skill-related qualities were obvious; environmental and economic improvements, far less so. Indeed, the effects of larger scale decisions about transmigration and oil palm development in the region contributed to a *decline* in environmental quality;[33] though the people were far more involved in determining the conditions under which these decisions were implemented than had been the case before ACM. About the same time, Moeliono (2006) conducted an assessment for the donor (DFID), which noted the high cost of the Jambi effort. He, like others, noted the comparative impotence of the CIFOR team and its NGO partners vis-à-vis the governmental bureaucracy. Though impressed with the vibrancy of the multi-stakeholder learning forum the ACM team assembled at the district level, he was disappointed in its inability to act, due partly to its lack of an institutional home.

Mutimukuru-Maravanyika (2010) provides a longer term view of the Zimbabwe outcomes (discussed in Chapter 6), stressing what she considers her team's political naïveté, expecting too much from participation, when (in retrospect) she felt the emphasis should have been more on making political changes. Still, I agree (as indeed, we intended) that attention to the distribution of power and structural elements in any context are central issues that need to be addressed. Subsequent work has included more explicit attention to politics, with some valuable insights; there's definitely more work to be done.

Since the 2003 assessment, a number of sites have managed to move ahead with ACM-like activities and related spinoffs. In Bolivia, for instance, the team expanded its work to the northern area, Pando, where they have worked with local communities to map trees, trails and other key features of the local Brazil nut harvesting landscape (Cronkleton *et al.* 2010). In this study (similar to one carried out in Borneo; Anau *et al.* 2005), the authors addressed the more 'political' issue of tenure.[34] Evans *et al.* (2006), building on their field experience, developed a field guide, showing users how to facilitate four kinds of future scenarios – a technique still used in the early stages of much ACM work. Another group explored the strengthening of grass roots networks (Taylor *et al.* 2010).

The work in Indonesia has had a variety of incarnations from the first ACM phase, funded by ADB, to a second one with DFID, expanding to a

GTZ/CAPRi-funded effort that aimed specifically at implementing PAR simultaneously in both community groups and among district officials in two districts in Jambi, Sumatra. That work (Komarudin *et al.* 2008; Komarudin *et al.* 2011) provided insights into ways to link the interests of local people and district officials (e.g., land tenure and access by local people with land use planning by bureaucrats; or village women's desires for empowerment with district level concerns to follow new legislation requiring more involvement of women in politics). It also produced a manual in Indonesian for use by district officials about gender (Wiliam-de Vries 2006); and a large number of policy briefs, produced jointly by the CIFOR team and district officials, in both English and Indonesian. District officials, struggling with how best to implement decentralization, were active in this process. This work was then taken up by the LM project, under ICRAF's direction, in the same district where ACM began; however, their interest in ACM was nil – they feared the approach was too time consuming. We did, however, conduct a pre-test of a 'governance assessment tool' there, designed to evaluate the quality of local governance (Colfer and Feintrenie 2011) – one step that could contribute to making ACM more politically savvy.

Nepal, like Indonesia, has managed to maintain ACM funding. The government has been interested from ACM's beginnings in improving its record on equity issues. As in Indonesia, one concern has been to improve collaboration between community groups and the formal, lower level government structures on which they depend for certain services. The team recently produced one summary of ACM conclusions for policymakers (McDougall *et al.* 2010) and another, longer guidebook for practitioners (McDougall *et al.* 2009). Banjade *et al.* (2008) describe a second round of PAR-like activity (called PAL, participatory action learning) in four new communities, focused on equity issues; the collections by Ojha *et al.* (2008a, 2008b) represent additional related spinoffs of relevance for ACM practitioners.

I give these examples simply to illustrate the locally relevant evolution of the method along with the useful spinoffs that have been partially stimulated by it. These outcomes and spinoffs have been diverse and thus difficult to capture succinctly; yet the wealth of materials that have been produced satisfies me that the research outcomes have been significant. From an empowerment perspective, there is ample evidence of significant cognitive and attitudinal change at community and higher levels from the ACM work. I believe for concrete, local level conservation and development outcomes (in terms of incomes and sustainability), longer time frames are needed.

Conditions for successful ACM

In the preceding text, I have shown that ACM can add value both in deepening learning and collaborative process, and through this, in generating better outcomes on social and environmental systems. That does not mean it can be applied successfully in every situation, and without challenges. Here, I return to one of ACM's original central questions: Under what conditions does ACM work? As a team, we continuously monitored our ACM process to identify and characterize the conditions and the critical processes required to move ahead.

Armitage *et al.* (2009), perhaps more ambitious than I, list 10 'conditions' for successful ACM. I see their conditions more as desirable goals that are in fact rarely met. It would be sad if researchers rejected the approach because, for instance, property rights are not 'reasonably clear': they rarely are in forests. I see many of their conditions as subject to change with the *use* of ACM.

Here, I first identify four conditions I consider essential for ultimate ACM success.

1 A problem must be recognized that is of sufficient gravity to motivate and mobilize participants to act voluntarily. Yet, conditions should not be so chaotic that prediction becomes impossible.
2 Facilitation, building on skills at social analysis, group process, diplomacy, and networking (with consistent, sustained concern for equity), must be regularly available and used.
3 The facilitator must operate within an institutional context that grants him/her significant freedom (including 'freedom to fail' and learn from such failures) and access to actors at various scales.
4 There must be a sufficient, funded time frame to follow up on actions planned, implemented, monitored and revised.

The five sections below expand on aspects of these conditions as experienced by our teams: community and district requirements, institutional issues, attitudes, the process itself, and the crucial issue of sufficient time.

Community and district requirements

One can begin an ACM process in almost any community, if one properly analyzes the local power structure, and attends to the realities and priorities of both marginalized members (women, lower castes, despised ethnic groups) and elites.

- It is typically necessary to either constitute PAR groups based on shared interests (e.g., increasing production in a garden area) or characteristics (e.g., young women or members of a hamlet); or fit in with existing groups (e.g., governmentally instituted cooperatives).
- Would-be participants (such as community sub-groups or, less commonly, district officials) must express some interest in the process. Participating in an effective and sustainable ACM process is probably necessarily voluntary.

- A sufficiently stable political context is important. There must be reasonable freedom of movement and no outright warfare, though problems and conflicts often actually contribute to ACM success by strengthening local will to resolve them.

Conducting complementary ACM processes with district personnel as PAR groups is desirable, but adds considerable complexity, particularly relating to the greater power of the participants (vis-à-vis the facilitators) and participants' greater geographical and administrative mobility as compared to most community members (see Moeliono 2006; Komarudin *et al.* 2008; Colfer and Pfund 2010). It can, however, open new doors for longevity of, and resources for, the more local-level processes – as seems to have happened to some extent in Indonesia (CAPRi), Madagascar (LM) and Nepal (ACM). I discuss some ambivalence about such government involvement below.

Institutional issues

Implementing an ACM project requires a high level of flexibility. This can vary from no pre-determined outputs to a number of them, but the fewer, the better. The approach stipulates that communities decide what they want and how to get it, within the constraints of their contexts (also subject to change). This resistance to pre-determined outputs is based on the conviction and experience that the ACM process will only continue to function, over time, to the degree that there is PAR group buy-in.

I attribute some of the failings of the LM project (in regard to ACM) to CIFOR's more stringent financial and time management policies. In the late 2000s, a 'time tracking' system was initiated, which required researchers to allocate each activity to a particular budget code (reducing various supervisors' time available to provide planned supervision). Where previously research and support personnel had been able to cooperatively determine the performance of needed tasks (allowing considerable responsiveness), new policies required pre-planning and attribution of every cent to a budget code – all reduced institutional flexibility.

The ACM team or facilitator must have the ability to take the institutional risks needed for implementation of PAR. His/her job must not be on the line when plans do not materialize quite as envisioned. There must be freedom to fail and to learn from such failures. This was not the case at CIFOR in the late 2000s.

There are additional requirements within the ACM team:

- Team members need good, constructive collaborative relationships.
- Ongoing self-evaluation and some source of continued motivation are also needed.
- Those field researchers best integrated into the global ACM team had the most success.

A facilitator's ability to draw on multiple sources of expertise, ideally in and out of his/her host institution and in the community context, is key. The process of changing local conditions – as ACM attempts to do – usually requires some external information, assistance and support. Part of the task we undertake when we begin ACM is to help local communities gain such support. This can be partially accomplished through formal partnerships, partly through a viable and vibrant network of resources.

I consider our distributed leadership approach to have had positive impacts, in the sense that field researchers felt comparatively free to experiment and follow leads that emerged in their particular contexts. That led to greater than average creativity and motivation, with positive outcomes, both for the site and for the global team. On the downside, there was a definite reduction in routine institutional support for our program when Bogor-based, formal program leadership ceased.

Attitudes

An ACM approach requires certain attitudes from the people who implement it. They must have sincere respect for community members (or other participants) and understand the utility of all participants' knowledge. This suggests an attitude of humility.

A facilitator must be sensitive to political realities, including the micro-politics within and among PAR groups and others in their contexts. There can be significant dangers (even death) to local people if care is not taken. Dangers to the continuation of project activities are more common.

ACM researchers must be comfortable with the risk of temporary failures (they happen in any event); and researchers must recognize that ultimately the existence and direction of change are in the hands of participants. This means they must be willing to acknowledge the ultimate decision-making power of the participants (not themselves or their institutions); they must relinquish much of their own (often false) perceptions of control.

The process

Facilitators' abilities and understanding of the ACM approach are among the most central elements of a successful ACM process. Facilitators must have the ability to catalyze true participation (in the sense of 'deliberation'), which usually also implies abilities to look at local situations holistically and analytically and understand cultural issues. Such skills are sometimes intuitive, but often they must be taught. Without such skills success is unlikely.

Facilitators must be comfortable with 'speaking truth to power'. This is true in their interactions within communities, but also in their own institutions and with other power-brokers in their field contexts (e.g., the timber industry, conservation projects, government officials). They have a unique position that combines the advantages of a perception of their neutrality (at least of their fairness) with the necessity to strengthen the voices of the marginalized (discussed in McDougall *et al.* 2007a; see also Greenwood and Levin 1998).

A deliberate process of handing over the reins of facilitation to participants should be built in from the beginning. Just as leadership is devolved within ACM teams, the ACM process is likely to work best if leadership is shared within the PAR groups. This is important as a capacity building mechanism for long-term sustainability, as a way to maintain the cycles of learning, improved equity, and links to outside actors (networking).

Time required for an effective ACM process

We have never had the test of ACM that we envisioned; serious livelihood and environmental impacts will take 10–15 years – a length of time donors so far have been unwilling to give. I have found the short term results we witnessed encouraging enough to still consider ACM a viable and valuable approach. But the length of time available for facilitation has been a crucial element in success. I believe with longer term funding it could be far more valuable and transferable.

One's expectations have to be geared to the time available. Our own funding of two to four years has been universally decried as inadequate. Much more can be accomplished when the momentum can be maintained. Startup costs, in terms of time and sustained effort much more than cash, are considerable. Over time the costs diminish and the payoff increases dramatically (assuming there is some periodic source of continued motivation and support). Without further input, though, the process is likely to grind to a halt, because of the lack of some resource that could be easily provided or a small problem that could be rectified, with small external inputs. With a short time frame, the human resource capacity building remains (a notoriously difficult 'impact' to track); action on the ground is likely to cease.

Discussion: expanding ACM approach

ACM's further expansion and consolidation depends on how ACM researchers and practitioners handle four critical issues: a) working among scales, b) linking with the political sphere of governance, c) building effectively on the positive elements of local people's cultural systems (which entails overcoming deep bureaucratic inertia that works against ACM), and d) persuading donors and sponsors to invest in ACM-like approaches.

Political and cultural aspects of ACM deserve some additional discussion.

ACM, scale and political action

The original ACM concept included clear attention to politics, both in its emphasis on the empowerment of community members/groups and in its explicit attention to multiple scales (the 'vertical' element). Turning first to the question of scale, our acceptance of the Cameroon team's decision to focus initially on the 'meso' level, for instance, rather than beginning at the local level, was clear acknowledgement that both levels were seen as important to the program and to the effort. Looking back on the evolution of ACM, we have moved progressively toward more explicit linking of levels. The Cameroon team began to incorporate more attention to local actions; and a number of other teams (Nepal, Indonesia, Bolivia; and the more recent efforts in Madagascar, Laos and Tanzania) increased – or at least planned to increase – attention to district or landscape levels. This was in recognition of the potential for local communities to monitor governmental actors and the possibility of additional resources and support available from governments.

When collaborative action begins at higher levels, there is a corresponding trade-off, a loss, in PAR in communities and with their sub-groups. In the original ACM program, PAR at the village level or below was attempted in all but one country; and seven of the 10 countries[35] included sites with high PAR 'grades' – see Table 3.4. In the Landscape Mosaics project, which began with an intended two-level focus (community and landscape/district), only two of the five country teams managed to conduct PAR, and then only at the community level (with these beginning late in the project). Very little was accomplished with regard to community empowerment (Colfer *et al.* 2011). More obvious results have been achieved beginning at the 'bottom' and working up; and of course, longer time frames have helped as well (e.g., Nepal; ACM and CAPRi in Indonesia).

A related issue is the degree to which the teams were able to follow up on our intention to work toward empowerment of forest communities and groups. Some informal external commentators have concluded that ACM was based on a generous portion of political naïveté; and recently Mutimukuru-Maravanyika (2010) reiterated this charge. ACM team members included a wide range of expertise, with several contributing political ecological sophistication: David Edmunds and Lini Wollenberg, for instance, had worked on devolution in forests for several years prior to ACM, and were thus interested in ACM's political dimensions. I myself began work on national-level governance issues, partly in response to needs identified under ACM.

There was considerable discussion within the global team about the degree to which we should focus on negotiations to reach agreements (as emphasized by one of our long-serving advisers, Robert Fisher (1995; and later by Jean-Laurent Pfund, the LM project leader); or on the more overtly political approaches proposed by Edmunds and Wollenberg:

> We suggest that rather than try to *eliminate* or *temporarily neutralise* political differences with negotiations to achieve broad agreement, practitioners should use negotiations to build alliances, gather information and test ideas strategically, with the explicit goal of increasing decision-making power of disadvantaged groups.
>
> (Edmunds and Wollenberg 2001: 232)

Table 3.4 Scale/starting points in ACM and longer term spinoffs (post-2005)

Country	*ACM Phase I (1999–2002)*	*ACM Phase II (2003–2006)*	*CAPRi (2004–2006)*	*LM (2007–2010)*	*2003 'grade'* (ACM only)*	*More recent examples of ACM-related contributions/spinoffs*
Bolivia	Local	Local	–	–	H	Collaborative mapping in conflict resolution (Cronkleton *et al.* 2010); development of participatory tools (Evans *et al.* 2006)
Brazil	Local	Local	–	–	L,M,H	Use by NGO, CTA, for community timber management in other communities
Cameroon	Meso	variable, splintered	–	Local–District–National	L,M,H	Continued use through Model Forests; collaborative use of COAIT[1] (Bonis-Charancle *et al.* 2009); links with GIS[2] (Robiglio *et al.* 2009); use as a conflict management tool (de Koning 2009)
Ghana	Local	–			L	?
Indonesia	Local, National	Local–District–National	Local–District–National	Local–District–National	H	Continuing ACM effort – on collective action (Komarudin *et al.* 2008); on gender issues (Wiliam de Vries 2006; Indriatmoko *et al.* 2007); on conflict management (Yasmi 2007); on links to national policy (Feintrenie and Martini 2010); on Bungo District (collection by Adnan *et al.* 2008); district level policies/issues (set of CIFOR policy briefs); on forest stewardship agreements, village forests, and REDD[3] (Akiefnawati *et al.* 2010); on conservation (EU Project, Y. Laumonier, L. Yuliani; Yuliani *et al.* 2008a, 2008b)
Kyrgyzstan	Local	–	–	–	L	?

Madagascar	Local	–	–	Local–District–National	L	Some follow through with National KoloAla program, use of criteria and indicators for sustainable forest management at district level, Swiss grad student looking at non-timber forest products used by women
Malawi	Local	–	–	–	L,M	?
Nepal	Local	Local–District	–	–	H	Development of policy and training materials (McDougall *et al.* 2009, 2010); ForestAction adaptation of approach (Ojha 2002, 2008); use of indicators (Paudel and Ojha 2007)
Philippines	Local	–	–	–	M,H	?
Zimbabwe	Local–District	–	–	–	H	Publications indicate decreasing sense of accomplishment over time (Mutimukuru 2005; Mutimukuru-Maravanyika 2008, 2010); comparative analysis of different ACM approaches (Mandondo *et al.* 2008); Prabhu's work with a pan-African network on ACM, based in ILRI[4] 2007–2009
Laos	–	–	–	Local–District–National	–	–
Tanzania	–	–	–	Local–District–National	–	–

* H = high success; M = medium success; L = low success

Notes:

1 COAIT: Community Options Analysis and Investment
2 GIS: Global Information System
3 REDD: Stewardship agreement to reduce emissions from deforestation and degradation
4 ILRI: International Livestock Research Institute

Wollenberg continued her ongoing work in Malinau, building on their ideas. Wollenberg *et al.* summarize some differences this sub-team perceived:

> Different from the other [Asian ACM] cases . . . this project team did not seek to test ACM approaches to understand their effectiveness. Thus we did not facilitate communities' development and use of indicator-based monitoring systems with different stakeholders or promote collaborative management per se. We sought instead to help communities create systems for settling differences among themselves . . . and to negotiate with external parties to reach agreements that were more just.
>
> (Wollenberg *et al.* 2007: 138)

The last sentence above would be accepted by most ACM teams as a central goal.

There were various preferences among teams and individual team members about the appropriate level of political engagement. My own view was that different contexts would permit different approaches. With good facilitation, some of the PAR groups in Jambi, for instance, combined these two strategies effectively – as indeed, I suspect, did Wollenberg's team.

Wollenberg and colleagues (2004) provide an outstanding analysis of how they have dealt with the chaos that Indonesia's decentralization process fomented in East Kalimantan (see also Rhee 2006) – a strategy she terms 'muddling through', which could perhaps have been helpful in the chaos of Nepal and Zimbabwe. These authors emphasize the inherent unpredictability in chaotic environments and argue for timely responsiveness to changing conditions, and the value of simply being there.

Some effort was consistently devoted throughout the ACM program and beyond, to looking at forest decentralization policies (e.g., collections by Colfer and Resosudarmo 2002; Edmunds and Wollenberg 2003; Contreras 2003; Colfer *et al.* 2008; German *et al.* 2009; Moeliono *et al.* 2009; Colfer and Pfund 2010) – as a complement to the PAR style efforts, and in recognition of both our political context and 'vertical' political concerns. The LM project had the most explicit attention to national level politics, in its linking of all community and landscape activity with a relevant national policy thrust in each country (e.g., KoloAla, an effort to develop co-management of forests, in Madagascar; resettlement and the development of a new intermediate level governmental institution, the *Kumban*, in Laos; etc.). But this is a different kind of politics from that urged by political ecologists.

Turning to macro-level political implications of our work, Rhee (2006) specifically notes CIFOR's facility at turning local level 'failure' into global success, in the sense of providing globally relevant prescriptions based on learning from local level experience. He writes of CIFOR's ability to

> [lend] an extraordinary amount of credibility, legitimacy and indisputability to their argument, regardless of whether CIFOR's applied research is 'successful' or not. There is a slippage here – by virtue of being an applied research institute, even if the application of their research does not result in the intended outcome, e.g., improved local livelihoods or secure tenure does not result, CIFOR's research is still successful because they can write about the 'lessons learned.' By being an applied research institute, CIFOR can toggle between being scientists and practitioners and hence elide 'failure' of applied research interventions through the idiom of science.
>
> (Rhee 2006: 211)

It does seem probable that our 'home' within CIFOR reduced our capacity to act in the overtly political manner called for by Edmunds, Wollenberg, some Wageningen professors, and most recently, Mutimukuru-Maravanyika. We, like most actors who would like to implement an ACM-like approach, worked within an institutional context, and were, at some level, dependent on the good will of the government in the countries where we worked to maintain our effort. An additional constraint on our political activity was the potential implications for the people living in the communities where we worked. Overt political activity can be dangerous – in our experience, particularly with the Zanu-PF regime in Zimbabwe and the Maoist and governmental conflicts in Nepal, ACM teams were cognizant of these dangers.

ACM and 'culture'

As an anthropologist, one of my interests has been to expand our understanding of people's ways of life. I have felt that successful 'international development' would actually mean the flowering of many different lifestyles and values. I saw ACM as a means to expose our team members to the wealth, complexity, beauty of existing forest socio-cultural systems, and to hopefully link the teams' knowledge with the knowledge, interests and motivations of local communities. I thought such a process could strengthen local communities' voice and power, particularly when tied to the multi-scale networking we also envisioned. I saw ACM as a mechanism to resist and reduce the cultural discontinuities that occur when fast-paced and externally induced changes are forced on local forest communities.

As we begin to look at different scales of action (e.g., a community PAR group with one at district level), we see other enduring mindsets that affect people's behavior and that are somehow resistant to change. Ojha (2008) identified some important ones in many forested contexts, in his analysis of governance in Nepal. He writes of feudalistic, bureaucratic, scientific doxa that result in serious symbolic violence. But his excellent and insightful analysis highlights an important dilemma that interferes with my vision of integrating cultural analyses into the mundane work of trying to improve human wellbeing. The 'deeper' an analysis of cultural or symbolic systems, the less accessible to people from other disciplines (or to policymakers, development workers, etc.).

Within the ACM project, the depth of cultural analysis I hoped for was rare – partly because we had few people trained in such holistic analysis, but also because those who were trained sent less qualified people to the field. Time constraints also mattered. Perhaps we simply tried to do too much within one program.

I confronted another problem – both terminological and conceptual – when I tried to analyze what was happening at community and district levels within the LM project. Looking at events from the community's perspective, I felt the oppression of government, the ways local people were marginalized, the disadvantages of being governed (á la Scott 2009). When I altered my viewpoint, to look at things from the district's perspective, the emphasis magically changed to one of improving the ability of the government to act effectively. If people are to be less marginalized, improvements are needed in governance. But if governments are improved, will they simply become more effective at marginalizing the citizenry? This question strengthens the rationale for the empowerment emphasis in ACM, and perhaps gets at what was troubling Mutimukuru-Maravanyika (Chapter 6).

Access to funds

The significance of long periods in the field for the successful implementation of an ACM process has been emphasized. The lack of funding to support such long-term rural residence/engagement has been identified as a crucial shortcoming of our efforts. As time goes by, and difficult issues such as climate change mitigation and adaptation raise their ugly heads, the need for sustained, collaborative effort in rural areas is likely only to increase. This assumes a commitment to democracy, a belief in all people's right to influence decisions about their own environment and future way of life. A capable, empowered rural population is key to ensuring effective management of the environment and continued improvements in people's living standards and quality of life.

Given that, and the clear evidence that such empowerment and capacity building are long-term projects, the importance of finding ways to access the funds to do this is obvious. But serious changes in donors' standard operating procedures are needed. Particularly in need of change is the requirement for rigid, pre-determined planning, longer term compliance with initial plans, and insistence on short term, quantifiable gains.

But not all the onus should be on donors. ACM advocates need to develop reasonable indicators of our own success. Some have argued for the use of what Robert Fisher has called 'plausible causal connections' as the best option for demonstrating impacts; but so far these have not been sufficiently convincing to donors. The fact that ACM itself includes ongoing self-evaluation and monitoring – as an integral part of the process – should be a potent argument in favor of such an approach. Yet, so far, most donors seem stuck on pre-defined, quantifiable indicators. More work is needed on both 'sides' (donors and ACM researchers/practitioners) of this dilemma.

Ways forward

ACM is an evolving idea, with many people around the world re-inventing, using and adapting it – which is as it should be. At local levels, our work to date has resulted primarily in intangibles – increased trust and self-confidence; skills in analysis, negotiation, proposal writing, networking, equity, collective action. These results are harder to 'sell' than would be clear increases in income or numbers of hectares sustainably managed, but they are critically important in empowerment and governance. I continue to believe that ACM-like approaches are ultimately the most likely to result in improved human wellbeing and healthier environments – through the actions of community members in concert with (or in protest against) actors working on broader scales. This is centrally important now, as we begin to cope more seriously with the challenges posed by climate change. Most agree that the people we work with are likely to be among the most adversely affected.

Some have wondered how best to encourage these approaches, seeing its loss at CIFOR as a harbinger of further discouragement. Although the formal ACM program disappeared, elements have continued vibrantly within the institution. The current Director General has strengthened research on human rights issues (e.g., Campese *et al.* 2009); considerable effort has been expended looking at land tenure (e.g., Sunderlin 2008; or a 12 country action research project reported in Larson *et al.* 2010); CIFOR's climate change adaptation 'domain' has shown serious interest in participatory approaches, particularly in Africa (cf. Brockhaus and Djoudi 2008, Locatelli *et al.* 2008; German *et al.* 2010); work in Asian protected areas has built directly on ACM (Yuliani *et al.* 2008a, 2008b), as well as a new project coordinated by Yves Laumonier; and of course those efforts discussed herein. One reason for the openness of the institution to elements of ACM has undoubtedly been the productivity of the ACM researchers, who have written extensively. As Rhee noted, within the research context, the number of research outputs can substitute to some extent for local level increases in tangibles.

Perhaps more important – and encouraging – is the expanding interest in ACM-like approaches globally. This volume describes valuable experience with some of these approaches. The divergent names, PAR, PAL, Farmer Field Schools, innovation systems approaches, and more, represent somewhat different emphases among like minds developing related approaches that recognize the value of rural peoples' perspectives, interests, goals and potential contributions to making the world a better place. A recognition of complexity, dynamism, systemic interconnections and holistic analyses are likewise features of them all. Here I put forth a few notions about future direction (bearing in mind the optimistic idea that the best way forward is to strengthen local peoples' positions within the larger systems of which they form a part):

- We need to maintain our concern with multiple scales, increasing our efforts to bring actors at these various scales together benignly.
- We need to strive even more intensely to work with the marginalized. We've learned how difficult it is, we've identified some of the barriers (time and energy constraints, lack of shared language, fear of the unknown and unfamiliar, political roadblocks). Now we need to overcome them.
- We need to remain open to continual improvements in our methods and in our understandings of the contexts in which ACM practitioners and communities are operating. Kusumanto (2007) uses the analogy of various understandings being spun off a spinning wheel, to reflect the evolving and widening changes that occur with social learning. A similar process characterizes the evolution of thinking about ACM; it should continue.
- Mutimukuru-Maravanyika has argued that our work was too politically naïve in Zimbabwe, that our approach had not adequately confronted the governmental power structure. Although I am skeptical that this team had any other choice, I agree with her that future ACM efforts should continue to take into account, and more forcefully engage, things political. Doing so will require care and sensitivity.

From a more institutional perspective, crucial elements include the need for significantly longer-term funding, yielding more time in the field. Strong, supportive leadership (and strong institutional support), mutually supportive teamwork, including effective intra-team communication of results, and holistic and deep understanding of local political and cultural realities are all important components of an effective ACM effort. The expanding use of variations on this method remains an important key to improving the wellbeing of both people and forests.

Notes

1 CIFOR, headquartered in Bogor, Indonesia, is part of the global Consultative Group on International Agricultural Research (CGIAR; referred to 'fondly' as the 'CG system').

2 CIFOR's work on this approach has included two to seven year projects in 30 sites in 11 countries under the auspices of the adaptive collaborative management (ACM) project (beginning in 1998); four sites in Jambi province in Indonesia under a Collective Action and Property Rights System wide Initiative (CAPRi) project (2004–2006); small projects in Danau Sentarum National Park in West Kalimantan

Province, Indonesia (periodically from 1996–present); and 16 sites in five countries within the Landscape Mosaics (LM) project (2007–2010); as well as related projects on policy and decentralization (1999–2010), and in the management of CIFOR's Malinau research program in East Kalimantan, Indonesia (1995–present).

3 As Rhee (2006) describes, these were common concerns at CIFOR, at all levels, by the early 2000s, a change from the late 1990s.

4 Rhoades (2006), for instance, comments on CIP (the International Potato Center): 'The all-important "show your stuff" annual review meeting began with prime-time presentations by the breeding programme and ended later with a few rushed minutes by the social scientists. Everyone knew exactly where one stood in the disciplinary hierarchy. Tongue in cheek, beleaguered CIP social scientists used to compare the social dynamics to the Hindu caste system' (2006: 406).

5 The first social scientists were quite junior, some from interdisciplinary backgrounds. In 1995, some of the 'social science' research was planned by an interested young forester. I witnessed one early seminar where his sensible suggestions were vituperatively attacked by senior biophysical scientists whose critiques actually showed more ignorance than the scientific rigour they purported to demand. Without the strong social science background, the young forester could not confidently defend his good ideas to such attackers. Prestige within the institution at that time also favored the established biophysical scientists, though that gradually changed.

6 Paraphrasing, these were 1) a rejection of grand theory, 2) social fluidity whereby the ever-changing social conditions require people to adapt (and they do), 3) an emphasis on the local and particular, 4) the existence of multiple groups with different concerns and voices, and 5) attention to signs and texts, with meanings contingent on social relations.

7 See Colfer (2005) for a thorough examination of ACM's conceptual underpinnings from my point of view.

8 See Greenwood and Levin (1998) for a reasoned response to such concerns.

9 See Rhee (2006) for a description of the growing acceptability of many of these ideas within CIFOR.

10 My own intention, using ethnographic research as a pattern, was for the facilitator to remain most of the time in the community. This rarely happened.

11 Such freedom for an anthropologist cannot be taken for granted within the CGIAR (cf. Cernea and Kassam 2006). Nor is such freedom and respect common in typical forestry institutions, where social science is even rarer, often less valued.

12 I led the program from 1998 to 2002, when I went on sabbatical and Ravi Prabhu took over. By my return, our funding had ended, there had been another restructuring, and our formal program no longer existed. We were part of the Forests and Governance Program, one of three CIFOR programs.

13 See Campbell and Sayer (2003), especially Hagmann *et al.* (2003), for evidence of the degree to which our views were eventually accepted.

14 In a subsequent ACM-type research project on landscape mosaics, several influential team members and leaders resisted allowing communities to dictate the teams' activities. The team leader particularly felt responsible to the donor to focus on biodiversity conservation, a central element in the proposal, and did not trust communities to take sufficient account of this issue (also argued by Sayer *et al.* 2006). I thought that the sustainability of effort *depended* on a strong voice by the communities.

15 Significantly, key members of the Bogor-based team moved closer to the field: Ravi Prabhu to Zimbabwe, and Chimère Diaw to Cameroon. The closer Asia sites were coordinated from Bogor: Cynthia McDougall in Nepal; Herlina Hartanto in the Philippines; and Linda Yuliani, Yanti Kusumanto and Lini Wollenberg coordinated the Indonesian sites. In the early days, we were still struggling with what to

do about Latin America, though our links with the University of Florida were growing.

16 Our ACM field sites were Guarayos in lowland Bolivia; Acre and Pará in Amazonian Brazil; Campo Ma'an, Ottotomo, and Akok in Central Cameroon and Lomié and Dimako in eastern Cameroon; Adwenaase and Namtee in central Ghana; Rantau Layung, Rantau Buta and Bulungan in East Kalimantan and Baru Pelepat in Jambi, Sumatra, in Indonesia; Achy, Arstanbap-Ata, Ortok and Uzgen in southern Kyrgyzstan; Ranomafana in southeastern Madagascar; Chimaliro and Ntonya Hill in Malawi; Bamdibhir, Deurali-Baghedanda in central Nepal and Andheribhajana and Manakamana in eastern Nepal; San Rafael-Tanabag-Concepción, Palawan and Basac, Mindanao, in the Philippines; and Mafungautsi in central Zimbabwe. The CAPRi field sites were all in Jambi Province, Sumatra, Indonesia. The Landscape Mosaic sites were in Takamanda-Mone, in western Cameroon; Bungo District in Jambi; Viengkham District in northern Laos; Manompana, in eastern Madagascar; and the East Usambaras in central Tanzania.

17 The conditions proposed by Armitage *et al.* (2009), in an excellent if idealistic set of 10, would predict such a problem.

18 The project structure involved the Cameroon team leader supervising the work in Ghana. Time demands in Cameroon – which took on the most ambitious set of activities and number of employees – reduced the time available to effectively integrate the Ghana team into the program.

19 The congruence in time between the various ACM efforts and Indonesia's decentralisation policy sparked many actions and analyses at the local level (as described in Kusumanto *et al.* 2005 or Komarudin *et al.* 2007, among many others; and in collections by Adnan *et al.* 2008; Fisher *et al.* 2007; Guijt 2007; Yuliani *et al.* 2007; Moeliono *et al.* 2009; at the national level, see Colfer and Resosudarmo 2002; Barr *et al.* 2006).

20 On the later LM project, it was only at the very end of the two-year project that the team leader and I understood that we had been using the phrase 'participatory action research' in different ways. This occurred in a context where we had a good relationship, frequent interaction, and extensive interdisciplinary experience and good will. We both naïvely assumed we knew what the other meant.

21 Besides the impact issue, there is a substantive, disciplinary issue here (see Denzin 2010). Ethnographic and other descriptive materials – such as clarification of ACM processes – are much more difficult to convey succinctly; books are more central in many people-oriented fields. The emphasis on journal articles within CIFOR reflects both the dominance of biophysical and deductive science within the broader CG system and a shift in that direction within CIFOR, over time. A serious and longstanding effort on my part, beginning in 2003, to get CIFOR backing for an ACM book series (which we did manage to produce, partly via RFF and Earthscan) was routinely stalled in the Publication Committee. The fact that there was never a firm 'no' reflects the divergence of opinion within the committee. One administrator (who represented a strange combination of antagonism and similar interests) sabotaged the effort by insisting that one of our researchers convert his book plans to CIFOR Occasional Papers and journal articles, arguing that these would be more prestigious for CIFOR.

22 This experience also precipitated a recurring ambivalence. I led an interdisciplinary and multi-cultural program, and wanted to be culturally sensitive. I knew that my formal position gave me 'power over' others (in some people's eyes, a great deal). I sometimes struggled with questions about the degree to which I should enforce principles related to human rights and equity in cultural contexts where people's values differed markedly from my own.

23 One case was related to the hierarchical expectations of Indonesian NGO staff: The original, older (half Dutch) site coordinator left the site to pursue further

education in 2004. We selected a capable and generally well-liked, junior, Indonesian staff member with long experience on site, and elevated him to the site-based leadership role quite innocently. The vituperative response from the NGO leaders was dramatic and unexpected. The newly promoted CIFOR leader's youth was seen as problematic probably because of the high prestige granted to CIFOR (notwithstanding the continued antagonism to the institution) and the NGO sense that a young CIFOR scientist (without an advanced degree) with leadership responsibilities diminished their own positions in the shared ACM endeavor.

24 Even with excellent training, an ACM facilitator can run into problems. In the recent LM project, our ACM efforts in Cameroon were derailed by our NGO partners' suspicions about CIFOR motives, despite having a facilitator well-versed in PAR methods.

25 The more academic NGO was New Era; the more action-oriented ones included Watch (briefly), Norms, and ForestAction. ForestAction appears to have strengthened its academic focus in recent years, with some exceptionally capable thinkers.

26 One disadvantage we had to live with was that we were prohibited from conducting ACM in the well-forested *terai* (an area later researched by Nepali ForestAction researchers, e.g. Ojha *et al.* 2008), and instead were required to work in the middle hills where community forests were small, not very profitable, and managed more for conservation than for production. On the other hand, we had strong support from some high level and very intelligent government officials, with unusually keen interest in improving equity.

27 These included, for instance, *Human Ecology, Ecology and Society, Agriculture and Human Values, Environmental History, Development and Change, ODI Natural Resource Perspectives, International Journal of Agricultural Resources, Governance and Ecology, Sustainable Development, International Forestry Review, CAPRi Working Paper Series, ACM News, CIFOR News, Small-Scale Forest Economics, Management and Policy.*

28 I have 24 endnote entries on books and theses developed by the original ACM team (not exhaustive). In 2008, I invited all LM team members (including officials, NGO staff, researchers, interns and graduate students) to contribute to an edited book (Colfer and Pfund 2010), documenting our governance work in the LM project. I promised to help anyone not comfortable with English to beautify their language and help with technical editing. In the end, graduate students or soon-to-be-graduate students were lead authors on all chapters not led by myself (seven of 11); and the same groups were included as coauthors, with others, in all chapters but one – reflecting a serious self-selection process.

29 Mutimukuru-Maravanyika (2010), in her analysis of what she considers ACM's long-term failure in Zimbabwe, attributes some of these longer term problems to the dependence the local team had encouraged, by paying community facilitators, since the team could not spend long periods in the field (cf. Zimbabwe's political chaos) – see Chapter 6.

30 When Rhee (2006) assessed ACM in Malinau, East Kalmantan, he concluded, in comparable fashion, that villagers understood CIFOR's institutional constraints in interaction with local government and related difficulties providing concrete livelihoods improvements.

31 And that evaluation was, we felt, tainted by the political maneuvring of some of our CIFOR administrators who preferred the donor to allocate funds to another program.

32 The LM teams expressed similar frustration in their interactions with local governments. The CAPRi team was more successful, but it devoted considerable ongoing effort to cultivating these relationships, under conditions that had become less volatile by the mid-2000s.

33 See Feintrenie and Martini (2011), for an analysis of the immediate positive effects of oil palm development on people's livelihoods in Jambi.
34 Tenure has been taken up more seriously by CIFOR recently, using a modified version of 'action research' in 12 countries (Larson *et al.* 2010).
35 Madagascar was included as an ACM country primarily because action research there had served as a prototype of the ACM program. I did not include it here, since activities there did not benefit from any training or advice from the ACM program itself; rather their earlier activities contributed to the development of our thinking.

References

Adnan, Hasantoha, Djuhendi Tadjudin, Linda Yuliani, Heru Komarudin, Dicky Lopulalan, Yuliana L. Siagian, and Dani Wahyu Munggoro, eds. 2008. *Belajar Dari Bungo: Mengelola Sumberdaya Alam di Era Desentralisasi.* Bogor, Indonesia: CIFOR.

Akiefnawati, Ratna, Grace B. Villamor, Farid Zulfikar, Iman Budisetiawan, Elok Mulyoutami, Asep Ayat, and Meine van Noordwijk. 2010. Stewardship agreement to reduce emissions from deforestation and degradation (REDD): Lubuk Beringin's hutan desa as the first village forest in Indonesia. *World Agroforestry Center Working Paper* 102:29.

Anau, Njau, Ramses Iwan, Miriam van Heist, Godwin Limberg, Made Sudana, and Eva Wollenberg. 2005. Negotiating more than boundaries: conflict, power and agreement building in the demarcation of village borders in Malinau. In *The equitable forest: diversity, community and resource management,* edited by C. J. P. Colfer. Washington, DC: Resources for the Future/CIFOR.

Arda-Minas, Lani. 2002. Cut flowers in Lantapan. *ACM News* 3 (2):5–6.

Armitage, Derek R., Ryan Plummer, Fikret Berkes, Robert I. Arthur, Anthony T. Charles, Iain J. Davidson-Hunt, Alan P. Diduck, Nancy C. Doubleda, Derek S. Johnson, Melissa Marschke, Patrick McConney, Evelyn W. Pinkerton, and Eva K. Wollenberg. 2009. Adaptive co-management for social–ecological complexity. *Fontiers in Ecology and the Environment* 7 (2):95–102.

Arnstein, S. R. 1969. Ladder of citizen participation. *Journal of American Institute of Planners* 35:216–224.

Banjade, Mani Ram, Harisharan Luintel, and Hari P. Neupane. 2008. Action research experience on democratising knowledge in community forestry in Nepal. In *Knowledge systems and natural resources: management, policy and institutions in Nepal,* edited by H. Ojha, N. Timsina, R. B. Chhetri and K. Paudel. Ottawa: Cambridge University Press.

Barr, Christopher, Ida Aju Pradnja Resosudarmo, Ahmad Dermawan, and John McCarthy, eds. 2006. *Decentralisation of forest administration in Indonesia: implications for sustainability, economic development and community livelihoods.* Bogor, Indonesia: Center for International Forestry Research.

Bogor ACM Team. 1999. Local people, devolution and adaptive co-management of forests: strategy document. Bogor, Indonesia: CIFOR.

Bonis-Charancle, Jean Martial, Anne-Marie Tiani, Michael Brown, George Akwah Neba, Zephirin Mogba, Guillaume Lescuyer, Rees Warne, and Brian Greenberg. 2009. Strengthening local analytical capability: Community Options Analysis and Investment (COAIT). In *In search of common ground: adaptive collaborative management in Cameroon,* edited by M. C. Diaw, T. Aseh and R. Prabhu. Bogor, Indonesia: CIFOR.

Brockhaus, Maria, and Houria Djoudi. 2008. Adaptation at the interface of forest ecosystem goods and services and livestock production systems in Northern Mali. *CIFOR InfoBrief* 19:4.

Buck, Louise, Charles C. Geisler, John Schelhas, and Eva W. Wollenberg, eds. 2001. *Biological diversity: balancing interests through adaptive collaborative management.* Boca Raton, FL: CRC Press.

Campbell, Bruce, and Jeffrey Sayer, eds. 2003. *Integrated natural resource management: linking productivity, the environment and development.* Oxon, UK: CABI Publishing.

Campese, Jessica, Terry Sunderland, Thomas Greiber, and Gonzalo Oviedo, eds. 2009. *Rights-based approaches: exploring issues and opportunities in conservation.* Bogor, Indonesia: CIFOR/IUCN.

Cernea, M. M., and A. H. Kassam, eds. 2006. *Researching the culture in agri-culture: social research for international development.* Oxon, UK: CABI.

Chambers, Robert, Arnold Pacey, and Lori Ann Thrupp, eds. 1993. *Farmer first: farmer innovation and agricultural research.* London: Intermediate Technology Press.

CIFOR. 1999. *C&I toolbox.* 9 vols. Bogor, Indonesia: CIFOR.

CIFOR. 2008. Adaptive collaborative management can help us cope with climate change. In *Infobrief 13*, edited by CIFOR. Bogor, Indonesia: CIFOR.

CIFOR News. 2002. Collaborating on collaboration. *CIFOR News* 29:11.

Colfer, Carol J. Pierce. 2005a. *The complex forest: communities, uncertainty, and adaptive collaborative management.* Washington, DC: Resources for the Future/CIFOR.

Colfer, Carol J Pierce, ed. 2005b. *The equitable forest: diversity, community and natural resources.* Washington, DC: RFF/CIFOR.

Colfer, Carol J. Pierce. 2008. From understanding to action: building on anthropological approaches to influence policy making. In *Against the grain: the Vayda tradition in human ecology and ecological anthropology*, edited by B. B. Walters, B. J. McCay, P. West and S. Lees. Lanham: Altamira Press.

Colfer, Carol J. Pierce, and Yvonne Byron. 2001. *People managing forests: the links between human well-being and sustainability.* Washington, DC: Resources for the Future/CIFOR.

Colfer, Carol J. Pierce, and Ida Aju Pradnja Resosudarmo, eds. 2002. *Which way forward? people, forests, and policymaking in Indonesia.* Washington, DC: Resources for the Future/CIFOR.

Colfer, Carol J. Pierce, Ganga Ram Dahal, and Doris Capistrano, eds. 2008. *Lessons from forest decentralisation: money, justice and the quest for good governance in Asia-Pacific.* London: Earthscan/CIFOR.

Colfer, Carol J. Pierce, and Jean-Laurent Pfund, eds. 2010. *Collaborative governance of tropical landscapes.* London: Earthscan, CIFOR.

Colfer, Carol J. Pierce, Etienne Andriamampandry, Stella Asaha, Emmanuel Lyimo, Endri Martini, Jean Laurent Pfund, and John Watts. 2011. Participatory action research for catalysing adaptive management: analysis of a 'fits and starts' process. *Journal of Environmental Science and Engineering* 5:28–43.

Colfer, Carol J. Pierce, Edmond Dounias, Marina Goloubinoff, Citlalli Lopéz, and William Sunderlin. 2006. Understanding forests–people links: the voice of social scientists. In *Researching the culture in agri-culture: social research for international development*, edited by M. M. Cernea and A. H. Kassam. Cambridge, MA: CABI Publishers.

Colfer, Carol J. Pierce, and Laurène Feintrenie. 2011. A dozen indicators for assessing governance in forested landscapes. In *Collaborative governance of tropical landscapes*, edited by C. J. P. Colfer and J. L. Pfund. London: Earthscan/CIFOR.

Contreras, Antonio P., ed. 2003. *Creating space for local forest management in the Philippines.* Manila, Philippines: LaSalle Institute of Governance and Antonio Contreras.

Cooke, Bill, and Uma Kothari, eds. 2002. *Participation: the new tyranny?* London: Zed Books.

Cronkleton, Peter. 2005. Gender, participation and the strengthening of indigenous forest management in Bolivia. In *The equitable forest: diversity, community and resource management*, edited by C. J. P. Colfer. Washington, DC: Resources for the Future/CIFOR.

Cronkleton, Peter, Marco Antonio Albornoz, Grenville Barnes, Kristen Evans, and Wil de Jong. 2010. Social geomatics: participatory forest mapping to mediate resource conflict in the Bolivian Amazon. *Human Ecology* 38:65–76.

Cunha dos Santos, Magna, Samantha Stone, and Marianne Schmink. 2007. Creating monitoring with rubber tappers in Acre, Brazil. In *Negotiated learning: collaborative monitoring in forest resource management*, edited by I. Guijt. Washington, DC: Resource for the Future.

Danau Sentarum ACM Team. 2005. Danau Sentarum National Park: The abandoned paradise. In *ACM Series*. Indonesia: CIFOR.

Dangol, Sushma. 2005. Participation and decisionmaking in Nepal. In *The equitable forest: diversity, community and resource management*, edited by C. J. P. Colfer. Washington, DC: Resources for the Future/CIFOR.

de Koning, Ruben. 2009. ACM as a tool for conflict management. In *In Search of common ground: adaptive collaborative management in Cameroon*, edited by M. C. Diaw, T. Aseh and R. Prabhu. Bogor, Indonesia: CIFOR.

Denzin, Norman K. 2010. *The qualitative manifesto: a call to arms.* Walnut Creek, CA: Left Coast Press.

Diaw, Mariteuw Chimère, and Trikurnianti Kusumanto. 2005. Scientists in social encounters: the case for an engaged practice of science. In *The equitable forest: diversity, community and resource management*, edited by C. J. P. Colfer. Washington, DC: Resources for the Future/CIFOR.

Diaw, Mariteuw Chimère, Ravi Prabhu, and Tony Aseh, eds. 2009. *In search for common ground: adaptive collaborative management of forest in Cameroon.* Bogor, Indonesia: CIFOR.

Edmunds, David, and Eva Wollenberg. 2001. A strategic approach to multistakeholder negotiations. *Development and Change* 32:231–253.

Edmunds, David, and Eva Wollenberg, eds. 2003. *Local forest management.* London: Earthscan.

Evans, Kristen, Sandra J. Velarde, Rocio P. Prieto, Sheila N. Rao, Sandra Sertzen, Karina Dávila, Peter Cronkleton, and Wil De Jong, eds. 2006. *Field guide to the future: four ways for communities to think ahead.* Nairobi, Kenya: Center for International Forestry Research (CIFOR), ASB, World Agroforestry Centre.

Feintrenie, Laurène, and Endri Martini. 2011. Role of the district government in directing landscape dynamics and people's futures: lessons learnt from Bungo District, in Jambi Province. In *Collaborative governance of tropical landscapes*, edited by C. J. P. Colfer and J.-L. Pfund. Bogor, Indonesia: CIFOR.

Fennella, David, Ryan Plummer, and Melissa Marschke. 2008. Is adaptive co-management ethical? *Journal of Environmental Management* 88:62–75.

Fisher, R. J. 1995. *Collaborative management of forests for conservation and development.* Gland, Switzerland: IUCN and WWF.

Fisher, R. J., and W. J. Jackson. 1998. Action research for collaborative management of protected areas. Paper read at Collaborative Management of Protected Areas in the Asian Region, at Royal Chitwan National Park, Sauraha, Nepal.

Fisher, Robert J., Ravi Prabhu, and C. McDougall, eds. 2007. *Adaptive collaborative management of community forests in Asia: experience from Nepal, Indonesia and the Philippines.* Bogor, Indonesia: CIFOR.

Foucault, M. 1980. *Power/knowledge.* New York: Pantheon Books.

German, Laura, Alain Karsenty, and Anne-Marie Tiani, eds. 2009. *Governing Africa's forests in a globalised world.* London: Earthscan/CIFOR.

German, Laura A., Anne-Marie Tiani, Ali Daoudi, Tendayi Mutimukuru-Maravanyika, Edward Chuma, Nathalie Beaulieu, Henri Lo, Cyprian Jum, Nontokozo Nemarundwe, Edward Ontita, Giselle Yitamben, and Victor Orindi. 2010. *The application of participatory action research to climate change adaptation: a reference guide.* Bogor: IDRC/CIFOR.

Greenwood, Davydd J., and Morten Levin. 1998. *Introduction to action research: social research for social change.* London: Sage Publications.

Guijt, Irene. 2007. *Negotiated learning: collaborative monitoring for forest resource management.* Washington, DC: RFF/CIFOR.

Hagmann, Jurgen, Edward Chuma, Kuda Murwira, Mike Connolly, and P. Paolo Ficarelli. 2003. Success factors in integrated natural resource management R&D: lessons from practice. In *Integrated natural resource management: linking productivity, the environment and development,* edited by B. Campbell and J. Sayer. Oxon, UK: CABI Publishing.

Hailu, Michael. 2002. The power of learning from peers. *ACM News* 3 (1).

Hartanto, Herlina. 2007. Monitoring with strong interests and weak incentives in Palawan, the Philippines. In *Negotiated learning: collaborative monitoring in forest resource management,* edited by I. Guijt. Washington, DC: Resource for the Future.

Hartanto, Herlina, with Lani Arda-Minas, Linda M. Burton, Asuzena Estanol, Maria Cristina Lorenzo, and Cecil Valmores. 2002a. *Planning for sustainability of forests through adaptive co-management: Philippines country report.* Bogor, Indonesia: CIFOR.

Hartanto, Herlina, Maria Cristina B. Lorenzo, and Anita L. Frio. 2002b. Collective action and learning in developing a collaborative monitoring system. *International Forestry Review* 4 (3):184–195.

Hartanto, Herlina, Ma. Cristina Lorenzo, Cecil Valmores, Lani Arda-Minas, Linda Burton, and Antonio Frio. 2003. *Learning together: responding to change and complexity to improve community forests in the Philippines.* Bogor, Indonesia: CIFOR.

Indriatmoko, Yayan, Elizabeth Linda Yuliani, Hasantoha Adnan, Eddy Harfia Surma, and Novasyurahati. 2006. *Executive summary: reassessment of criteria and indicators of sustainable and equitable forest management in Baru Pelepat Village.* Bogor, Indonesia: CIFOR.

Indriatmoko, Yayan, Linda Yuliani, Yunety Tarigan, Farid Gaban, Firkan Maulana, Dani Wahyu Munggoro, Dicky Lopulalan, and Hasantoha Adnan, eds. 2007. *Dari Desa ke Desa: Dinamika Gender dan Pengeloloaan Kekayaan Alam.* Bogor, Indonesia: CIFOR.

Jambi ACM Team. 2006. *The people of Baru Pelepat: learning to change.* Indonesia: CIFOR.

Kusumanto, Trikurnianti. 2007. Learning to monitor political processes for fairness in Jambi, Indonesia. In *Negotiated learning: collaborative monitoring in forest resource management,* edited by I. Guijt. Washington, DC: Resources for the Future.

Kusumanto, Trikurnianti, Linda Yuliani, Phil Macoun, Yayan Indriatmoko, and Hasantoha Adnan. 2005. *Learning to adapt: managing forests together in Indonesia.* Bogor, Indonesia: CIFOR.

Komarudin, Heru, Yuliana L. Siagian, Carol J. Pierce Colfer, with Neldysavrino, Yentirizal, Syamsuddin, and Deddy Irawan. 2007. *Collective action to secure property rights for the poor: a case study in Jambi Province, Indonesia.* Bogor, Indonesia: CIFOR for CAPRi (Collective Action and Property Rights).

Komarudin, Heru, Yuliana L. Siagian, Carol J. Pierce Colfer, with Neldysavrino, Yentirizal, Syamsuddin, and Deddy Irawan. 2008. *Collective action to secure property rights for the poor: a case study in Jambi Province, Indonesia* Vol. 90, *CAPRi Working Paper.* Washington, DC: Collective Action and Property Rights Systemwide Initiative.

Komarudin, Heru, Yulia Siagian, and Carol J. Pierce Colfer. 2011. The role of collective action in securing property rights for the poor: a case study in Jambi Province, Indonesia. In *Collective action and property rights for poverty reduction*, edited by E. Mwangi, H. Markelova and R. Meinzen-Dick. Philadelphia: University of Pennsylvania Press.

Landscape Mosaics Team. 2008a. *Integrating livelihoods and multiple biodiversity values in Landscape Mosaics: research guidelines version 1.* Bogor, Indonesia: CIFOR, ICRAF, Joint Biodiversity Platform.

Landscape Mosaics Team. 2008b. *Integrating livelihoods and multiple biodiversity values in Landscape Mosaics: field methods.* Bogor, Indonesia: CIFOR, ICRAF, Joint Biodiversity Platform.

Larson, Anne, Deborah Barry, Ganga Ram Dahal, and Carol J. Pierce Colfer, eds. 2010. *Forest for people: community rights and forest tenure reform.* London: Earthscan/CIFOR.

Leach, Melissa, and Robin Mearns, eds. 1996. *Lie of the land: challenging received wisdom on the African environment.* Oxford, UK: International African Institute.

Lee, Kai N. 1993. *Compass and gyroscope: integrating science and politics for the environment.* Washington, DC: Island Press.

Locatelli, Bruno, Markku Kanninen, Maria Brockhaus, Carol J. Pierce Colfer, Daniel Murdiyarso, and Heru Santoso. 2008. *Facing an uncertain future: how forests and people can adapt to climate change.* Bogor, Indonesia: CIFOR.

Malinau ACM Team. 2005. *Our forests: voices from Malinau about the future of one of Asia's richest rainforests.* Indonesia: CIFOR.

Mandondo, Alois, Ravi Prabhu, and Frank Matose, eds. 2008. *Coping amidst chaos: studies on adaptive collaborative management from Zimbabwe.* Bogor, Indonesia: CIFOR.

McDougall, C., H. Ojha, M. Banjade, B. H. Pandit, T. Bhattarai, M. Maharjan, and S. Rana. 2010. *Forests of learning: experiences from research on an adaptive collaborative approach to community forestry in Nepal.* Bogor, Indonesia: CIFOR.

McDougall, Cynthia, Chiranjeewee Khadka, and Sushma Dangol. 2007a. Using monitoring as leverage for equal opportunity in Nepal. In *Negotiated learning: collaborative monitoring in forest resource management*, edited by I. Guijt. Washington, DC: Resources for the Future.

McDougall, Cynthia, Ravi Prabhu, and Robert Fisher. 2007b. Discussion and conclusions. In *Adaptive collaborative management of community forests in Asia: experiences from Nepal, Indonesia and the Philippines*, edited by R. Fisher, R. Prabhu and C. McDougall. Bogor, Indonesia: CIFOR.

McDougall, Cynthia, Bishnu Hari Pandit, Mani Ram Banjade, Krishna Prasad Paudel, Hemant Ojha, Manik Maharjan, Sushila Rana, Tara Bhattarai, and Sushma Dangol. 2009. *Facilitating forests of learning: enabling an adaptive collaborative approach in community forest user groups, a guidebook.* Bogor, Indonesia: Center for International Forestry Research.

Milne, M. with Cynthia McDougall, Lal Uprety, and Yulia Siagian. 2001. *A participatory research and gender analysis impact assessment: case studies from the local people, devolution and adaptive and collaborative management of forests research program*. Bogor, Indonesia: CIFOR, as a part of the Small Grants Programme of the CG Systemwide Programme on Participatory Research and Gender Analysis and Asian Development Bank RETA 5812.

Moeliono, Ilya. 2006. Laporan Pengkajian Program ACM di Kabupaten Bungo, Jambi [Report on Assessment of the ACM Program in Bungo District, Jambi]. Bogor, Indonesia: CIFOR/DFID.

Moeliono, Moira, Eva Wollenberg, and Godwin Limberg, eds. 2009. *The decentralisation of forest governance: politics, economics, and the fight for control of forests in Indonesian Borneo*. London: Earthscan/CIFOR.

Mutimukuru-Maravanyika, Tendayi. 2010. Can we learn our way to sustainable management? Adaptive collaborative management in Mafungautsi State Forest, Zimbabwe, CERES Research School for Resource Studies for Development, Wageningen University, Wageningen, the Netherlands.

Mutimukuru, Tendayi, Richard Nyirenda, and Frank Matose. 2005. 'Learning amongst ourselves': towards adaptiveness by stakeholders in forest management through social learning, in Mafungautsi. In *The equitable forest: diversity, community and resource management*, edited by C. J. P. Colfer. Washington, DC: Resources for the Future/CIFOR.

Mutimukuru-Maravanyika, Tendayi, Ravi Prabhu, Frank Matose, Richard Nyirenda, and Witness Kozanayi. 2008. Facilitating adaptive collaborative management in forested landscapes: the Mafungautsi case study. In *Coping amidst chaos: studies of adaptive collaborative management from Zimbabwe*, edited by A. Mandondo, R. Prabhu and F. Matose. Bogor, Indonesia: CIFOR.

Ojha, Hemant R. 2008. *Reframing governance: understanding deliberative politics in Nepal's Terai Forestry*. New Delhi, India: Adroit Publishers.

Ojha, Hemant, Bharat Pokhrel, Krishna Paudel, and Cynthia McDougall. 2002. Comparative case studies on adaptive collaborative management: a review of eight community forestry sites in Nepal. Bogor, Indonesia: CIFOR, as a part of ADB RETA 5812, 'Planning for Sustainability of Forests through Adaptive Co-Management' Research Project in Nepal, a Collaborative Initiative of the Center for International Forestry Research (CIFOR), Indonesia and Ministry of Forests and Soil Conservation, Nepal.

Ojha, Hemant, Netra Timsina, Kumar Chetan, Mani Ram Banjade, and Brian Belcher, eds. 2008a. *Communities, forests and governance: policy and institutional innovations from Nepal*. New Delhi: Adroit Publishers.

Ojha, Hemant, Netra Timsina, Ram B. Chhetri, and Krishna Paudel, eds. 2008b. *Knowledge systems and natural resources: management, policy and institutions in Nepal*. New Delhi: Cambridge University Press.

Pandey, Raj Kumar. 2002. Self-monitoring as an effective tool to enhance adaptive and collaborative management (ACM) in community forestry in Nepal. *ACM News* 3 (4):9–10.

Paudel, Krishna, and Hemant Ojha. 2007. Imposing indicators or co-creating meanings in Nepal. In *Negotiated learning: collaborative monitoring in forest resource management*, edited by I. Guijt. Washington, DC: Resource for the Future.

Poffenberger, Mark, ed. 1990. *Keepers of the forest: land management alternatives in Southeast Asia*. New Hartford, CT: Kumarian Press.

Pokorny, Benno, Guilhermina Cayres, and Westphalen Nunes. 2007. Testing the limits of criteria and indicators in the Brazilian Amazon. In *Negotiated learning: collaborative monitoring in forest resource management*, edited by I. Guijt. Washington, DC: Resources for the Future.

Prabhu, Ravi, Carol J. Pierce Colfer, Chimère Diaw, Cynthia McDougall, and Robert Fisher. 2009. Action research with local forest users and managers: lessons from CIFOR's research on adaptive collaborative management. In *Farmer first revisited*, edited by I. Scoones and J. Thompson. Warwick, UK: Practical Action Publishing.

Rhee, Steve. 2006. *Brokering authority: translating knowledge, policy and practice in forestry institutions in Indonesia*. Yale University, New Haven, CT.

Rhoades, Robert E. 2006. Seeking half our brains: constraints and incentives in the social context of interdisciplinary research. In *Researching the culture in agri-culture: social research for international development*, edited by M. M. Cernea and A. H. Kassam. Cambridge, MA: CABI Publishing.

Robiglio, Valentina, William Armand Mala, and Mariteuw Chimère Diaw. 2009. Integrating participatory mapping and GIS to build local information systems. In *In search of common ground: adaptive collaborative management in Cameroon*, edited by M. C. Diaw, T. Aseh and R. Prabhu. Bogor, Indonesia: CIFOR.

Rudel, T. K, and J. M. Gerson. 1999. Postmodernism, institutional change, and academic workers: a sociology of knowledge. *Social Science Quarterly* 80 (2):213–228.

Ruitenbeek, Jack, and Cynthia Cartier. 2001. The invisible wand: adaptive co-management as an emergent strategy in complex bio-economic systems. *CIFOR Occasional Paper* No. 34.

Sayer, Jeffrey, and Bruce Campbell. 2003. Research to integrate productivity enhancement, environmental protection, and human development. In *Integrated natural resource management: linking productivity, the environment and development*, edited by B. Campbell and J. Sayer. Oxon, UK: CABI Publishing.

Sayer, Jeffrey, Bruce Campbell, Lisa Petheram, Mark Aldrich, Manuel Ruiz Perez, Dominique Endamana, Zacharie-L. Nzooh Dongmo, Louis Defo, Stephen Mariki, Nike Doggart, and Neil Burgess. 2006. Assessing environment and development outcomes in conservation landscapes. *Biodiversity Conservation*. DOI 10.1007/s10531-006-9079-9. http://portals.wi.wur.nl/files/docs/landscapes/Sayer_Assessing%20environment%20and%20development%20outcomes_paper.pdf.

Scott, James C. 2009. *The art of not being governed: an anarchist history of upland Southeast Asia*. New Haven, CT: Yale University Press.

Scott, James C. 1990. *Domination and the arts of resistance*. New Haven, CT: Yale University Press.

Shaner, W. W., P. F. Philipp, W. R. Schmehl, and Consortium for International Development. 1982. *Farming systems research and development: guidelines for developing countries*. Boulder, CO: Westview Press.

Sithole, Bevlyne. 2001. *Uncovering shrouded identities: institutional considerations and differentiation in an adaptive co-management project in Zimbabwe*. Bogor, Indonesia: CIFOR.

Sithole, Bevlyne. 2002. *Where the power lies: multiple stakeholder politics over natural resources: a participatory methods guide*. Bogor, Indonesia: CIFOR.

Sunderlin, William D. 2008. From exclusion to ownership? Challenges and opportunities in advancing forest tenure reform. Washington, DC: Rights and Resources Initiative.

Taylor, Peter J. 2005. *Unruly complexity: ecology, interpretation, engagement*. Chicago: The University of Chicago Press.

Taylor, Peter Leigh, Peter Cronkleton, and Deborah Barry. 2010. Learning in the field: using community self studies to strengthen forest-based social movements. *Sustainable Development* (wileyonlinelibrary.com).

Tiani, Anne Marie, Joachim Nguiebouri, and Chimère Diaw. 2001. Criteria and indicators for collaborative monitoring and adaptive co-management of tropical forests. *ACM News* 2 (3):7–8.

Vanclay, Jerry. 2003. Special issue on ACM modelling. *Small-Scale Forest Economics, Management and Policy* 2 (2).

Vanclay, Jerry, Ravi Prabhu, and Fergus Sinclair. 2006. *Realizing community futures.* London: Earthscan.

Wiliam-de Vries, Dede. 2006. *Gender Bukan Tabu: Catatan Perjalanan Fasilitasi Kelompok Perempuan di Jambi [Gender is not taboo: notes on the facilitation of women's groups in Jambi].* Bogor, Indonesia: CIFOR.

Wollenberg, E., and A. Ingles, eds. 1998. *Incomes from the forest: methods for the development and conservation of forest products for local communities.* Bogor, Indonesia: CIFOR.

Wollenberg, Eva, David Edmunds, and Louise Buck. 2000. *Anticipating change: scenarios as a tool for adaptive forest management (a guide).* Bogor, Indonesia: CIFOR.

Wollenberg, Eva, David Edmunds, Louise Buck, Jefferson Fox, and S. Broch, eds. 2001a. *Social learning in community forests.* Bogor, Indonesia: CIFOR.

Wollenberg, Eva, Jon Anderson, and David Edmunds. 2001b. Pluralism and the less powerful: accommodating multiple interests in local forest management. *International Journal of Agricultural Resources, Governance and Ecology* 1 (3/4):199–222.

Wollenberg, E., R. Iwan, G. Limberg, M. Moeliono, Steve Rhee, and I. Made Sudana. 2004. Muddling towards cooperation: a CIFOR case study of shared learning in Malinau District, Indonesia. *Currents* 33:20–24.

Wollenberg, Eva, Jon Anderson, and Citlalli Lopez. 2005. *Though all things differ: pluralism as a basis for cooperation in forests.* Bogor, Indonesia: CIFOR.

Wollenberg, Eva, Ramses Iwan, Godwin Limberg, Moira Moeliono, Steve Rhee, and Made Sudana. 2007. Muddling towards cooperation: spontaneous orders and shared learning in Malinau District, Indonesia. In *Adaptive collaborative management of community forests in asia: experiences from Nepal, Indonesia and the Philippines*, edited by R. Fisher, R. Prabhu and C. McDougall. Bogor: CIFOR.

Yasmi, Yurdi. 2007. Institutionalisation of conflict capability in the management of natural resources: theoretical perspectives and empirical experience in Indonesia, Wageningen University, Wageningen, the Netherlands.

Yuliani, Elizabeth Linda, Hasantoha Adnan, and Yayan Indriatmoko. 2008a. The use of appreciative inquiry as a tool for enhancing adaptive capacity in natural resources management. Paper read at International Association for the Study of the Commons, at Cheltenham, England.

Yuliani, Elizabeth Linda, Yayan Indriatmoko, Valentinus Heri, Seselia Ernawati, Leon Budi Prasetyo, and Zul MS. 2008b. *Promoting good governance in managing Danau Sentarum National Park: through adaptive collaborative management approach.* Bogor, Indonesia: CIFOR/GemConBio.

Yuliani, Linda, Djuhendi Tadjudin, Yayan Indriatmoko, Dani Wahyu Munggoro, Farid Gaban, Firkan Maulana, and Hasantoha Adnan, eds. 2007. *Multistakeholder forestry: steps to change.* Bogor, Indonesia: CIFOR.

Zimbabwe ACM Team. 2004. Learning by doing: a role play for learning about community forest management. In *ACM series*, edited by R. Prabhu. Zimbabwe: CIFOR.

4 Learning in the social wild

Farmer Field Schools and the politics of agricultural science and development in Ecuador

Stephen Sherwood, Marc Schut and Cees Leeuwis

Introduction

As a result of its impressive success as a knowledge-based, community-led approach for change in Integrated Pest Management (IPM) in Southeast Asia and elsewhere, Farmer Field School methodology (here referred to as Farmer Field Schools or FFS) was introduced in the Andes, initially to help communities overcome pesticide-health concerns. Eventually, the approach was adapted to address other concerns in agriculture and natural resource (ANR) management, including the sustainable management of small and large animals, local seed systems, soil fertility, and water for food production and climate change adaptation. Beyond helping to solve technical concerns, FFS was explicitly inserted as a political device for shifting the designs of ANR management away from technology- to people-centred approaches. In this paper, we examine the arrival and rise of FFS in Ecuador, followed by encounters with the socio-technical regime organized around agricultural modernization and subsequent transformations and counter-movements. The exercise sheds light on the conflicts between present institutional designs and needed re-direction towards more adaptive agricultural science and development practice.

Following initial success, FFS became widely recognized as 'best practice' in Ecuador, leading to the proposal of new policies for agricultural development. During processes of 'scaling', when FFS was released into the 'social wild' of institutional collision, collusion and coercion, the approach became systematically transformed in the hands of researchers, extensionists and farmers and their organizations and projects to the point where FFS no longer represented a serious threat to established ways of thinking, organizing, and doing in science and development practice. In the process, the original idea of FFS as a means of adaptive collaborative management was lost. The experience provides rare insight into the politics of institutional continuity and change in ANR, in particular over how actors involved in agricultural development become organized around prestigious symbols and active in translation of meanings, truth construction, enrolment and other processes of social networking in favour of certain agendas at the cost of competing interests.

We begin with an explanation of the theoretical approach underpinning this inquiry, followed by a description of 'FFS-by-design' and summary of its rise to the status of best practice in Ecuador. In continuation, we examine ex-post the transformation of FFS in the social wild. The data that is presented brings together reflective practice based on the practitioner experience of Sherwood as a FFS Facilitator, Master Trainer and Project Coordinator in the Andes from 1998–2009,[1] as well as the action-research of Schut, Sherwood and Leeuwis[2] in Ecuador to explore the challenges and opportunities of implementing a methodology aimed at strategically questioning existing forms of knowledge production and institutional practice in rural development. In addition, we draw on complementary studies (Paredes, 2001 and 2010; Barrera *et al.*, 2001; Mendizabel, 2002; Barrera *et al.*, 2004; Borja, 2004; Maurceri, 2004), project documents (Jiggins *et al.*, 2001), field visits and interviews with trainers and project participants. Due to the large quantity and complexity of the material as well as our interest in grounding an analysis and examining central institutional features, we summarize our empirical data as three, highly contrasting case studies of FFS facilitators, followed by evidence on the institutionalization of field schools in their contrasting organizational environments. After analyzing and discussing the data, the chapter ends with conclusions on how methodology-based, niche-level interventions may survive the social wild, though perhaps in new forms, and contribute to more fundamental institutional change.

Farmer Field Schools: a niche-level intervention for socio-technical change

In this section, we describe and analyze the introduction and performance of FFS as a strategic niche-level intervention for shifting a paradigm of agricultural science and development practice towards more adaptive, collaborative agriculture and natural resource management. To better understand what happened to FFS in the social wild, we draw on two theoretical concepts. First, we explore the ideas of Gibbons (1994) and Nowotny *et al.* (2003) on extreme 'modes' of scientific knowledge production to position the relevance and ambition of the introduction of FFS in the Andes. Second, in order to explain the spontaneous social processes of appropriation and transformation of FFS in the hands of different organizational actors and their networks, we analyze our experiences in the context of socio-technical regimes, as described in Geels (2004).

In his book *Cognition in the Wild*, Edwin Hutchins (1996) explored the performance of cognition in its 'natural habitat' – i.e., in the realm of spontaneous, culturally constituted human activity. Borrowing from Hutchins' metaphor, we embarked to study the performance of FFS in the context of the 'social wild' of science and development practice. Our research interest was to examine how, over time, FFS, as a niche-level intervention for broader institutional change, could become diversely translated and utilized in the hands of different actors, sometimes against the explicit purposes of the methodology.

Modes of scientific knowledge production

Although agricultural modernization has led to increases in food production and economic growth in many places, the recent International Assessment of Agricultural Knowledge, Science and Technology for Development (IAASTD)[3] as well as an exhaustive study by the US Academy of Sciences (NRC, 2010) concur that the benefits often have not been equally distributed and gains have come with severe social and environmental costs that place into question the sustainability of past contributions. As a result, the IAASTD concluded with a call for 'a fundamental shift in science and technology policy and practice that maintains and enhances environmental and cultural services, while increasing sustainable productivity, and safeguarding nutritional quality and the diversity of food and farming systems'.

Gibbons (1994) and Nowotny *et al.* (2003) recommend an institutional shift in science and development from abstract, specialized and externally based knowledge production ('Mode 1') to more localized, applied activity ('Mode 2'). In contrast to Mode 1 science, Mode 2 aims at knowledge production that is produced in the context of application and thus locally embedded and reflexive (Nowotny *et al.*, 2003; Maasen and Lieven, 2006). Mode 2 knowledge production emerges from people's processes of daily activity, problem solving and organization. This 'people-centred' approach is heterogeneous and transdisciplinary in bringing together diverse knowledge, experiences and skills. Mode 2 knowledge production operates in heterarchical structures, with interactions between mixed and dissimilar organizations – each with unique and often contradictory and competing codes and rules. Instead of blind, external review – the gold standard in Mode 1 science – the hallmark of quality assurance in Mode 2 depends on local relevance of outcomes and impacts. Thus, Mode 2 knowledge production demands scientific and development processes based on open, transparent and flexible negotiations and decision-making processes that are responsive to localities. Table 4.1 contrasts Mode 1 and Mode 2 knowledge production.

Institutions are collective sets of formal and informal rules in the form of routines, protocols, norms, values, interpretative schemes, incentive structures and regulations that ultimately shape problem definitions, inform perspectives, engineer practices, and exploit niches of development opportunity. During processes of endless formation and re-formation (i.e., institutionalization),

Table 4.1 Comparison of Mode 1 (generalized/expert) and Mode 2 (local/lay) knowledge production (after Gibbons *et al.*, 2000)

Criterion	*Mode 1: Knowledge produced in the context of abstraction*	*Mode 2: Knowledge produced in the context of application*
Focus	Technology-centred	People-centred
Nature of knowledge production	Theoretical – produced from within a disciplinary community	Practical – produced from within a problem context
Bias – rules that govern conduct	Disciplinary and multi-disciplinary – single or multiple system of rules governing conduct	Transdisciplinary – dynamic, multiple systems of rules collide and collude
Problem-solving – experience and skills employed	Homogeneous – focused, well defined experience and skill set	Heterogeneous – diverse experiences and skills involved
Organization structures	Centralized and hierarchical – well-established; graded and top-down	Diverse and heterarchical – loose, flexible, and fluid structures; mixed and dissimilar constituents
Negotiation and consensus – resolution of differences	Closed and static – conditioned by pre-established norms and rules	Open and transient – conditioned by context of application and evolves with it
Nature of knowledge	Generalizeable and cumulative	Context specific and dependent on locality
Social accountability and reflexivity	Low – offer oriented, exclusive and low sensitivity to impact of outcomes; preoccupied with internal criteria and priorities	High – demand oriented, inclusive and high sensitivity to impact of outcomes; preoccupied with relevance
Quality control – enforcement of 'good science'	Self-referential – 'Peer review' judgements; peers selected based on compliance with norms; emphasis on individual creativity from within disciplinary bounds	Broadly based – composite and multi-dimensional; dependent on social composition of review system; emphasizes 'group think', socially extensive and accommodating
Theory of knowledge spread	Spontaneous diffusion based on merit	Repeated processes of generation

socio-technical networks or regimes spontaneously arise, fall and re-organize around symbols and bodies of knowledge as well as the generation and distribution of resources to challenge existing institutions, creating rules and implementing regulatory frameworks that affirm certain ideals (best practices), while marginalizing others (Rip and Kemp, 1998). A socio-technical regime, such as that which became organized around a global project of agricultural modernization during the second half of the twentieth century, can be seen as both a factory and a storehouse of institutional perspectives and arrangements that enable and regulate the use, development and survival of a particular value system, sets of rules and technical processes and products. Thus, over time a socio-technical regime is likely to become entrenched, focusing on the defence and reproduction of its own self-referential purposes (Schot and Geels, 2007). Ironically, over time the very success of a regime can lead to intransigence that ultimately reduces its adaptability to overcome the inevitable challenges of interests organized around competing purposes. A capacity to open up pathways of innovation with existing socio-technical networks is understood as central to overcoming the urgent socio-environmental challenges confronting modern-day society.

Geels and Raven (2006) describe social spaces as 'niches' that open up as actors experiment with and organize around particular ideals, interests and purposes. As emerging groups of actors drive radical change, they generate tensions with the existing order, commonly leading to spirals of counter activity and ensuing responses. People and their organizations may raise doubts over whether or not established norms and rules are being followed, or they may enlist in emerging networks. As a result of endlessly creative developments from within a niche environment, over time a certain network converges around increasingly appropriated and acceptable rule sets or novelties, thereby displacing existing orders.

In the late 1990s, a group of scientists based at International Potato Center (CIP) and organized around ideals of agroecology and collaborative natural resource management proposed the introduction of Farmer Field School methodology in the region (Sherwood *et al.*, 2000). In practice, their utilization of FFS represented a radical departure from an established socio-technical regime built upon externally based, Mode-1 knowledge and technology. In this study, we describe the introduction of FFS in Ecuador as a niche-level intervention intended to promote institutional transition towards more locally led, Mode-2 science and development practice.

FFS-by-design: transition towards more adaptive, collaborative science and development practice

There is growing concern about how science and development can more effectively bring forth more sustainable solutions to complex social, economic and environmental problems. Many of today's most impenetrable problems are embedded in past 'solutions', such as agricultural modernization, and require innovative restructuring of supply chains, fundamental institutional change, adaptive and collaborative management and decision-making structures towards sustainable development. At the heart of the anxiety are questions whether science and development are capable of organizing and implementing such change, or have become locked into dominant non-adaptive, non-collaborative and self-destructive paradigms. To address these questions we present empirical data on the introduction and performance of Farmer Field Schools in Ecuador – a strategic niche-level intervention intended to advance a transition in agricultural science and development towards more adaptive, collaborative practices.

As a result of the harmful consequences of agricultural modernization in Ecuador – in particular, severe health problems associated with pesticide exposure, degrading soils and declines in productivity (Crissman *et al.*, 1998), during the 1990s a growing number of people became concerned over industrial era technology, leading to rising waves of protests and questioning of public policy. Impressed by the success of FFS in Integrated Pest Management (IPM) in Southeast Asia and elsewhere, colleagues at the International Potato Center (CIP) invited the FAO's Global IPM Facility to help it introduce FFS in the Andes (Sherwood *et al.*, 2000). As a high-order, interactive lay-expert learning approach based on well-established principles of adult education (e.g., discovery-based learning), ecological literacy (filling knowledge gaps on the existence of beneficial organisms), and social learning (adaptive, collaborative learning in heterogeneous contexts), FFS has proven capable of helping communities to address immediate productivity, human health and environmental concerns (Pontius *et al.*, 2002). It was proposed that FFS also could provide an entry point for profound institutional change: a shift from expert-led, technology-centred agricultural science and development practice towards more people-centred designs, emphasizing local knowledge and community-based innovation (Luther *et al.*, 2005). Nevertheless, providing its fundamental conflict with the agenda of a conflict with the well established norms of science and development practice, it remained to be seen whether FFS could overcome existing power interests and spark (or contribute to) broader institutional change.

FFS methodology explicitly emerged in response to the adverse consequences of modern, industrial era rice farming in Asia, especially the health and environmental effects of pesticides (Kenmore *et al.*, 1987). The methodology is based on the premise that farmers are motivated and capable innovators who, provided insights from biology and ecology, could devise effective solutions to their pest management problems themselves. FFS accommodates typical learning styles of lay people, such as hands-on, in-the-field learning that is closely linked with the priority crop of interest (Pontius *et al.*, 2002). Over time, however, the methodology diversified to include new crops, animals, and other aspects of rural life (e.g., market interactions, specialized soil and water management, human health, and – more recently – climate change adaptation) across different geographies and cultures (for examples and further information, see *www.farmerfieldschool.net*).

The didactics of FFS are explicitly based on Dewey's concepts of discovery-based learning (i.e., adaptive, open-ended, self-directed learning through solving problems) and learning-by-doing (practice and repetition).[4] Instead of answers, the facilitator employs lively experiments where participants collaboratively discover not just new technical content, but also how to work together in finding answers to pressing problems. Rather than providing answers to pre-determined questions, FFS brings together farmers in groups and involves them in discussions over production concerns. The methodology strategically aims at limiting information to: what farmers do not know, but need to know to improve their agriculture. FFS seeks to shed new light on previously hidden agricultural phenomena, for example, over insect ecology and especially pest-beneficial interactions. It employs learning experiments to help farmers 'see' what was previously hidden, for example, over the life cycle of a particular insect of concern. Agroecosystem analysis (AAE) lies at the heart of the FFS learning experience (Pontius *et al.*, 2002; Pumisacho and Sherwood, 2005). It involves teaching farmers how to 'read' the health and ecology of a crop through field observations, to sample relevant crop health indicators – plant growth, disease incidence and severity, pest-beneficial insect populations, presence of weeds, and moisture conditions – and to report and openly discuss findings with the group. Rather than rely on external, expert-based advice, the negotiated outcomes of the AAE lead to management decisions based on the experience of the participants. Box 4.1 summarizes the central design features of FFS.

By design, FFS addresses adaptive management through group problem solving that is grounded in the realities of a farmer's field as well as in his or her community (Pontius *et al.*, 2002). Rather than promote specific technologies, groups of men, women and children from a particular community systematically test their ideas through comparative trials. The learning–action process is transdisciplinary in that it responds to a broad interpretation of agriculture – the agronomy of soils, plants and pest management, marketing as well as social concerns, such as work in groups and the formation of associations. FFS actively employs both farmer and expert knowledge in open-ended discovery. It emerges from organizational processes that are socially mixed and demand extensive collaboration, negotiations and consensus building among participants. The end product is not based on a pre-determined package of solutions, for example, on how to implement IPM in potato. Instead, FFS seeks to promote critical thinking and collaborative problem-solving skills that enable participants to independently find solutions through farm-led experimentation. Successful graduates deepen their knowledge of biological principles of agriculture and specific local problems as well as possible alternatives that they themselves have devised and that have been subjected to individual and group performance criteria. As a result, spread of the FFS experience is not simply a matter of repetition, but rather each group must develop its own curriculum and learning history in light of nuanced local experiences.

Thus, FFS can be viewed as a strategic departure from the expert system and its modernization project based on the 'extension' of pre-conceived solutions (Table 4.2). Instead of depending on formally trained specialists who provide answers and solutions, FFS employs 'facilitators' – either formally trained professionals or lay volunteers – who guide participants in process design and open-ended learning. According to this model, participants take responsibility for conducting experiments and finding

Box 4.1 Learning principles of Farmer Field Schools (based on Gallagher, 1999 as referenced in Pumisacho and Sherwood, 2005)

- Non-formal adult education – adults have substantial life experience and are largely independent, self-directed learners.
- Interactive group learning – people learn best in groups, where they can interact and exchange ideas based on their diverse life experiences.
- Content linked to crop/animal stages/life cycle – farmers address technical content immediately, as it emerges with the development of the crop or animal and the labour and cultural activities associated with its production.
- The field is the best place for learning – rural people prefer applied knowledge and learn best during hands-on interaction with their crops and animals in the field.
- Basic scientific concepts – every farm and every season is different; farmers need to manage general concepts and apply them to specific local contexts.
- Continual learning and experimentation – agriculture is a highly dynamic enterprise, and as a result, farmers never stop learning and innovating.
- Farmers become 'experts' – as a result of the above, participants become critical thinkers, highly capable of independently solving problems and taking on the challenges of their agricultural development.

answers. These features make FFS methodology an explicit application of Mode 2 knowledge production and people-centred development.

Based on its success, the FFS methodology has been proposed as a promising novelty for transforming agricultural practice (Luther *et al.*, 2005). The methodology is radically different from predominant development discourses in positioning farmers as highly capable innovators, emphasizing group learning, and utilizing real-time, iterative learning–action processes to address a multitude of social and environmental problems. Numerous studies rigorously demonstrate the contributions of FFS (summarized in van den Berg and Jiggins, 2007), leading an international call for 'scaling up' the methodology (LEISA, 2003a and b). As a result of growing popularity and its demands for radical change in development practice, however, FFS has become vulnerable to competing interests from existing institutional power interests. Taking into account the required changes to programme staffing,

Table 4.2 Comparison between conventional extension and FFS (translated from Pumisacho and Sherwood, 2005)

	Expert-led extension: technology-centred development	*Farmer Field Schools: people-centred development*
Underlying logic	Instrumental and directed	Organic and interactive
Who chooses content/ themes?	Expert/specialist	Participants themselves
Learning content	Narrow and well-defined	Broad and open to questioning
Teaching/learning methods	Formal lectures Demonstration Controlled experiments	Dialogue Discovery-based activities Open-ended experiments
Role of participants	Passive subjects: collect and memorize what is taught	Active subjects: question, learn and teach, find solutions
Role of teacher/trainer	Active subject: talk, teach, discipline, determine relevant content, provide answers	Active subject: facilitate, raise questions, learn and teach, provoke discovery
Primary source of experience	Expert/specialist	Everyone
Who is knowledgeable?	Expert/specialist	Everyone
Communication style	Uni-directional: from expert to participants, limited and controlled	Bi-directional: between facilitator and participants, open and free
How do people learn?	Collecting and memorizing what is taught	Reflection over personal experience
Desired effect on participants	Learn to adopt and manage technology	Learn to analyze and solve problems independently

resource distributions, administrative procedures and field practice, it could be expected that efforts to promote and institutionalize FFS would provoke strong reactions.

Arrival of FFS in Ecuador

By the early 1990s, the promotion of FFS became a major thrust in Southeast Asia (Kenmore, 1991). The studies of Peter Kenmore, a pioneer of FFS, revealed that by killing off beneficial organisms such as parasitic wasps, insecticide use was largely responsible for pest problems in rice. Over time, an FFS in IPM movement contributed to a broad ban on pesticides for rice in Indonesia. Kenmore leveraged this experience at the Food and Agricultural Organization (FAO), and he convinced numerous countries to develop national FFS programmes on rice IPM, including: the Philippines, Indonesia, Cambodia and Vietnam. By the end of the decade, Pontius *et al.* (2002) estimated that over two million farmers in Southeast Asia had graduated from FFS. Subsequently, Kenmore and colleagues established the Global IPM Facility (GIF) at FAO headquarters in Rome, in part as a means of exporting FFS to other parts of the world, especially Africa and Latin America. The International Potato Center and in particular Sherwood was charged with coordinating the introduction of FFS in the region (Luther *et al.*, 2005).

Establishment of a pilot programme

By the mid-1990s, word of the success of FFS in Southeast Asia and similar people-centred approaches in Latin America (e.g., Bentley, 1992) had reached people at the Ecuadorian Instituto Nacional Autónoma de Investigación Agropecuaria (INIAP) and the International Potato Center through publications, conferences and increasingly donor demands. Meanwhile, governments throughout Latin America were well into 'structural adjustment' and 'economic modernization', which, in Ecuador, became articulated as the 1993 Law of Modernization of the State (Beckerman and Solimano, 2002; Gallardo-Zavala, 2003). For public agriculture policy, this generally involved the dismembering of public extension and research services and placing the responsibility of 'capacity building' and 'technology transfer' in the hands of private consultancy companies and the agrochemical industry. Despite major reorganizations and budget cuts in the public system, a dominant logic of the expert and technology-based change continued to shape and enforce agriculture development (Sherwood, 2009).

Agriculture professionals and their organizations in the Andes commonly were resistant to ideas from other regions, but they were willing to explore common experience among successful IPM work and to adapt local methods (Thiele *et al.*, 2001; Luther *et al.*, 2005). 'Participatory approaches' became a passageway for FFS. In 1997, CIP and its partners in Bolivia and Peru started to experiment with participatory training (Torrez *et al.*, 1999a and b), incorporating some elements of the FFS approach, but not the methodology defining practice of the Agroecosystem Analysis.

In 1999, the Global IPM Facility and CIP led a three-month Training of Trainers (ToT) course on FFS methodology in potato IPM for professional extensionists and farmer promoters from Ecuador, Bolivia and Peru at the Guáslan training centre in the Ecuadorian highlands. The organizers subsequently charged the graduates with the task of establishing FFS in their respective countries. Between 1999 and 2004, the GIF and CIP raised funds, implemented new ToTs, and organized graduates into support networks. Moreover, they advocated and lobbied the methodology in government circles. By design, FFS in the Andes sought to conserve the central features of FFS in Asia. Nevertheless, in the Andes FFS emerged in unique contexts of agricultural modernization in the region (Box 4.2), which posed new challenges for institutional transition from Mode 1 to Mode 2 modalities.

Box 4.2 Obvious differences between FFS in Asia and the Andes

In 1999, when researchers and development practitioners began to test FFS in the Andes, the main focus was on potato and the crop's major phytosanitary concern: the disease late blight. Moreover, potato was the principal economic crop and the one that demanded the greatest use of external inputs. FFS in potato IPM centred technical content on the organisms that tended to have a comparative advantage in higher altitude production systems: plant pathogens and resulting diseases. As a result, pathologists and host resistance came to play a central role in FFS in the Andes. Additionally, in Ecuador researchers quickly found that mechanized tillage on steep hillsides was a major cause of soil loss, so FFS focused much attention on soil erosion, which was not a concern for lowland rice farmers. Also, potato farmers often over-applied synthetic fertilizers. As a result, FFS often emphasized how to increase productivity through the introduction of resistant varieties and reducing fungicide and fertilizer use.

Rural demographics were very different between the regions. In Southeast Asia, farmers commonly cultivated much less than a hectare. Meanwhile, in Carchi, they tended to farm 3 to 5 hectares. Further, rural people in Southeast Asia tend to live in concentrated communities of thousands of families. The communities of the Highlands Andes are commonly made up of fewer than 10–50 families spread out across large areas. This demographic created challenges in bringing together farmers as well as difficulties with communication between sessions. Often participants asked for fewer, more intensive meetings.

While public investment in agriculture continued in Southeast Asia, agriculture modernization in Latin America and the different project and institutional contexts placed unique demands on FFS. Andean countries underwent sharp cuts in public investment in agriculture and rural development during the 1990s. Because of a vacuum of public support for rural development, non-governmental actors in Central and South America, particularly non-governmental organizations (NGOs) and community-based organizations (CBOs), were called on from the beginning to play a lead role in coordination and facilitation of FFS. As a result, governments generally were not amenable to large national-level programmes such as those that championed FFS in Southeast Asia. Increasingly, short-term consultants began to dominate community development and NGOs and these operated time-bound, project-based contributions. As a result, a number of NGO partners claimed that their competitiveness depended on project-based results and that staff had to 'pay for themselves'. They were not commonly willing to send staff to intensive three-month Training of Trainers (ToT) courses. This led us to shift to less intensive trainings based on bi-weekly sessions over periods of six or more months. It also led to an emphasis on market-oriented approaches to financing FFS and farmer-led FFS.

Becoming 'best practice'

As a result of the startling research outcomes on health, environmental and productivity problems of modern potato production in Carchi (Crissman *et al.*, 1998; Yanggen *et al.*, 2003), CIP and INIAP's agenda in Ecuador became centred around the improvement of farm productivity and rural health through decreasing dependence on agrochemicals. The first goal in establishing FFS was to demonstrate the potential of the methodology, leading in 1999 to a series of pilots, followed by internal evaluations and commissioned impact studies. As a growing number of FFS were implemented and completed in the country, ex-post studies assessed longer-term considerations, such as technology adoption and adaptation (Borja, 2004; Barrera *et al.*, 2004; Mauceri, 2004; Schut, 2006). Results from 'ballot box' tests showed that FFS helped farmers to significantly increase knowledge of relevant ecological principles (Barrera *et al.*, 2004). Economic data from FFS comparison plots demonstrated how farmers could substantially decrease agrochemical use (including synthetic pesticides and fertilizers) without decreasing production by area. Cost-benefit analysis demonstrated that FFS graduates could nearly eliminate dependence on highly toxic pesticides while decreasing production costs for a tonne of potatoes by about 30 per cent (from $104 to $80), leading to arguments that widescale application of FFS could generate substantial gains in production efficiency that would carry on year after year.

The arrival of additional FAO project resources brought on board the Minister of Agriculture (MAG), who assigned members of the National Directorate of Investigation and Technology Transfer (DITTE) to provide government certification of FFS in Carchi and elsewhere. This involved the creation of minimum standards for FFS graduates and facilitators as well as the establishment of a database on approved field schools (Borja, 2004). As part of the FAO project, in 2001, Sherwood and colleagues organized an intensive 400-hour ToT for farmers and *técnicos* from the northern Ecuadorian provinces of Imbabura and Carchi. Most of the participants were from public organizations (INIAP, MAG, municipal governments and technical schools), national NGOs and community-based organizations (communities and farmer cooperatives). A handful came from private consulting firms and agrochemical companies. The central technical theme was widened from pest management to 'integrated crop management' (ICM), which permitted attention to other crop concerns, such as soil fertility and marketing. Similar to the Asian approach to IPM, ICM potentially included: soil fertility through managing soil life, limited tillage, cover crops and green manures, multi-cropping, rotation schemes, and minimal use of synthetic fertilizers. The FFS in ICM were primarily applied to potato, but to deepen the penetration of the methodology, partners were encouraged to apply FFS to other crops and animal systems, including: field beans, tomato, Andean grains (quinoa and lupine), onion, pasture/milk production, poultry and guinea pigs.

By the end of 2001, results from longer-term evaluations arrived, showing favourable results. Studies found that a high percentage of FFS graduates adopted IPM technology, especially resistant varieties and insect traps, leading to substantially less pesticide use (Paredes, 2001, 2010 and Barrera *et al.*, 2004). Farmers said they welcomed the methodology as a refreshing change to the 'theory' and 'impractical' recommendations of *técnicos*. A number of graduates said that they had felt 'abandoned' by the government and were eager to work with people who were willing to spend time with them in the field and learn about the difficulties they faced, which were not always limited to production agriculture. While some farmers were impatient with the open-ended, discovery-based learning of FFS (i.e., having to find answers on their own), they favoured how the experience exposed them to 'new technologies', such as seed and varieties as well as information on pests. Participants liked that FFS facilitators treated them as equals. As a result of their enthusiasm, FFS graduates became the best salespeople of the methodology.

By 2003, hundreds of farmer groups in the country were requesting support in implementing FFS. Mayors and provincial governments in Carchi and elsewhere regularly mentioned field schools in their public addresses. The regional head of MAG cited that through FFS his office was promoting the 'cutting-edge of agriculture extension', and the Ministry formally requested assistance in commandeering international financing for field schools. Representatives from national and international pesticide companies contacted the project leaders to see how they could support FFS. Sherwood (2009) observed that by this time, the methodology had become part of the spontaneous political discourse of central players of the socio-technical regime in Ecuador. Arguably, within five years of its introduction, FFS had obtained the prestigious status of best practice.

Results: FFS in the social wild

In 2003, CIP and INIAP ended its project to introduce FFS, representing a relinquishment of control over FFS to its partners in the field – the release of FFS into the social wild. To illustrate the diverse expressions of FFS after their release in the social wild, three highly contrasting experiences of FFS facilitators are presented: Joel, an agronomist from a public research institution (INIAP), Lenin, a forester from a conservation NGO (EcoPar), and Eduardo, a leader from an organization of voluntary farmer promoters (APRODIC). The names of the facilitators have been changed to preserve anonymity.

Tales of three facilitators

Case 1: Joel at INIAP: the national agriculture research service

Joel was from a small rural town that was taken over by Quito's urban sprawl during the oil boom of the 1970s. He grew up in a rapidly urbanized

Kichwa-speaking community. His grade school teacher was proud of the agricultural roots of the town and taught the students about gardening and fruit trees. Joel applied much of his learning at home. His family raised crops and vegetables and reared chickens and pigs to feed themselves, while their income came from his father's irregular employment as a factory worker or construction labourer in the city. Joel said with conviction: 'That teacher and at home is where my passion for agriculture was born.'

Joel was a good student and became the first in his family to reach university, where he studied animal science and conducted a thesis on pasture systems in the Central Highlands. Following graduation, INIAP hired him for a permanent extension-researcher position, and he was sent to support the Unit of Validation and Transfer of Technology in Carchi. As Carchi was the most intensive potato production region of the country, INIAP's activities there centred on the crop's production problems. Joel conducted applied research on varieties resistant to late blight. He researched IPM of the Andean weevil and the leafminer fly. He also conducted research on crop response to chemical fertilizers. In 1996, he was sent to CIAT in Colombia, where he was trained in a participatory research methodology known as Comites de Investigación Agrícola Local (CIAL). In 1999, Joel, who had become a leader in the EcoSalud project on pesticide health effects, was selected for the initial FFS training at Guáslan.

Joel described the ToT as an 'awakening'. He was particularly impressed by the Master Trainer at Guáslan, a Cambodian and Head of the National IPM Programme who had studied in Cuba and spoke Spanish. Joel said that the trainer's 'commitment and passion for IPM impressed the entire group'. Joel became a champion of FFS and earned the fame of being one of the country's best facilitators.

As the provincial head of INIAP-Carchi, Joel came to coordinate many projects. Each included FFS as the lead intervention platform. In 2001, Joel organized and led the first ToT for *técnicos* and farmer leaders of the north. In the evaluations, the participants gave Joel a perfect five out of five star ranking in four categories: enthusiasm, teaching, example and friendship.[5]

In a workshop leading up to the Ecuadorian methodological guide (Pumisacho and Sherwood, 2005), Joel expressed concerns over the 'erosion of FFS'. Based on his experience he had the growing feeling that participants did not possess clear knowledge of the agroecosystem analysis and that their experiments lacked creativity. He emphasized that a field school was not a field school if it did not include agroecosystem analysis, learning plot, insect collections and zoos, and experiments. He lobbied for the creation of a certification system to make sure that facilitators and graduates met minimum standards in both the technical aspects of IPM and the process management aspects of 'participation'. Among his colleagues in the FFS movement, Joel earned a reputation as a staunch advocate of FFS-by-design.

Joel actively resisted collaboration with the pesticide industry. When he left for two years of graduate study, however, his supervisor at INIAP took advantage of his absence to establish a new project with CropLife, an international organization that was created to collectively represent the agrochemical industry's policy interests, which involved placing an industry representative inside the field office in Carchi. Thereafter, INIAP's approach to FFS underwent major transformations. Following his return in 2004, Joel felt that he had been forced to implement a watered-down 'hybrid' of the FFS methodology. The new approach involved five modules that centred on getting farmers to adopt an 'IPM technology package'. One farmer FFS facilitator who collaborated with Joel argued that INIAP's re-designed field schools had little similarity with the original methodology. He explained, 'The content is pre-determined, and there is no time for learning plots, agroecosystem analysis, and experiments.' Despite Joel's early clarity and enthusiasm over 'FFS-by-design', organizational constraints eventually led him to implement technology-centred FFS.

Case 2: Lenin at EcoPar: an environmental NGO

Lenin grew up in a small city in southernmost Ecuador. His parents had rural roots, but like many youths of his era, Lenin's aim was to become a professional and move out of the rural area. The nearby Universidad Nacional de Loja had a highly demanding forestry programme, known as the best programme of its kind in Ecuador, which immediately attracted Lenin. Later, FAO's Community Forestry Program (DFC in Spanish) hired him as an agroforestry extensionist to work in the Central Highlands. He was trained in what he described as 'cutting edge participatory methodologies' and became part of the FAO's largest programme in the Andes – one of its most lauded initiatives.[6] One of his supervisors at DFC described Lenin as: 'Young, smart, and hardworking – one of our best community extensionists.' The FAO eventually ended the project in the early 2000s, and when the private consulting firm EcoPar won a grant to implement a biodiversity conservation project in Carchi, many former DFC employees were hired, including Lenin.

Midway through his career, Lenin completed an MSc in Community-based Natural Resource Management at the Catholic University in Ibarra. Of the subjects presented in the diverse curriculum, Lenin favoured economics, where he excelled as a student. His thesis on environmental services argued for forest conservation through different forms of resource privatization and administrative procedures to monetize relationships between urban- and rural-based people. He felt that, given the financial incentives, farmers would protect their land and watersheds for the people of the cities.

In 2004, the coordinator of capacity building at EcoPar – an avid proponent of people-centred development and social learning – enrolled Lenin in an FFS ToT in Otavalo, which focused on 'Integrated Potato Management'. The coordinator was concerned about Lenin's 'vertical' teaching style, and added that: 'Lenin likes to spend as little time as possible in the field, and he tells farmers what they should do', leading to little impact. Lenin explained he attended the ToT reluctantly, clarifying: 'The DFC prepared me in participatory methodologies. The [FFS] training was very basic and not always relevant for me.' Nevertheless, he said, 'The technical training in potato IPM is new and useful, since I am working in Carchi.' Further, EcoPar's donor agency, International Development Research Centre (IDRC), was enthusiastic over FFS, so he had decided to learn the methodology.

Following the ToT, EcoPar proposed a series of FFS as a means of building a grassroots initiative around the themes of biodiversity and conservation of the forest remnants on the Eastern Andean ridge of Carchi. Lenin was placed in charge of implementation. Technical themes included ecological potato production, soil conservation and forest management.

According to FAO project reports and his colleagues at EcoPar, Lenin immediately adapted FFS to fit his own priorities. He chose to work not with groups but rather individuals, and he made decisions on the dates and times of meetings. He did not involve participants in open-ended curriculum design. Instead, he pre-determined technical content and process. His FFS did not include the central process elements of the methodology. Participants summarized the course's technical content as lectures on home gardening and controlling pests. They had not participated in the open-ended, group learning normally associated with FFS. Independent of the ultimate contributions of his activity, Lenin's utilization of the methodology involved the transformation of FFS from people- to technology-centred designs.

Case 3: Eduardo at APRODIC: a community-based organization of farmer promoters

Eduardo's parents were resource-poor immigrants from Colombia who, in the 1960s, crossed the border into Ecuador in search of work. His father became a member of the '24 of May Cooperative' that, following a violent conflict, were awarded land rights under the national land reform policies in El Angel, Carchi. His father, his younger brother, and Eduardo himself were three of 18 participants in the first pilot FFS organized by Sherwood and colleagues in Ecuador. Over time, the creativity and enthusiasm of the brothers began to set them apart. They enjoyed experimentation and immediately made their own insect collections at home. The brothers reported back to the group impressive details, for example, on the mating practices of the Andean weevil, the number of eggs laid and hatched, and the results of their studies on the insect's life cycle. Similarly, they implemented their own detailed studies on the Andean weevil and on resistance of different local potatoes to late blight.

Since he did not own land of his own, Eduardo applied his skills to his family's farm. Within a year of the FFS his family stopped using highly toxic pesticides, and they decreased by half their use of fungicides for late blight. His learning was not limited to the potato crop. Eduardo set up a tree nursery of native species that was used to reforest the boundaries and steep hillsides of the family farm. They set up contracts with the municipality and NGOs to sell trees for local reforestation programmes. Eduardo and Luis experimented with irrigation and pasture varieties and mixtures, leading to substantial improvements in milk production. One of his cows produced about 24 litres of milk per day, nearly three times the average of his neighbours.

Joel invited Eduardo to participate in the 2001 ToT. Joel explained that while the técnicos struggled under the applied technical content of the FFS, Eduardo's intelligence and his earlier FFS experience enabled him to stand out among the group of 26 trainees. Often, Master Trainers would draw on Eduardo to explain the process aspects of the methodology as well as technical content, for example, on insect and plant ecology or the practical aspects of managing a healthy potato crop in Carchi. Eduardo had grown confident standing up in front of a group and defending his ideas, but he was best known as an expert producer. A potato specialist at INIAP told Sherwood, 'No one knows how to grow potatoes better than Eduardo.'

Prior to the end of the ToT, Eduardo set up an FFS on IPM with primary school children in his community. Later, he organized FFS with adults on Integrated Potato Management as well as on how to manage pastures. FAO evaluators observed that Eduardo effectively applied the central elements of the methodology: the content was determined by the participants, there were comparison plots, people conducted experiments, and they regularly used agro-ecosystem analysis to inform decision-making.

Eventually, Eduardo went beyond merely implementing FFS. He became involved in the Humanist Farmers Movement and helped to set up an organization of farmer promoters and researchers in Carchi – the Asociación de Promotores de Desarrollo Integrado de la Comunidad (APRODIC). Eduardo summarized that the purpose of APRODIC was: 'To make a difference.' Initially, APRODIC hoped to receive financial support from the second phase of the CIP-run EcoSalud project to run field schools, but for political reasons, the project channelled resources to the Ministry of Agriculture, municipalities and the Provincial Government, even though those entities had little to no experience in FFS. Eventually, the government organizations did not meet their commitments, so CIP resorted to directly contracting Eduardo and APRODIC. Eduardo was charged with running an FFS based on an imposed curriculum centring on pesticide safety. When Eduardo was asked about this, he responded that when hired by CIP, 'I have to follow their curriculum, which is Safe Use of Pesticides.' He added, however, 'When I run an FFS myself, I do it the right way – according to what the people want.' Table 4.3 summarizes how Joel, Lenin and Eduardo diversely applied FFS.

Encountering the social wild

According to established FFS practice, participants must selected the thematic platform. Then, through interactive needs assessment, the facilitator

Table 4.3 Three expressions of FFS

Characteristic	*Joel at INIAP*	*Lenin at EcoPar*	*Eduardo at APRODIC*
Professional background	Diligent técnico – locally-based agronomist with specialization in biological pest control	Development bureaucrat – urban-based forester with specialization in resource economics	Eccentric farmer – lifetime of practical experience with potato-milk farming in Carchi
Reason for learning/ applying FFS	Idealistic to pragmatic – 'improving lives' was later supplanted by supervisor's interests	Pragmatic – 'told to do so' by supervisor and donor	Idealistic – 'wanting to make a difference' and pragmatic when working with projects
Facilitation style	Open and interactive	Closed, top-down	Open and interactive
Preferred location of sessions	In the field and classroom	In the classroom	In the field
Source of funding for salaries and FFS	State and external donors (COSUDE, USAID, CropLife)	External donors (IDRC)	Self-financed by facilitator and participants; punctual contributions from municipalities and external projects (EcoSalud)
Technical platforms	Potato IPM (adoption of technologies), safe use of pesticides	Agroforestry, sustainable biodiversity, organic agriculture	Potato IPM (pesticide use reduction), pasture improvement/ animal management
Number of participants	10–15	1–5	15–30
Average number of sessions	Decreased from 15 to 6	5 informal meetings	15–20, depending on group (primary school students or adults)
Who decides technical content	Técnico, in line with project objectives, some consultation from participants on details	Técnico, in line with project objectives	Participants, with feedback from facilitator or employer
Use of agroecosystem analysis?	Yes and no; full discussion	No	Yes; full discussion and negotiation
Experimentation?	Yes and no; largely pre-planned and project determined	No	Yes; highly open-ended and spontaneous
Didactics	Lively and fun: dynamic but largely unidirectional	Formal: lectures, unidirectional learning	Practical: discovery-based learning, learn by doing in the field
Follow-up	Dependent on project funding	Dependent on project funding	Dependent on group enthusiasm

and the participants work together to determine technical content based on local needs, priorities and preferences. Over time in Carchi, however, external agents commonly pre-determined the themes and technical content of FFS. In Joel and Lenin's cases, their organizations' projects in IPM and conservation, respectively, conditioned the thematic platforms of the field schools. Similarly, when CIP's EcoSalud project hired Eduardo, the resulting field schools shifted to the employer's priority of safe use of pesticides. Only when Eduardo ran his own, self-funded FFS did he allow participants to select their own thematic platform and curriculum. Eduardo added that: 'When I run my own FFS, I do not have snacks or lunches to offer participants. If they [the FFS participants] do not like the content, why should they show up?' In contrast, Joel and Lenin argued that free lunches and other incentives were essential to maintaining the enthusiasm of participants.

By design, FFS requires participants to identify priority research concerns and to conduct comparative, participant-led experiments. Nevertheless, several studies (e.g., Jiggins *et al.*, 2001; Schut, 2006) observed that professional biases, projects and organizational priorities commonly influenced the content and design of experiments. In Lenin's case, no experiments were used. In Joel's case, experiments were used to 'sell' INIAP's IPM alternatives, especially resistant varieties, fungicide spray regimes, and Andean weevil traps as well as CropLife's priority: 'appropriate', 'correct' and 'safe' use of pesticides. In the EcoSalud-financed FFS, no farmer-led experiments were included. In contrast, in Eduardo's field schools the participants usually led their own experiments, for example, on themes such as potato varieties, planting distances, and organic versus inorganic fertilizers.

The extent to which the critically important component of agroecosystem analysis was used differed considerably between cases. Initially, Joel applied AAE in all of his field schools, and he asked elective questions that forced participants to take good measurements and to address contradictions between their findings. Nevertheless, the hybrid FFS that he came to design and implement de-emphasized AAE. In some cases, the activity was no longer applied. According to one farmer participant, 'The goal [of Joel's FFS] is to get farmers to adopt his improved practices.' Similarly, Lenin used an AAE activity in one session and then never repeated it. His explanation was that: 'It took too much time', and that he could 'already predict the outcome beforehand.' In contrast, Eduardo drew heavily on AAE in all of his FFS, and he challenged participants to look for ways of decreasing outlays, especially of highly toxic pesticides and synthetic fertilizers. Joel's original utilization of AAE was more nuanced and sophisticated, with biological details, such as the name of certain insects and their habits, but Eduardo's group paid considerable attention to measurements and time for discussion and negotiation. When asked why he emphasized participant decision-making in his FFS, Eduardo responded, 'They [the participants] have invested a lot of money in the [FFS] plots. I'd be "lynched" if the experiments failed.'

In the place of theoretical, classroom lectures, FFS emphasizes in-the-field exercises where participants learn through farming practice. Joel invented numerous creative, discovery-learning activities. On multiple occasions he employed creative activities, such as 'What is this?' – an elicitive approach to helping farmers find answers to their own questions. He continued to use lively teaching in his hybrid FFS, but rather than biological and ecological principles, the content shifted towards the use of externally based technologies. Lenin's colleagues argued that, 'He [Lenin] refused to apply discovery-based activities, that he preferred to give lectures, and that he does not like to dirty his hands.' Eduardo, on the other hand, was uncomfortable writing and learning in a classroom. As a result, his FFS took place almost entirely in the field. Eduardo was judicious with the use of discovery-based learning: 'When a farmer is testing me, I'll give him an answer so that he knows that I am a true potato farmer. Otherwise, he has to find it [the answer] for himself. Sometimes my answers will not be his answers. Each potato field is different. That's where the ingenieros make mistakes.'

Joel, Lenin and Eduardo utilized FFS methodology for diverse purposes, involving consequent adaptations. Of the three, only Eduardo continued to allow participants to determine content, to use AAE, farmer-led experiments, discovery learning and learn-by-doing, but this only occurred when he was free of project constraints. In process management and didactics, both Joel and Lenin excluded participants from determining the thematic platform, content design, decision-making and independent learning. Be it due to project and organizational constraints or individual preferences, in the hands of these *técnicos*, over time FFS became transformed from a people- to a technology-centred endeavour.

Multiple appropriations of FFS by regime actors

Mendizabel (2002) explored how researchers at CIP and its national partners drew on FFS for advancing science-based contributions to agricultural development in the Andes and Southeast Asia. In so doing, FFS became hybridized to 'Farmer Field Schools – Farmer Participatory Research' (FFS-FPR). This involved shifting technical content around institutional research priorities, such as pesticide-use efficiency (Torrez *et al.*, 1999a and b) or selection of late blight resistance varieties (Nelson *et al.*, 2001; Ortiz *et al.*, 2004). Operating under the demands of scientific institutes, researchers commonly increased the complexity of single variable demonstrations to the point where FFS began to include dozens of variables and other subtleties. As a result, the outcomes of FFS experiments could be seen only through sophisticated statistical analysis. As described in Nelson *et al.* (2001), researchers commonly conserved the use of discovery learning exercises, but participants commonly were excluded from the selection of thematic platforms and curriculum content. In certain cases, researchers strategically dropped central FFS decision-making tools, such as the agroecosystem analysis, and their feedback into activity. In terms of thematics, content and process, researchers systematically transformed FFS beyond its original designs.

The priorities of researchers, however, were not the only factors influencing FFS. As a *técnico* from a conservation-oriented NGO, Lenin never fully appreciated the time demands of FFS. He argued that participatory research offered a better balance between the 'onerous' requirements of participation and his time limitations as an urban-based professional. His field schools included smaller groups, fewer encounters and no agroecosystem analysis. With regard to pedagogy, the expression of the methodology shared the features of FFS-FPR. While disciplinary demands often constrained CIP and Fundación Promoción e Investigación de Productos Andinos (PROINPA) researchers, Lenin altered FFS design for reasons associated with the 'high costs' of participation. Despite encouragement from his employer, Lenin's worldview and personal priorities kept him from implementing people-centred FFS.

Charged with both applied research and extension activities at INIAP, Joel faced very different constraints. His supervisor in the capital – an economist like Lenin – commonly evoked the work of economists at the World Bank (Feder *et al.*, 2004), which likewise criticized the 'high costs of participation'. He often referred to the primacy of INIAP and CIP, arguing that extension was secondary to research. Further, decreases in public funding as a result of government restructuring produced a subsequent need for public–private collaboration (e.g., between INIAP and CropLife), which worked against Joel's immediate interests in Carchi. On multiple occasions, central government policies intended to combat corruption interfered with FFS. For example, on one occasion, the police arrested Joel and impounded the project vehicle for working over the weekend when he travelled to support an FFS session in a nearby village. Concerned about corruption, the national government had passed a rule forbidding the use of public vehicles on weekends – exactly when most FFS groups preferred to hold meetings. Diverse constraints, such as the preferences of a supervisor, organizational mandates, private funding and public policies, combined to work against Joel's personal commitment to farmer-led FFS.

Meanwhile, Eduardo, an independently financed farmer who lived in a rural village, escaped many such constraints. This was particularly true when he was able to finance FFS through local resources. Eduardo's challenges were more localized, such as how to work together with neighbours to advance what were perceived as common interests. In the expert-led examples, particularly when Joel was coerced into running technology-centred FFS, the central objective became: 'What people did not do' (e.g., adopt certain agroforestry or IPM practices), and 'How to get them to do something differently' (Table 4.3). This primacy often had little to do with the priorities of farmers. In Eduardo's example, the preoccupation was: 'What people did and why', and how to keep participants interested in wanting to learn and collaborate. When Eduardo conducted the externally funded FFS on pesticide safety, he could no longer conserve this orientation. As a result, he could no longer motivate FFS participants to make financial contributions, and he had to draw on new incentives to get people to attend meetings, such as attractive lunch breaks or baseball caps. Through the freedom of self-employment, however, Eduardo owned a substantial degree of manoeuvrability allowing him to implement people-centred FFS and, to a large degree, conserve the methodology's central design feature: adapting to local realities and local relevance.

Viewed as part of an institution, the technical experts in Ecuador never seriously entertained the broader proposal of FFS as a people-centred approach and potentially part of a broader farmer-led movement for social change. During a meeting, one extensionist complained to Sherwood: 'If farmers begin to lead FFS, then what are we [the técnicos] supposed to do?' The impact studies in Ecuador de-emphasized the methodology's knowledge and social contributions, while emphasizing economic priorities, in particular cost-benefit and net return (Barrera *et al.*, 2004; Mauceri, 2004). In such cases, the objective of FFS shifted from 'the empowerment of individuals and communities' to 'profit in the market'. In the process, participants moved from the source of innovation to imitating the experts. In practice, FFS shifted from a pathway for human and social development to a means of technology transfer and economic development.

In the hands of experts operating in their organizations, FFS became diversely packaged and sold. It appeared in proposals as a means to 'organic' or 'clean production' (e.g., by INIAP and EcoPar), 'pesticide-use reductions' (CIP), and 'increases in productivity' (IPM-CRSP: Integrated Pest Management–Collaborative Research Support Program, Ecuador). The expected outputs of FFS became part of an individual or institutional marketing strategy. Researchers, técnicos and their institutes reduced FFS from a participant-led, multi-faceted and iterative learning–action methodology to a relatively pre-determined and standardized means of technology transfer. The methodology became a simplified object: a packaged course. In the process of creating the object, FFS was re-shaped into a new form: one essentially in line with the norms of the expert system.

In the social wild, FFS was vulnerable to competing forces involved in knowledge production. This was particularly true during processes of going to scale with the methodology. Diverse institutional factors – including professional thinking and preferences, project-based constraints, organizational culture, and centralized decision-making – combined to create obstacles for FFS. Through processes of objectification and commoditization, the methodology became de-localized and de-humanized, moving away from the essential adaptive and collaborative qualities of the methodology. The institutionalization of FFS largely transformed the methodology into an approach that no longer seriously threatened established ways of thinking and doing. Nevertheless, the transformation was not complete. Certain FFS facilitators, such as Eduardo, enjoyed conditions shaped by their geography and institutional culture that better positioned them to conserve the original people-centred designs of FFS.

Scaling-up in name, but not in meaning

After visiting the first wave of FFS throughout Central and South America, Sherwood and Thiele (2003) expressed concern over the 'methodological erosion' of FFS (Box 4.3). With the growing utilizations of FFS in Ecuador (scaling up), the methodology and many of its defining principles – interactive design, open-ended discovery-based learning, learning by doing, AAE and farmer-led experimentation – became vulnerable to translation. As Paredes (2001), Mendizabel (2002), Borja (2004), Schut (2006) and Schut and Sherwood (2007) illustrate, FFS progressively acquired diverse and even contradictory meanings.

Between 2005 and 2006, Schut (2006) conducted research on the diverse expressions of FFS in northern and central Ecuador. Visiting the FFS in Carchi, he explored the nuances of these developments. He found that previously competent facilitators (i.e., those who were acknowledged in project documents as fully applying the methodology) no longer applied FFS in its original, people-centred form. This led him to conclude that the

Box 4.3 Early observations on the transformation of FFS in the Andes (summarized from Sherwood and Thiele, 2003)

- Group establishment – women were systematically excluded. FFS leaders were chosen on the outset based on criteria external to the FFS experience. In most cases, FFS changed leaders over time.
- Technical content – facilitators commonly pre-packaged technical content, as per their professional backgrounds and project priorities.
- Learning plots – often competitive between farmers and facilitators. Sometimes there was no comparative 'traditional' plot. Often limited to demonstration of 'improved technologies'.
- Discovery-based learning – facilitators commonly 'cut corners' on learning. Instead of facilitating open-ended learning experiments, facilitators often gave lectures in classrooms.
- Experiments – facilitators often pre-determined content and design. Researchers often introduced complicated designs.
- Follow-up – project priorities determined follow-up. This central feature was commonly neglected.

Box.4.4 From Farmer Field Schools to Agrarian Revolution Schools

In 2010, Ecuador's Ministry of Agriculture, Aquaculture and Fisheries (MAGAP) created a national farmer training programme, explicitly transforming Farmer Field Schools around new purposes. The Ministry of Agriculture re-named FFS 'Agrarian Revolution Schools' (in Spanish, ERAs), broadened the thematic platform around 'integrated farm management' (from soils, water, crops, pests, animals, markets), and introduced political indoctrination in President Correa's call for a 'Citizen Revolution'. Urban-based professionals entirely pre-determined the curriculum and several hundred recent university graduates and young professionals (by policy, under 30 years of age) underwent an intensive, five-day training programme as facilitators. During the first five months of implementation, they were charged with organizing 977 ERAs involving 25,047 participants from each of Ecuador's 24 provinces. From pedagogical and political perspectives, FFS underwent a fundamental transformation to the point where the original design of the methodology was no longer recognizable.

fundamental shift of FFS in the hands of research and development practitioners was no mere oversight. Instead, he viewed it as the result of patterned behaviour. Researchers and técnicos expressed appreciation for how the methodology 'reached' into rural villages as well as the platform it provided for enabling technologies to arrive in communities. Nevertheless, other elements, particularly those associated with independent learning and decision-making, and open-ended and iterative design, proved more problematic. Over time, it was the interactive, discovery-based features of FFS (i.e., those elements that made FFS a people-centred methodology) that captured the attention of the experts, as they organized to re-work and re-design FFS for new purposes.

Within five years, CIP, INIAP and the FAO had largely abandoned FFS in its original form. The largest and most explicit transformation of FFS, however, came in the hands of the Ministry of Agriculture. Drawing on President Correa's call for a 'Citizen Revolution' in 2009, the Ministry of Agriculture re-designed FFS methodology around pre-determined technical training of rural people in modern technologies and indoctrination in the Government's political ideology (Proaño, 2011). The end product was a national-level campaign called the *Escuelas de la Revolución Agraria* or Agrarian Revolution Schools. While borrowing certain attractive features from FFS, they fundamentally undermined the people-centred intentions of FFS (Box 4.4). While FFS certainly survived within certain circles, in particular in community-based organizations and the agroecology movement, the methodology no longer influenced the activity of the larger organizations in Ecuador and thus was less threatening than previously.

Discussion

The FFS methodology became diversely appropriated. Initially, the adaptation in the Andes – in terms of the themes of interest, field conditions and cultural practices – was generally consistent with the methodology's original proposals in Asia. Nevertheless, over time changes to process management and didactic design – i.e., the erosion of people-centred elements – revealed conflicting expressions of FFS. The experience of FFS in the social wild illustrates how competing processes of Mode 1 (expert-led/technology-centred) and Mode 2 (lay- or people-centred) knowledge production can lead to conflicts during the negotiated processes of social change. Viewed as a social construction, knowledge is not produced in a political vacuum. New claims of knowledge and knowledge production are judged in reference to established forms. This is particularly true for interventions that represent entirely new ways of thinking and organizing. Those actors that do not conform may be subject to public criticism and sanction. Gibbons *et al.* (2000: 2) argue: 'It seems to be a recurrent historical pattern that intellectual innovations are first described as misguided by those whose ideas are dominant, then ignored, and, finally, taken over by original adversaries as their own invention.' FFS in Ecuador faced a similar fate.

The process orientation of FFS led people from INIAP, MAG, the pesticide industry and elsewhere to declare the methodology as expensive. After the benefits of FFS became overwhelmingly clear and the methodology became legitimized as best practice, many of the very same actors began to claim ownership of it. In the process of taking over, however, these actors systematically changed FFS around new purposes. The facilitation of open-ended discovery learning became specialized top-down lectures. Questions became answers. The content and processes of FFS were simplified to the point where differences between individual FFS were lost. Consistent with the design features of expert systems, FFS underwent degrees of homogenization. Rather than broaden Mode 1 production of knowledge to accommodate the necessary conditions for Mode 2, we observed that the experts and their organizations commonly sought to transform FFS in line with their competing priorities. Within that process, FFS as Mode 2 methodology was pulled into the Mode 1 paradigm: knowledge production through abstraction. Was the transformation of FFS due to a simple oversight or questions of competence, or were more subtle forces at play?

Drawing on diverse literature, we continue to explore three underlying social processes that influenced whether or not FFS as a niche-level intervention could function as a catalyst for broader socio-technical change:

the reshaping of farmer perceptions and expectations, confrontations with the regime over the creation of innovation pathways, and the generation of unexpected spin-offs.

Challenging farmers' perceptions and expectations

Andean farmers do not operate alone in isolated farming domains. Rather, they are part of unique socio-technical networks in which other actors, organizations and entities collectively inform and define courses of action and development opportunities (van der Ploeg, 1993). As a result, for farmers as well as *técnicos*' perceptions of the meaning and value of FFS respond to a certain way of defining relevant problems and solutions within a socio-technical network. Paredes (2010) explains that certain farmers operate in networks in which pesticides have come to be accepted as obligatory or unavoidable elements of good potato farming – despite the risk they represent for a family's health. For this group, agrochemicals are not seen as a problem, but rather a solution. In other farming styles, less agrochemical use is desired to avoid risks, and, in such cases, 'good farming' has come to mean safe farming, which partly explains why different actors may find different meanings and opportunities in FFS and how those influence its insertion into social life.

In retrospect we can see that the arrival of FFS in Ecuador challenged prevailing perceptions and expectations. FFS-by-design helped to create awareness and enabled communities to address latent pesticide concerns, particularly the health effects due to chronic exposure to highly toxic pesticides. As described in Sherwood *et al.* (2005), through the interactive learning–action process of FFS, the communities of Carchi increasingly uncovered unwanted environmental, health and economic products associated with modern farming. We found that people needed to discover for themselves the 'bads' of their agriculture before they could begin to understand the shared responsibility that farmers, communities and broader society held in co-producing those consequences.[7] Many communities no longer blindly accepted the recommendations of the técnicos from NGOs, the government and agrochemical stores. They began to demand non-chemical, ecologically based means for improving their agriculture. In several cases, agrochemical salespeople were chased out of town.

Nevertheless, as we experienced with Lenin and with Joel's supervisor at INIAP, many experts did not share the vision of people-centred FFS and its impacts, demonstrating that changing community-level perceptions was not enough. It showed the need to also reach the institutions of professional brokers of externally based knowledge and technology, which involved unseating established seats of power and influence. Despite such constraints, the emergence of new expectations through niche-level learning (see also Geels and Raven, 2006) is a necessary entry point for further developments.

Confrontations with the regime over innovation pathways

Attempts to institutionalize FFS have centred on scaling-up – i.e., increasing the methodology's geographical and institutional spread, usually meaning its organizational reach (LEISA, 2003a and b). To support this, people have proposed greater investment in the ToT, better impact studies (usually through 'more rigorous' quantitative means), and better (i.e., more) communication as a means to convince researchers and policy makers in their own language. Nevertheless, despite a wealth of extraordinarily specialized economic, environmental and health studies and substantial investment in communicating results in Ecuador (Crissman *et al.*, 1998; Yanggen *et al.*, 2003), the institutionalization of FFS turned out to be a far less rational process. While FFS scaled in name, its fundamental principles often were lost in translation. Instead of enabling desired institutional change, scaling exposed vulnerabilities that led to the transformation of FFS (Schut and Sherwood, 2007).

Over time, FFS methodology was pulled back into the dominant institutional paradigm it was supposed to challenge (Schut, 2006; Schut and Sherwood, 2007). Supporting collaboration amongst farmers in local innovations became top-down technology-transfer, and the farmer-led, demand-driven character was replaced by externally driven extension and development. Reasons can be found in the hierarchical and formal organization of national research and development institutes such as CIP and INIAP, where disciplines, procedures, protocol, mandates and responsibilities were clearly formulated, respected, and defended. Moreover, funding structures, time-constraints and donor-demands often did not provide sufficient space to adequately respond to the needs and interests of farmers.

On the surface, it appeared that organizations such as CIP, INIAP and MAG were in favour of Mode 2 knowledge production including: growth of multi-functionality, more pluralistic science and development, increasingly fluid communication and practice, and a preoccupation with relevance (Gibbons *et al.*, 2000). Nevertheless, the experience of FFS in the social wild suggests that in practice these actors continued to enforce Mode 1 knowledge production (Table 4.4). In essence, we see that a Mode 2 methodology was scaled-up in an environment where institutional pre-conditions for Mode 2 work (e.g., conducive incentive structures and professional attitudes) were largely lacking or incapable of advancing interests against competing perspectives.

In hindsight, we found our initial proposals of regime change towards people-centred designs ambitious. It implied transformation of assumptions about the underlying causes of poverty and environmental degradation, best practice and 'good agriculture', and how learning and development should be supported and facilitated. The performance of FFS in the social wild exposes the limitations of methodology-based contributions and especially

Table 4.4 Institutional conditions favourable to Mode 1 (expert-led) and Mode 2 (lay or people-led) knowledge production

Factor innovation	*Mode 1: expert-led science*	*Mode 2: lay/people-led*
Knowledge	Specialized – produced in the context of abstraction	Distributive – produced in the context of application
Role of science	Produce basic knowledge and spin-offs, set priorities – separation between science and technology	Facilitate innovation – little difference between science and technology
Preoccupation of science	Conformity to discipline and truth	Local relevance and accountability
Expected products	Solutions – answers to questions	Process management – management of interchanges
Criteria for recognition	Disciplinary knowledge, compliance to norms	Multi-functionality, adaptability and responsiveness
Communication	Formal, controlled, limited sources, and coherent	Informal, open, multiple sources, and complex
Access to information	Achieved through position	Achieved through social connections and networking
Organizational structure	Rigid, with stable boundaries	Fluid, fuzzy and porous boundaries
Role of government	Promote competition	Promote competition and collaboration

the utility of going to scale with such approaches as a means to change existing, unsustainable techno-centric regimes.

At different moments of our experience, especially when we managed considerable project resources and we held tight control over FFS as a best practice, we succeeded in convening a growing array of actors. Nevertheless, as FFS left the protected margins of the project agenda and entered the social wild, the meaning of FFS became vulnerable to translation and cooption by the competing interests of the dominant socio-technical regime. It appears that FFS was released before its people-centred form took full social hold. Schut (2006) found that by 2005 the FFS network of actors appeared to have divided between two ideological lines: expert vs. locally led development. In Figure 4.1 we draw on the interaction of these two factors to describe competing stages of socio-technical development, useful for describing the evolution of FFS in Carchi from multiple forms of IPM as a novelty to IPM in practice.

FFS was introduced as a niche-level intervention with the intention of making it an attractive novelty that would contribute to broader institutional change towards people-centred development (from quadrant IV to III).

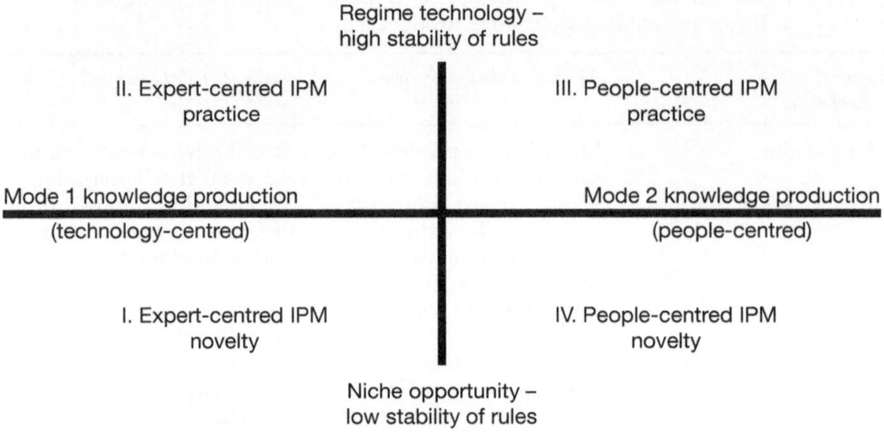

Figure 4.1 Competing stages of socio-technical change.

While FFS was successful at becoming a popular novelty and best practice, conserving its adaptive and people-centred qualities during processes of diffusion or 'massification' of the methodology became a struggle. At first, it seemed a question of competency (e.g., as a result of the quality of the ToT), but after observing previously competent FFS facilitators, such as Joel, lose hold of the methodology, it became obvious that more subtle social forces were at play. FFS became transformed into an expert-led IPM approach (movement from quadrant IV to I) that quickly scaled-up across heterogeneous groups of actors (movement from quadrant I to II) as a result of the structural alignment of Mode 1 knowledge production that still dominated the established socio-technical system.

How could regime change towards people-centred IPM have been enabled? Spontaneous regime level movement between expert- and people-centred development is not expected, since by definition the two regimes are organized around conflicting sets of formal and informal institutions, rules and rationalities. Change tended to happen at the moment of attempting to scale-up the methodology, when the priorities of the project leaders shifted from FFS implementation to diffusion of the methodology. At that point, the boundaries of influence became more porous, making it vulnerable to competing influences.

Institutional movement towards people-centred IPM practice requires a discontinuity with the established socio-technical regime that generally must come from 'below'. Since there is little to no alignment with the rules of the regime, change cannot happen through extending reach or scaling up but rather through expanding the depth and breadth of social embrace or scope. Influencing regime transformation requires that FFS broadly penetrate institutional spaces, transforming norms and values and organizing a new power centre around the enforcement of emergent niches and their rules.

Bypassing the regime: spin-offs and new movements

So far, this chapter largely described the diverse and often contradictory expressions of FFS that evolved when the methodology was introduced within the regime. While some space for manoeuvre was created within the regime, many institutional constraints and forms of resistance occurred. The fact that thousands of farmers gained insight into the biological and ecological processes of agriculture has had a number of spin-offs. Substantial evidence suggests that, for some, this contributed to fundamental improvements in farming, especially with regard to reducing the reliance on external inputs, whilst maintaining or increasing their productivity (Barrera *et al.*, 2004). Although FFS in Ecuador did not appear to have achieved the sort of organization described in the 'Community IPM' literature (Pontius *et al.*, 2002 and *www.communityipm.com*), it had initiated collaboration and collective action in a handful of communities, local organizations (APRODIC) and networks (MACRENA – Red de Manejo Comunitario de Recursos Naturales), with diverse and, it appears, unpredictable consequences.

By 2011, the few remaining examples of people-centred FFS continued to operate on the margins, but their strong internal organization, self-financing and diversifying activity suggested that they had gained a social foothold in northern Ecuador and beyond. Seeds of change were planted, but it did not appear that FFS as people-centred proposal would continue to grow into an increasingly coherent body of knowledge capable of defining and enforcing rules of 'good agriculture' and 'good development practice'. Instead, a related spin-off is the emergence of Canastas Solidarias – largely self-financed consumer groups from marginal neighbourhoods of Quito – that began to meet with groups of FFS graduates from the north to negotiate new consumer-grower arrangements around 'healthy food', a concept preoccupied with not just the end product of commodities but also with the production process itself.[8] Creative social networks promised to shorten the production–consumption chain in ways that could improve efficiencies through cutting out exploitative intermediaries and improving returns to both producers and consumers. Through strengthening linkages and interactions between growers and producers, such activities also increased accountability at both ends. This appeared to generate new social benefits, such as opportunities for the children of the Canastas Solidarias to be able to escape the city and live and work on the farms of FFS graduates during breaks in the school year. In addition, many of the former FFS collaborators operating at the grassroots had become reorganized around ideals of 'food sovereignty'. Acting as part of the Colectivo de Agroecología, these actors played an important role in provoking new public debates during the 2008 Constitutional Convention and subsequent legislation, including legislation to outlaw highly toxic pesticides[9] and proposals for food sovereignty.[10] Thus a sector of the original FFS participants had migrated to become part of diversifying networks of actors capable of producing and enforcing, at least to a limited degree, innovative policy reform, though in a form distant from what was expected at the onset of introducing FFS a decade earlier.

Conclusions: lessons for institutional change

In this chapter, we have drawn on the extremes of Mode 1 and Mode 2 knowledge production to describe the multiple epistemological expressions of FFS. Additionally, we have drawn on theories of institutionalization and socio-technical change to assess how organizations appropriated the methodology – e.g., the evolution of FFS as a niche-level innovation to its utilization by the dominant socio-technical regime, being pulled back in the research and development paradigm it was supposed to challenge (Schut, 2006). Although FFS scaled-up in name, it did not scale-up in meaning (Schut and Sherwood, 2007).

Key for understanding the transformation of FFS as well as similar methodology-based interventions is that such interventions are not politically neutral. In their public demonstrations of new desirabilities and possibilities, the FFS movement threatened established ways of thinking, doing and ordering. In the process, it became vulnerable to competing interests. Following its release into the social wild, FFS became prone to the ordering processes of the established socio-technical regime. Instead of shaping new rationalities and processes of change, commonly the design of FFS was re-organized to fit existing institutional mandates and objectives, donor-demands and funding structures, and administrative procedures.

The performance of FFS in the social wild exposes limitations of methodology-based interventions, and it suggests that calls for scaling-up innovations such as FFS are overly simplistic. Just like the relevance of FFS was tailored to the expectations and objectives of its farmer-participants, we could conclude researchers, development practitioners, the private sector and policymakers did the same. FFS were adapted and modified to fit political, economic and institutional realities, harmonizing it with existing norms, values and established rules and power relations, rather than challenging them. As this chapter demonstrates, this meant the loss of key features of the methodology, in particular its threatening people-centred character. In all, the experience indicates that one cannot realistically hope to achieve people-centred adaptive collaborative management of agriculture and natural resources through the mere development, documentation and scaling-up of a specific methodological approach, such as FFS.

Nevertheless, in some cases FFS as a methodological intervention arguably opened up unexpected spaces and pathways for subsequent social change. Certain FFS facilitators and graduates diversified activity further towards people-centred purposes, leading to innovative activities, such as the Canastas Solidarias and Food Sovereignty movements that were largely self-financed and self-organized and capable of producing and enforcing new 'rules of the game'. Over time, the FFS experiences contributed to the emergence of social networks capable of creating new relationships between producers and consumers, which led to the construction of collaborations around the production, consumption and circulation of food, ultimately shifting the investment of existing resources to new purposes. In essence, we see that the methodology and its underlying principles could be useful outside the sphere of influence of the regime. In practice, this represented a self-organized 'bypassing' of the regime in favour of people-centred development.

The question remains how to deal with the challenges facing methodology-based interventions in the social wild? In order to re-direct the present trajectory, beyond scale, FFS would need to influence the *scope* of activity. In order to respond to those expectations, niche-level interventions and counter movements must provide attention to how they can influence or address established norms, values and power relations, as well as reflect on issues of desirability and feasibility seen from the perspectives of different stakeholders. Most likely, growth and institutionalization of emergent proposals would require an increasingly influential social network composed of assemblages of actors from both the old regime as well as previously marginalized sectors capable of creating and enforcing new rules on the production and re-production of methodology-based interventions among the heterogeneous set of actors involved in agricultural science and development.

In the light of present institutional challenges it should be questioned whether fundamental changes can be designed and planned by means of specific methodological devices. Our analysis suggests that it makes more sense to approach fundamental change as a pathway, in which FFS as well as other methodology-based interventions can create space for innovation, new social networking and momentum for learning that could provide the foundations for exploring sustainable solutions to the complex problems facing us. We found that such a process also includes and requires forms of struggle and coalition formation towards changing dominant meanings, norms and incentive structures within organizations and establishments in order to create better conditions for people-centred development approaches.

We conclude that one cannot realistically hope to achieve people-centred adaptive management of agriculture and natural resources through the mere development, documentation and scaling-up of a specific methodological approach such as FFS. This needs to be complemented with context-specific strategies aimed at supporting the emergence of new networks and coalitions for changing dominant meanings, norms and incentives within organizations that may employ such methodological approaches. Alternatively, one could choose to bypass such establishments altogether and embed FFS within networks that have self-organizing capacity and genuine interest in people-centred development. Over time, such networks may well evolve into relevant movements for change that cannot be ignored by competing interests.

Acknowledgements

The studies reported here were originally financed by the Rockefeller Foundation. Subsequent investigations were conducted with the support of the USAID Soil Management and IPM Collaborative Research Support Programs as well as the United Nations' Food and Agriculture Organization, the Ecosystem Health Program of the Canadian International Development Research Council, the Dutch Fund for Eco-Regional Research, and a CEPIP-W Research Grant. The authors would like to thank Niels Röling for his comments and suggestions. We also would like to thank colleagues at CIP, INIAP, EkoRural, Carchi Consortium, Grupo Randi-Randi, EcoPar, MACRENA, the Canastas Solidarias, the Colectivo de Agroecología, and other individuals and communities that participated in different aspects of the research and development efforts in Ecuador.

Notes

1 In collaboration with the FAO's Global IPM Facility, the International Potato Center and others, Sherwood played a central role in introducing Farmer Field Schools in Latin America as of 1998, as summarized in Sherwood *et al.* (2000), Luther *et al.* (2005) and Pumisacho and Sherwood (2005).
2 The authors' research of FFS is summarized in Schut (2006) and Sherwood (2009), M.Sc. and Ph.D. studies respectively that were supervized by Leeuwis.
3 www.agassessment.org.
4 John Dewey's influential work on 'experiential learning' is nicely summarised at http://wilderdom.com/experiential/ (accessed 1 December 2010).
5 FAO, 2003. 'Development of Innovation Capacity for IPM and Agricultural Innovation for Greater Food Security in the Highlands.' Technical Cooperation Program/ECU/0067, 22 pp.
6 The Community Forestry Program approach in the Andes was amply described in Kenny-Jordan *et al.* (1999).
7 The 'reflexive modernization' and 'risk society' literature views the 'externalities' of modernity as an 'internality' or systemic product of societies. As such, socio-technical development in recent times is seen to lead to not just 'goods', but also 'bads' (see e.g., Giddens, 1990; Beck, 1992)
8 The Canastas Solidarias experience is summarized in Kirwan (2008). The broad concept of 'healthy food' was articulated in a national campaign document entitled 'Eat well, healthy, and in sovereignty'.
9 Official Congressional Registry of Ecuador No. 224, Tuesday, 29 June 2010.
10 The 2008 Ecuadorian Constitution calls for a shift in public policy towards 'food sovereignty', defined in the multi-dimensional context of agricultural production, emphasizing the 'social purpose' of land as a means of equitable, democratic social development and natural resource conservation and in favour of biodiversity (articles 276, 282, 334 and 400), equitable food distribution and commercialization (article 335), and ample access to culturally appropriate food and a healthy diet, in particular by means of native crops, animals, and other food sources (articles 13 and 281).

References

Barrera, V., L. Escudero, G. Norton, and S. Sherwood. 2001. Validación y difusión de modelos de manejo integrado de plagas y enfermedades en el cultivo de papa:

Una experiencia de capacitación participativa en la provincia de Carchi, Ecuador. *Revista INIAP*, 16: 26–28.

Barrera, V., L. Escudero, G. Norton, and J. Alwang. 2004. *Encontrando salidas para reducir los costos y la exposicin a plaguicidas en los productores de papa: experiencia de la intervencion en la provincia de Carchi, Ecuador*. CropLife, INIAP, IPM-CRSP, and FAO, Quito, Ecuador.

Beck, U. 1992. *Risk society: towards a new modernity*. Sage Publications, London, UK.

Beckerman, P. and A. Solimano (eds.). 2002. *Crisis and dollarization in Ecuador: stability, growth and social equity*. International Bank for Reconstruction and Development and World Bank, Washington, DC.

Bentley, J.W. 1992. Alternatives to pesticides in Central America: applied studies of local knowledge. *Culture and Agriculture*, 44: 10–13.

Borja, R. 2004. Documenting Farmer Field Schools in the Ecuadorian Highlands: a case study of the Province of Carchi. MPS thesis, International Agriculture and Rural Development, Cornell University, Ithaca, NY.

Crissman, C.C., J.M. Antle, and S.M. Capalbo (eds.). 1998. *Economic, environmental, and health tradeoffs in agriculture: pesticides and the sustainability of Andean potato production*. Kluwer Academic Publishers, Dordrecht, the Netherlands.

FAO. 2003. Development of Innovation Capacity for IPM and Agricultural Innovation for Greater Food Security in the Highlands. Technical Cooperation Program/ECU/0067, 22 pp.

Feder, G., R. Murgai, and J. Quizon. 2004. The acquisition and diffusion of knowledge: the case of pest management training in Farmer Field Schools, Indonesia. *Journal of Agricultural Economics*, 55(2): 217–239.

Gallagher, K. 1999. *Farmer Field Schools (FFS): a group extension process based on adult non-formal education methods*. FAO Global IPM Facility, Rome, Italy, 7 pp.

Gallardo-Zavala, J. 2003. Ecuador: lecciones de otra decada perdida y la dolarizacion. Poligrafica, C.A., Guayaquil, Ecuador.

Geels, F.W. 2004. From sectoral systems of innovation to socio-technical systems: insights about dynamics and change from sociology and institutional theory. *Research Policy*, 33: 897–920.

Geels, F. and R. Raven. 2006. Non-linearity and expectations in niche-development trajectories: ups and downs in Dutch biogas development (1973–2003). *Technology Analysis & Strategic Management*, 18(3): 375–392.

Gibbons, M. 1994. Transfer sciences: management of distributed knowledge production. *Empirica*, 21(3): 259–270.

Gibbons, M., C. Limoges, H. Nowotny, S. Schwartzman, P. Scott, and M. Trow. 2000. *The new production of knowledge: the dynamics of science and research in contemporary societies*. Sage Publications, London, UK.

Giddens, A. 1990. *The consequences of modernity*. Polity Press, Cambridge, UK.

Hutchins, E. 1996. *Cognition in the wild*. MIT Press, Cambridge, MA.

Jiggins, J. et al. 2001. *Mid-term review of the Global IPM Facility*. FAO, Rome, Italy, 71 pp. plus annex.

Kenmore, P., J.A. Litsinger, J.P. Bandong, A.C. Santiag, and M.M. Salac. 1987. Philippine rice farmers and insecticides: thirty years of growing dependency and new options for change. In: J. Tait and P. Napompeth (eds.). *Management of pests and pesticides: farmers' perceptions and practices*. Westview Press, Boulder, CO, pp. 98–115.

Kenmore, P.E. 1991. *Indonesia's integrated pest management: a model for Asia*. Southeast Asia Regional Programme, FAO, Manila, the Philippines.

Kenny-Jordon, C., C. Herz, M. Anazco, and M. Andrade. 1999. *Construyendo Cambios: Desarrollo Forestal Comunitario en los Andes*. Rome, Italy, FAO.

Kirwan, E. 2008. Building an urban–rural platform for food security. *LEISA Magazine*, September, 24(3): 22–24.

LEISA. 2003a. Learning with Farmer Field Schools. *LEISA Magazine for Low External Input and Sustainable Agriculture*, March, 19(1).

LEISA. 2003b. Aprendiendo con las ECAs. *LEISA Revista de Agroecología* (Latin America edition), June, 19(1).

Luther, G.C., C. Harris, S. Sherwood, K. Gallagher, J. Mangan, and K. Touré-Gamby. 2005. Developments and innovations in Farmer Field Schools and the training of trainers. In: G.W. Norton, E.A. Heinrichs, G.C. Luther, and M.E Irwin (eds.). *Globalizing IPM*. Blackwell Publishing, Ames, Iowa, pp.159–190.

Maasen, S. and O. Lieven. 2006. Socially robust knowledge. Transdisciplinarity: a new mode of governing science? *Science and Public Policy*, 33(6): 399–410.

Maurceri, M. 2004. Adoption of Integrated Pest Management technologies: a case study of potato farmers in Carchi, Ecuador. MSc thesis, Department of Agricultural and Applied Economics, Virginia Tech University, Blacksberg, VA.

Mendizabel, V.G. 2002. The development of participatory research at the International Potato Center (CIP): an ethnographic study of the social construction of expertise. MSc thesis, Department of Communication and Innovation Studies, Wageningen University, Wageningen, the Netherlands.

Nelson, R., R. Orrego, O. Ortiz, J. Tenorio, D. Mundt, M. Fredrix, and N.V. Vien. 2001. Working with resource-poor farmers to manage plant disease. *Plant Disease*, 85(7): 684–695.

Nowotny, H., P. Scott, and M. Gibbons. 2003. Introduction: 'Mode 2' revisited: the new production of knowledge. *Minerva*, 41(3): 179–194.

NRC. 2010. Towards sustainable agriculture systems in the 21st century. Committee on Twenty-First Century Systems Agriculture. US National Research Council, National Academy Press, Washington, DC, 598 pp. Available at: www.nap.edu/catalog/12832.html.

Ortiz, O., K.A. Garrett, J.J. Heath, R. Orrego, and R. Nelson. 2004. Management of potato late blight in the Peruvian highlands: evaluating the benefits of Farmer Field Schools and farmer participatory research. *Plant Disease*, 88: 565–571.

Paredes, M. 2001. We are like the fingers of the same hand: peasants' heterogeneity at the interface with technology and project intervention in Carchi, Ecuador. MSc thesis, Department of Communication and Innovation Studies, Wageningen University, Wageningen, the Netherlands.

Paredes, M. 2010. Peasants, potatoes and pesticides: heterogeneity in the context of agricultural modernization in the Highland Andes of Ecuador. PhD dissertation, Wageningen University, Wageningen, the Netherlands, 344 pp.

Pontius, J., R. Dilts, and A. Bartlett. 2002. *Ten years of building community: from Farmer Field Schools to community IPM*. Community IPM Programme, FAO, Jakarta, Indonesia.

Proaño, M. 2011. *Escuelas de la Revolución Agraria. Instituto Nacional de Capacitación Campesina, Ministerio de Agricultura*, Ganadería, Acuacultura y Pesca, 66 pp.

Pumisacho, M. and S. Sherwood (eds.). 2005. *Escuelas de Campo de Agricultores en America Latina: guia para facilitadores*. INIAP, CIP and World Neighbors, Quito, Ecuador.

Rip, A. and R. Kemp. 1998. Technological change. In: S. Rayner and E.L. Malone (eds.). *Human choice and climate change*. Battelle Press, Columbus, OH, 2: 327–399.

Schot, J. and F. Geels. 2007. Niches in evolutionary theories of technical change. *Journal of Evolutionary Economics*, 17(5): 605–622.
Schut, M. 2006. A house does not make a home: challenging paradigms through Farmer Field Schools. MSc thesis, Department of Communication and Innovation Studies, Wageningen University, Wageningen, The Netherlands.
Schut, M. and S. Sherwood. 2007. FFS in translation: scaling up in name, but not in meaning. *ILEIA Magazine*, 4(23): 28–29.
Sherwood, S. 2009. Learning from Carchi: agricultural modernization and the production of decline. PhD dissertation, Wageningen University, Wageningen, the Netherlands, 284 pp.
Sherwood, S and G. Thiele. 2003. Facilitar y dejar facilitar: ayudemos a los participantes a dirigir las ECAs. *LEISA Revista de Agroecología*, 19(1): 80–83.
Sherwood, S., R. Nelson, G. Thiele, and O. Ortiz. 2000. Farmer Field Schools in potato: a new platform for participatory training and research in the Andes. *LEISA Magazine on Low External Input and Sustainable Agriculture*, 16(4): 24–25.
Sherwood, S., D. Cole, C. Crissman, and M. Paredes. 2005. From pesticides to people: improving ecosystem health in the Northern Andes. Chapter 10 in: J. Pretty (ed.). *The pesticide detox: towards a more sustainable agriculture*. Earthscan Publications, London, UK, pp. 147–164.
Thiele, G., E. van de Fliert, and D. Campilan. 2001. What happened to participatory research at the International Potato Center? *Agriculture and Human Values*, 18: 429–446.
Torrez, R., J. Tenorio, C. Valencia, R. Orrego, O. Ortiz, R. Nelson, and G. Thiele. 1999a. *Implementing IPM for late blight in the Andes*. CIP, Lima, Peru.
Torrez, R., A. Veizaga, E. Macías, M. Salazar, J. Blajos, A. Gandarillas, O. Navia, J. Gabriel, and G. Thiele. 1999b. *Capacitación a agricultores en el manejo integrado del tizón de la papa en Cochabamba*. Fundación PROINPA, Cochabamba, Bolivia.
van den Berg, H. and J. Jiggins. 2007. Investing in farmers: the impacts of Farmer Field Schools in relation to IPM. *World Development*, 35(4): 663–686.
van der Ploeg, J. 1993. Potatoes and knowledge. In: M. Hobart (ed.). *An anthropological critique of development*. Routledge, London, UK, pp. 209–227.
Yanggen, D., C. Crissman, and P. Espinosa (eds.). 2003. *Los plaguicidas: impactos en producción, salud y medio ambiente en Carchi, Ecuador*. CIP, INIAP, and Abya-Yala, Quito, Ecuador.

5 Learning through networking

Enabling an adaptive learning network of local communities for integrated floodplain management in Bangladesh

Parvin Sultana and Paul Thompson

Introduction

This chapter documents and analyses how over 250 existing community-based organizations (CBOs) managing natural resources in Bangladesh have improved their management through an adaptive learning network, and the challenges through this process of encouraging CBOs to adopt improvements in their practices that take a more system-wide view of the productivity of floodplains, which we call 'integrated floodplain management' (IFM). The chapter is based on an action-research approach. The authors have worked in development initiatives to establish CBOs and in research and evaluations in this field since 1996, and specifically since 2007 we have piloted adaptive learning among CBOs. Through this CBOs have taken up elements of IFM, and at the same time we have investigated the effectiveness of this and impacts in participating communities. The chapter is based on working with a partnership and network of CBOs – about 150 since 2007, which expanded to about 250 from the end of 2008. Our experience is reported and assessed in the context of recent contributions on adaptive collaborative or co-management. The main questions that we seek to address are whether and how adaptive learning can be effective between a set of independent and dispersed CBOs, and what are the outcomes from this approach.

The remainder of the paper is divided into: a conceptual framework, the context of Bangladesh floodplain resources and community management, the process followed and how the adaptive learning network approach evolved, some of the outcomes from this adaptive learning network, lessons and issues arising, and a brief conclusion.

Conceptual framework

Research has highlighted the resilience of successful traditional common property institutions and associated resource exploitation and livelihoods based on feedback and adjustment to environmental variability (Berkes 1994; Freeman 1989; Holling *et al.* 1998). Where such traditional institutions were weak and/or new local institutions and organizations have been created for natural resource management a comparable dimension has been developed: adaptive management – 'an approach based on the recognition that the management of natural resources is always experimental, that we can learn from implemented activities, and that natural resource management can be improved on the basis of what has been learned' (Borrini-Feyerabend *et al.* 2000). Adaptive co-management is now seen as a specific management approach that incorporates a hierarchy of institutional arrangements to share management responsibilities over scales of resource and an explicit commitment to learning (Armitage *et al.* 2009).

Adaptive learning is about systematically using the results of management and monitoring to test assumptions (Margoluis and Salafsky 1998) and thereby improve interventions. If the intervention did not achieve the expected results, it is because either the assumptions were wrong, the interventions were poorly executed, the conditions at the intervention site had changed, the monitoring was faulty, or some combination of these problems. Adaptive learning involves changing the assumptions and the interventions to respond to new information obtained through the monitoring efforts. Tyler (2008) argues that the key element of adaptive management is that it specifies a formal design to test resource management interventions experimentally, rather than waiting for scientific uncertainties around such actions to be resolved before acting. However, he also noted that much adaptive management is set within an existing resource management framework that does not promote wider shared learning and focuses on technical/scientific issues. In our case we see managers (CBOs) identifying knowledge gaps and what they will test, assess and compare as adaptive learning, although the participants may or may not design actual experiments.

In management through adaptive learning the existence of uncertainties is not only accepted but made a focus of management efforts which seek to reduce them at the same time as managing the resource. In such cases learning, and reducing uncertainties about the resource system being managed, becomes a vital and integral part of management itself. Learning is seen as a three-stage process comprising: information generation, information sharing and information utilization (Arthur and Garaway 2005). Learning is about systematically documenting the process that was followed and the results that were achieved. Armitage *et al.* (2009) stress four aspects of learning related to co-management:

1 Social interaction based on trust.
2 A diversity of learning strategies particularly experiential.
3 A model of learning that accounts for social context, reflection, system thinking or interconnectedness, and a diversity of approaches to adaptation.

4 Attention to who is learning and the linkages among learners – scaling up individual learning.

These are issues that have been neglected in Bangladesh floodplain co-management. While CBOs have been formed that built on and enhanced social trust within local communities, the projects that supported this process did not take a strategic view of learning or systems. The CBOs themselves try to adapt but without support to do this in a systematic way. Floodplains and waterbodies comprise contested and over-exploited resources, and the earlier initiatives discussed in the next section each focused on part of the natural resource system – fisheries management or water for agriculture, developing CBOs and supporting skills and investments to address that focus. Pressure on limited dry season water for irrigation and fish remained an issue that CBOs with these different original aims raised as a gap in management during adaptive learning forums. The integrated floodplain management approach supported by our team has offered options for CBOs to test, adapt and address this issue. In this regard CBOs that have a legacy of different objectives in their management of similar floodplain resources handed to them from past project support and were encouraged to take a wider view of their natural resource systems by sharing lessons in crop, water and fishery management, and to consider other system linkages not anticipated at the outset by the research team.

The adaptive learning network approach discussed in this chapter addresses several issues and challenges that arise in the literature; discussed in this section:

- networking to scale up learning across management units and its consequences;
- relative merits of horizontal and vertical linkages in learning;
- the scope for adaptive learning extending to agencies other than the local level and beyond natural resource management to institutional arrangements;
- participatory monitoring and indicators.

Olsson *et al.* (2006) and Fabricius *et al.* (2007) note that learning becomes even more effective when knowledge networks are formed. Previously adaptive learning networks have been proposed between individuals (Davidson-Hunt 2006), or involved villages but focused on a more technical aspect of resource management such as culture based management of small waterbodies (Arthur and Garaway 2005). This project attempted to scale up the learning units, by linking up CBOs. In Bangladesh community based management of floodplain resources has expanded to a considerable extent, but the uniqueness and isolation of each of these locally managed units limit their scope for adaptive management and learning. The solution attempted in this project is learning among a network of CBOs, in areas identified by those CBOs. We term this learning a 'multiplier effect' where the benefits and lessons generated among a network of similar units or CBOs is greater than the scope for learning separately by each individual CBO. It also brings an advantage neglected in the literature – that of peer pressure from CBOs following practices collectively recognized as good on those CBOs that may be lacking whether in resource management norms or in organizational procedures. The project was designed to test this approach by working with a range of CBOs in Bangladesh that no longer receive support from the projects that initiated them, but which continue to manage floodplain resources.

Hence we focused on a horizontal learning process between CBOs that are comparable in status and environment but have diverse experiences. While we accept the importance of 'bridging organizations' such as those in co-management, that can strengthen social capital, particularly trust, across a hierarchy of levels or scales of different types of organizations (Berkes *et al.* 2003), we observed that this is effective when poorer grass-roots stakeholders already have achieved a sufficient level of 'bonding' social capital. This minimizes the potential risk raised by Adger *et al.* (2005) that powerful stakeholders could use information from cross-scale learning to undermine trust and reinforce their own authority. This applies in establishing community based co-management – first helping the poor to organize and develop trust and a common platform and voice before they are thrown into the frame of formally interacting with traditionally more powerful actors (government and local elites). Similarly the adaptive learning network started among CBOs in the expectation that from strength within a network CBOs can subsequently bring knowledge and views collectively to be heard and respected by government agencies, and even enable CBOs to advocate and negotiate over policy constraints with the government system. In Bangladesh this is particularly relevant to the issue of secure access to waterbodies for communities, but as will be seen access to and influence over higher levels of government administration remain a challenge for a CBO federation.

Adaptation is also possible in natural resource governance systems – if organizations and institutions are static they are less likely to survive in a dynamic world (Jentoft 2007). But this raises a new set of challenges – under what conditions will community organizations and co-management forums adapt and change (for example in terms of membership and constitutions and linkages) as part of adaptive learning, or will they focus adaptive learning only on natural resource management? We have taken a bounded view, by facilitating adaptive learning among CBOs rather than also targeting government agencies. This was for two reasons. First, an action research project of NGOs and researchers could not easily work to change practices across government agencies within a few years. Second, the earlier projects had already focused on the government agencies but the main gap that we perceived from working with and reviewing CBO experience since 1996 was the lack of links and sharing among CBOs. Hence substantial change in or introduction of forums where government stakeholders interact with CBOs was beyond our scope. But as will be seen CBOs did compare their governance systems, and some adopted good governance practices that they had previously ignored, or expanded membership to include under-represented stakeholders, such as women.

Monitoring and assessment methods and scope have to be considered in adaptive learning and in assessments of co-management performance. For example, in the institutional assessment framework developed by Ostrom (2005) and Oakerson (1992) which focuses on contextual factors in terms of biophysical/material conditions, community attributes, and rules. More generally Armitage *et al.* (2009) recommend that context, uncertainties and objectives should be considered in deciding what outcomes to monitor. However, for learning by resource managers the process is at least as important as the information and its analysis, and participatory processes in both the choice of indicators and monitoring are vital (Garaway and Arthur 2004; Armitage *et al.* 2009). Our approach focused on the learning and participatory process, and we endeavoured to enable the CBOs to direct the route of this process. However, coordinating adaptive learning across and between many CBOs whose leaders had not previously thought about such interactions or how to assess the effectiveness of innovations and changes in practices was a challenge. We discuss how input from the research team was needed to advise CBOs on the practicality of some proposed indicators and to help CBOs from different regions standardize on the details of measurement and methods.

In this regard adaptive learning among CBOs also builds on a wider literature on learning within organizations (including businesses) that has emphasized how organizations can encourage learning to overcome gaps between expectations and practice (Argyris 1996). However, most work focuses on individual organizations with well-trained staff or members, and a space and opportunity for employees to take a role in learning processes (Senge 1990). The CBOs individually differ in having members, generally with little formal education, who volunteer time but depend for their livelihoods on use of floodplain resources. Working with each CBO separately to enhance learning would have been possible, but would have replicated the past way of facilitating and assisting CBOs. The challenge addressed by an adaptive learning network is how to empower in learning many independent collective action initiatives, each with a limited scope to learn on its own. The significance of learning among social actors sharing similar management contexts and challenges as well as networking among them has not been appreciated in the existing literature on adaptive management. We have focused on learning among similar organizations and within those organizations their leaders, the networking process was designed to help develop trust among peers, and learning was based largely on sharing experiences taking explicit account of the interconnections in resources and their uses in the floodplain. Project resources limited the scope to directly involve many members of each CBO in learning workshops, so sharing between CBO leaders and general members was encouraged.

Social learning and adaptive management are thus interlinked, but in the context of co-management will most often involve shared learning among divergent stakeholders (individuals and organizations) with different orientations and values, as a means of developing a collective understanding. Consequently one of the main challenges for a social learning approach is the application of participatory methods and communications skills in the face of social and political power differences (Beck and Fajber 2006; Armitage *et al.* 2008). While, as noted by Tyler (2008), social learning can foster innovation through deliberation, relationship building, better communications, shared power, and recognition of mutual interest; it can be time consuming and is vulnerable to conflicts among stakeholders. In this project, by focusing on adaptive learning among CBOs rather than links between CBOs and government stakeholders, we have sought to provide a framework favourable to social learning, and have adopted an iterative process led by the participant CBOs.

General context of the case study

Bangladesh has achieved notable development successes in recent decades, with average life expectancy increasing to 65 years, economic growth sustained at about 6 per cent a year, GDP increased to US$ 494 per person in 2008 (compared to US$ 326 per person in 2000), and the percentage of people living in poverty reduced to 40 per cent by 2005. However, Bangladesh has the highest population density of any country excluding city states concentrated in a deltaic environment. Consequently the environment is under severe and increasing stresses. Most notably, wetlands and rivers are dying, a considerable part of over 4.5 million ha of floodplain have been drained and degraded, siltation and flooding are growing problems, and native fishes and other aquatic life are in decline.

About two-thirds of Bangladesh may be classified as wetlands according to the Ramsar Convention definition. About 6–7 per cent of Bangladesh is always under water, and in the monsoon 21 per cent is deeply (>90cm) flooded and around 35 per cent experiences shallow inundation (FAO 1988) – these areas form the floodplains of Bangladesh. Within this system are a wide variety of ecosystems including natural lakes, freshwater marshes, reservoirs, oxbow lakes, *haors* (deep depressions in the north-east that coalesce to form a vast inland sea in the monsoon – rainy season), *beels* (permanent freshwater depressions), *baors* (oxbow lakes), estuarine waters, and extensive seasonally inundated floodplains.

Floodplain wetlands in Bangladesh provide local people, especially the poor, with food, most notably fish but also other aquatic animals and plants. Some 82 per cent out of about 125,000 households fishing for income in one survey were found to be poor (WorldFish Center 2003). Up to 80 per cent of rural people and about half of rural poor households living in the floodplains catch fish and use other aquatic resources, and up to 70 per cent of animal protein consumption in Bangladesh is derived from fish (Minkin *et al.* 1997; Thompson *et al.* 1999; Muir 2003; Toufique and Gregory 2008). Wetlands are sources of fodder, building materials, water retention to recharge groundwater and absorb floodwaters, and a means of transport. Hence local communities have a direct interest in maintaining floodplain productivity.

However, the complex, dynamic nature of floodplain systems is a challenge to managing these areas. Permanent waterbodies and rivers are state property, but the majority of land in the floodplain system is privately owned and cultivated with rice. When it is flooded private land becomes a seasonal commons where people catch fish and use a multitude of natural aquatic resources, all interlinked in an ecosystem connected through water. Bangladesh wetlands have ample water for half of the year, but the limited amount of surface water in the dry season drives productivity. In the dry season surface water is needed for agriculture, for domestic uses, and for the survival of fish. Agricultural development has largely focused on rice production. Although embankments and flood control have played an important part in draining wetlands to expand agriculture and protect crops from floods in the monsoon, rapid growth in agricultural production since the 1980s has been led by abstracting ground and surface water to irrigate high-yielding rice in the dry season (Sultana *et al.* 2008), and embankments have had negative impacts on fisheries (Ali 1997; Halls 1998).

There are already a large number of policy initiatives and strategies that are supportive of sustainable development in Bangladesh. However, their implementation and adoption into practice has been at best piecemeal and often constrained by contradictory policies for land management and revenue. Underlying these problems is an administration that is highly departmentalized; this calls for a high degree of coordination in tackling complex interrelated problems. The challenge is nowhere greater than in the management of floodplain natural resources. The Ministry of Fisheries and Livestock (MOFL) is responsible for all technical aspects of inland fisheries, but the ownership and management of waterbodies rests largely with the Ministry of Land (MOL). Meanwhile agricultural development in floodplains has largely been driven by water management infrastructure built by the Bangladesh Water Development Board and Local Government Engineering Department (LGED), although technical support to farmers is crop-based and comes from the Department of Agricultural Extension. Each Ministry and attendant departments has its own mandate and priorities and these are not compatible most of the time. For example, the aim of protection and sustainable use of fisheries sought by the MOFL comes into direct conflict with the aim of increased revenue earning from the inland fisheries by the MOL.

Despite changes in national policies that call for an end to drainage of remaining wetlands (MWR 1999), wetlands continue to be encroached for agriculture, industry, brickfields and aquaculture with no sign of abatement, and the views of poor floodplain users have continued to be unheeded in policy formulation. In addition, 'floodplain aquaculture', where better off people enclose seasonally flooded private land with bunds and stock carps, has rapidly expanded in the last decade at the expense of natural fisheries used by poor people (Sultana 2010). Aquatic common pool resources, such as fisheries, have been declining as a result of these factors. Fish consumption fell by 11 per cent between 1995 and 2000 (but by 38 per cent for the poorest households), and it is estimated that inland capture fisheries catches fell by 38 per cent between 1995 and 2002 (Muir 2003).

This chapter discusses how an adaptive learning approach has linked a large number of community based organizations (CBOs) to help them exchange information and ideas, adopt good practices and innovate; and also identifies and discusses various forms of challenges encountered in the process, and building on this, attempts by the CBO federation and project partners to inform and influence government practices and policies related to floodplains. It has been supported since 2007 by two projects supported by the Canadian International Development Research Centre and UK Department for International Development's Research Into Use programme (RIU) by a partnership involving Bangladesh Environmental Lawyers Association, Banchte Shekha, Center for Natural Resources Studies, Flood Hazard Research Centre, MRAG, and 250 CBOs linked together through a federation.

Existing 'traditional' resource management practices and institutions

In this section we discuss the existing institutional arrangements in floodplain management prior to the adaptive learning initiative in 2007. As will be seen, government policies and practices regarding floodplain natural resources are fragmented along sectoral lines, resulting in uncoordinated and even conflicting interventions, that have in general given preference to farmers and water management for rice at the expense of the poor dependent on other wetland products. However, there have been parallel moves in water and fishery management to enable greater local participation since the 1990s. In addition it should be noted that most traditional research in Bangladesh related to fisheries and agriculture has been driven by scientists, or at best has involved farmers in trials; little has been driven by the demands and priorities of rural people.

Although MOFL and Department of Fisheries (DOF) are responsible for conserving and enhancing fisheries, and have set policies, strategies and rules, these agencies do not directly control the use of waterbodies. The larger permanent waterbodies including rivers and *beels* (depressions in the deeper parts of the floodplain) that form the more valuable components of the overall fisheries are government property and have been divided up into about 12,000 *jalmohals* or fishery estates. The fishing rights in *jalmohals* have historically been managed by the Ministry of Land for revenue generation (although by now this makes a minuscule contribution to the national budget). They have been leased out on a competitive basis to the highest bidder for three years. Usually this means they are controlled by wealthy and influential lessees who then hire traditional professional fishers to catch fish for them or charge tolls from those fishers.

The government of Bangladesh has attempted to reverse this pattern. In the 1970s a preference for leasing *jalmohals* to fisher cooperatives was established, and from 1986 to the mid-1990s the New Fisheries Management Policy (NFMP) piloted licensing of individual fishers in about 270 *jalmohals*. However, these policy changes had little impact since fisher cooperatives tend to be under the patronage of moneylenders and *de facto* lessees who pay for the lease, while the decision on who received licenses was also controlled by the cooperatives and therefore indirectly their patrons (Ahmed *et al.* 1997) and gave no incentive for individual fishers to cooperate to conserve fish. In 1995 a government decision ended leasing of rivers or 'open waters', with the stated aim of freeing poor fishers from exploitation, and they are now open access. This change was influenced by a group of wealthy boat owners (Huda 2003) and opened rivers and their fishers to exploitation and capture by those who invest in brushpiles (tree branches submerged to create fish shelters as a traditional way of catching fish) and have local power (Thompson *et al.* 2003). This system creates a significant number of management-related barriers affecting fisheries.

Table 5.1 summarizes the various policy shifts and projects that have had a significant bearing on floodplain resource management since 1985. Since the mid-1990s through several projects the responsibility for around 300

Table 5.1 Key events and changes in floodplain natural resource management in Bangladesh 1985–2010

Year	*Type*	*Event/change and its significance*
1986	Policy	New Fisheries Management Policy pilots licensing of fishers in about 270 waterbodies
1990	Project	Start of Flood Action Plan – component studies in the next four years highlighted impacts of flood control on fisheries and need for participation (ends 1995)
1991	Project	Oxbow Lakes Project phase II, although it focused on stocking of carps in closed waterbodies, developed over time community management systems (ends 1997)
1993	Practice	Initial guidelines on participation in water sector developed under Flood Action Plan
1994	Practice	Community Managed Wetland Habitat Restoration project – pilots re-excavation of waterbodies to restore connectivity and fisheries
1995	Policy	'Open *jalmohals*', i.e. flowing rivers taken out of leasing and in effect made open access resources (effectively signalled end of NFMP)

Table 5.1 continued . . .

Year	*Type*	*Event/change and its significance*
	Policy	Environment Conservation Act passed establishing Department of Environment and designating some Ecologically Critical Areas – mainly wetlands
1996	Practice	Start of Community Based Fisheries Management (CBFM) Project (1st phase) – gradually focuses on community institutions in 19 pilot waterbodies (ends 1999)
	Practice	Start of Small Scale Water Resource Development Sector Project of Local Government Engineering Department – forms 250 CBOs for water management under a special provision for cooperatives and hands over ownership of infrastructure to CBOs (ends 2002, but 2nd and 3rd phases continue to present)
1997	Policy	Ministry of Land places leasing of closed *jalmohals* smaller than 20 acres (8 ha) under Ministry of Youth and Sport
1998	Practice	Start of Management of Aquatic Ecosystems through Community Husbandry (MACH) project – develops community based co-management in three large floodplains (ends 2008)
1998	Policy	National Fisheries Policy (NFP) – provides general policy guidance including for participation but does not specifically support a role for CBOs
1999	Policy	National Water Policy – for the first time water development sector aims to maintain wetlands rather than drain them for agriculture
2000	Practice	Influenced by CBFM-1, Ministry of Land and Ministry of Fisheries and Livestock agree a framework for reserving *jalmohals* for 10 years under DOF oversight as a means of enabling management by CBOs; this is implemented through a set of project-specific Memoranda of Understanding
	Policy	Bangladesh Water Development Act passed – makes stakeholder consultation compulsory at all stages of project cycle in water sector
2001	Practice	CBFM-2 and Fourth Fisheries Project start; over next five years they expand waterbody management through CBOs under inter ministerial MOUs
	Practice	Water sector (Ministry for Water Resources) issues updated consolidated 'Guidelines for Participatory Water Management'
2006	Policy	National Fisheries Strategy (of DOF), includes proposals for moving all *jalmohals* into community management by reserving leases for a nominal fee for 'well functioning' CBOs. Not implemented
2007	Practice	Adaptive learning network project helps over 150 floodplain CBOs link together
2009	Policy	Ministry of Land issues a revised *Jalmohal* Management Policy, ends competitive tendering for leases to waterbodies replacing this with a preference for nearest CBO composed of fishers, but leaves definitions and objectives imprecise and places considerable power in hands of Members of Parliament as advisors. Tenure of existing CBOs close to expiring
2010	Project	Department of Fisheries and World Bank start planning a major project proposed to expand community management to 1,000 waterbodies

jalmohals has been handed over for 10 years by the Ministry of Land to MOFL. Under MOFL the Department of Fisheries has reserved these *jalmohals* for use by CBOs formed through those projects. In most cases these CBOs are still functioning and many have established sanctuaries and closed seasons thereby restoring fishery productivity. Based on experience developed through these projects since the late 1980s the National Fisheries Strategy (DOF 2006) regarding inland fisheries aims to support sustainability based on community participation, leading to a more equitable distribution of benefits. It proposed gradually reserving *jalmohal* leases for supervised CBOs against nominal lease payments, and restoration of fisheries.

By comparison, the majority of *jalmohals* leased under the traditional system have experienced over-exploitation, declining catches, and a lack of conservation measures. Since fishers are usually poor and leases have to be paid at the start of the year, access for fishers is compromised. Even with a preference in the traditional system for leasing to fisher cooperatives, middlemen pay the lease and take effective control using lists of their 'fishers'. In 2009 the Ministry of Land introduced its latest 'Jalmohal Management Policy' which on paper may encourage conservation based management by CBOs and ends competitive leasing. Instead, a registered CBO formed of 'fishers' would receive a three-year lease, with the lease payment increasing by 5 per cent in each round. However, it is open to potential manipulation over which group ('CBO') gains access, including potential political interference from Members of Parliament, who have been given a role in advising on which CBO will get a lease. The continued adoption of such top-down policies reflects a lack of commitment in government agencies, particularly the land administration, to share responsibilities in decision making and to listen to grass-roots actors such as CBOs. The latter are keen to bridge this gap, but to do so depends on the willingness of policy makers to listen and also build a bridge.

In the water management sector decisions since the 1960s had been dominated by a top-down engineering perspective resulting in embankment construction. During 1990–1995 the Flood Action Plan (FAP) process saw considerable debate over participation in flood mitigation planning and implementation. One of the guiding principles of FAP was stated to be 'Encourage popular support by involving beneficiaries in the planning, design and operation of flood control and drainage works' (World Bank 1990, p. 3). NGOs and civil society groups raised questions in national and international media over negative effects of existing embankments on the environment and fisheries, and inequitable impacts that disadvantage poorer people. The growth in debate was correlated with the move to a democratically elected government in 1991, greater freedom of the press, and public openness in FAP itself. Although by the close of FAP in 1995 the donor consortium had concluded that they could not justify implementation of major embankments, Guidelines for Public Participation in water sector projects were initiated under FAP (Brammer 2000; Hanchett 1997).

Subsequently, a formal process within the framework of the National Water Policy formulated Guidelines for Participatory Water Management (MWR 2001). The guidelines emphasize participation of all stakeholders in decision making in all project cycle steps. This policy was reflected in the Bangladesh Water Development Act, 2000 which made stakeholder consultation and participation at all stages of the project cycle mandatory. The Local Government Engineering Department, which is responsible for constructing smaller scale water control works, was also involved in finalizing the guidelines and adopted a policy of formally handing over water control structures to be owned and managed by local communities. To do this it has implemented projects that formed local cooperatives (CBOs) with which it develops plans for infrastructure, and transfers ownership of that infrastructure after construction where the command area is under 1,000 ha.

Although these changes were the result of public debate within Bangladesh involving a wide range of stakeholders, they were also influenced by international recognition of the failure of top-down approaches to natural resource management, and promotion of community based management for sustainability, empowerment, efficiency and poverty reduction (Agrawal 2001; World Bank 1996). From the mid-1990s, the donors and government accepted a participatory development approach based on increased roles for NGOs and devolving responsibilities to local people for floodplain resources, including water management schemes and fisheries. This was influenced by international research on how local institutions regulate and manage common pool resources, such as fisheries and water, which gave rise to understanding of complexity and recommendations on the design of more effective bottom-up management systems (Stern *et al.* 2002).

Since the mid-1990s the Government of Bangladesh has, therefore, undertaken several projects to improve local fisheries management and water resources management through CBOs. In the water management projects community participation directly followed from the participation guidelines. In the fisheries projects community management was promoted by funding agencies, initially the Ford Foundation, and subsequently Danida, the UK's DFID, USAID and World Bank. Importantly the Department of Fisheries lacked any direct control over waterbodies until it gained a role through projects, and therefore saw a benefit for itself in community based co-management.

The various funded projects were all time bound, but had the intention of establishing community management of fisheries, wetlands or water resources structures. Floodplain resource users are either poor and weak or rich and influential. One of the main themes of community based management has been to encourage and empower the poor to take part in management and for the rich to support them. For example, *jalmohals* and water control structures have been transferred for long-term community management under government oversight, and poor fishing communities using these resources are expected to continue benefiting for generations to come. But sustainable use requires local coordination and collective action and for this local institutions and CBOs were established. In larger systems – larger wetlands and large water management projects – both CBOs and co-management forums that link CBOs and government have been established but are still at an experimental stage (Sultana and Thompson 2010). However, the task of establishing community management and introducing interventions in resource management dominated those projects, which did not support CBOs to build capacity in continually analyzing their performance and adapting to changing circumstances and opportunities. Nor did projects make much effort to help CBOs interact to learn from one another; at best they may have sponsored some meetings between CBOs under the same project, but usually to disseminate information. Ultimately there is the question of the sustainability of CBOs when intensive project support ends, but as will be seen many CBOs have continued to function.

Evolution of adaptive learning network

Adaptive learning network

In Bangladesh by the mid-2000s there were already several hundred floodplain CBOs. These are people's organizations comprising 50 to 400 or more members, mostly poor, and covering the several villages that depend on a defined floodplain or waterbody. They were formed to manage natural resources (water and fish) by government projects, and facilitated by NGOs. They have graduated from those projects and are registered as independent legal entities with recognized rights and responsibilities to their resources.

Typically each CBO has an executive committee of leaders representing the whole organization. These leaders have developed management plans, sometimes through discussion with the general members or wider community, but the planning process was most often not well structured in terms of systematically reviewing recent experience and seeking the opinions of different categories of stakeholders. Also, although every year the CBOs reviewed and revised their management plans, there was no sharing between CBOs because the projects that created the CBOs had treated each one separately and did not encourage cross-learning. The adaptive learning network process when proposed aimed to improve on this practice through:

- the multiplier effect of shared learning across CBOs based on experience;
- identifying common constraints and gaps in knowledge of what works;
- coordinated piloting of innovations and good practices that address a wider range of interlinked floodplain resource management issues than sectoral projects have usually addressed;
- networking and collective advocacy and negotiation of rights.

In early 2007, with IDRC support, we reviewed the status of existing CBOs that had been formed by projects that had ended or were just ending. Based on assessments from the projects CBOs were identified that were continuing to function and had scope to improve their management systems and resource base and therefore might gain during the project period from adaptive learning (hence some of the weakest and strongest CBOs were not considered). Just over 150 CBOs were contacted to see if they were interested to participate in the initiative (as discussed later a further 100 CBOs were added in 2008). Table 5.2 summarizes the origins of all of the participating CBOs. The CBOs were invited to attend workshops and informed that through this they could learn about and then try natural resource management actions that other CBOs in other regions were doing.

In May–June 2007 initial workshops were held with small groups of leaders from the CBOs from a given environment and region, so that sufficient time could be spent by the research team on discussions with each of the CBO leaders. We anticipated that meetings with large numbers of CBO leaders would result in confusions and frustrations at the initial stage, if all leaders did not get a chance to explain their own experiences and ask questions of the team. Here we discussed how the adaptive learning network process could function and understood the existing management plans of the CBOs, how those had evolved and what expectations, outcomes and gaps the CBOs had. These small group meetings were a starting point, but were not intended to be repeated as they would neither be cost effective nor address the aim of learning across a larger network of CBOs. In addition to identifying a number of lessons and successes, the CBOs also identified the following main gaps and uncertainties facing them:

- Lease period: Most CBOs managing *jalmohals* have access through DOF for 10 years, but after project support finished some Deputy Commissioners tried to auction out use rights ignoring the agreements made by the Ministry of Land, and after 10 years all face an uncertain future.
- Lease value: Per hectare lease costs vary between *jalmohals* according to previous competitive tenders, in this and other leasing issues the policy makers never involved the fishers. The CBO leaders in the workshops wanted to be involved in policy formulation, and this could more easily be done if they organize themselves.
- Access to rivers: Since 1995 access to fish in rivers became free, and influential people put brushpiles in the best fishing locations; the gap is how to establish fisher rights to limit use in rivers, for example licensing different types of use.
- Compliance with closed seasons: Appropriate closed seasons that balance higher long-term catches and short-term loss of income for professional fishers is an issue; alternative occupations and seasonal loans could be solutions.

Table 5.2 Completed projects that formed CBOs and distribution of those participating in adaptive learning in 2008–2010

Project		*Agencies (donor; government agency)*	*CBOs formed*	*No. of participating CBOs*				
				Total	*SW*	*NW*	*CE*	*NE*
AqDP	Aquaculture Development Project	IFAD; DOF	9	6	6			
CBFM	Community Based Fisheries Management (1&2)	Ford Foundation then DFID; DOF	107	74	20	17	29	8
FFP	Fourth Fisheries Project	World Bank and DFID; DOF	46	36	15	15	6	
MACH	Management of Aquatic Ecosystems through Community Husbandry	USAID; MOFL	16	16			16	
OLPII	Oxbow Lakes Project phase II	Danida and IFAD; DOF	22	8	8			
PBAEP	Patuakhali-Barguna Aquaculture Extension Project	Danida; DOF	9	3	3			
SEMP	Sustainable Environmental Management Programme	UNDP: MOEF	30	25	2		2	21
SSWRDSP	Small-Scale Water Resources Development Sector Project (1st phase)	ADB, IFAD and Dutch; LGED	462	87	34	39		14
Total			701	255	88	71	53	43

Key: ADB – Asian Development Bank; Danida – Danish international development agency; DFID – UK Department for International Development; DOF – Department of Fisheries; Dutch – Government of the Netherlands; IFAD – International Fund for Agricultural Development; LGED – Local Government Engineering Department; MOEF – Ministry for Environment and Forests; MOFL – Ministry for Fisheries and Livestock; UNDP – United Nations Development Programme; USAID – US Agency for International Development.

Note: all of the projects involved one or more national NGOs, and most involved international technical assistance.

- Siltation of waterbodies: Waterbodies are silting up and being encroached for agriculture and human settlements. CBOs raised the question whether different arrangements of brushpiles used as sanctuaries in rivers affect siltation.
- CBO registration: Most CBOs are registered as cooperatives or as social welfare organizations; CBOs that are not registered might be more vulnerable to losing access to waterbodies, and the type of registration affects CBO operations.
- Water pollution: CBOs reported worsening surface water quality and fish kills. Some CBOs had already lobbied for enforcement of regulations on industrial effluent; jute retting also causes water quality problems.
- Other threats: CBOs reported that fisheries officers demand cash and big fish – otherwise they threaten to cancel the access rights of the CBOs. The CBOs proposed to organize in one platform so that they can better resist malpractice.
- Increasing fishing pressure in rivers: Fifty years ago professional fishers were by tradition low caste Hindus, but with population growth and poverty Muslims started fishing, resulting in excess effort. Support for alternative livelihoods may reduce this.
- Demarcation of waterbodies and competing uses: The boundaries of most of their waterbodies are not demarcated resulting in conflicts with farmers.
- Mobilizing funds to maintain water control structures: CBOs established by LGED reported difficulty in raising their 50 per cent contribution to larger maintenance works but fisheries CBOs are able to raise funds for leases and management activities.
- Faulty design: Some CBOs reported that water control structures were originally faulty and proposed that they should be involved more in planning and design.
- Coordination among fishers and farmers: Although fishers and farmers all need water for their livelihoods, they only coordinated their activities in one or two sites. In the floodplain beels with sluice gates, influential farmers close and open the gates for their interest, which may not be beneficial for the fishery.

Figure 5.1 contrasts the typical planning-management action process of individual CBOs (Figure 5.1a) with the process that evolved in the adaptive learning network (Figure 5.1b). In three larger workshops (bottom large circle of Figure 5.1b) held in July 2007 the leaders of all the CBOs from each of three floodplain environment types (open floodplains, rivers, and smaller 'closed' waterbodies) came together from across Bangladesh and reviewed their management plans and practices, identified common issues and uncertainties, solutions that had already be proven by some of the CBOs/sites, potential changes in their draft management plans, and other aspects of their structure, decision making and operations that they wanted to improve, test or experiment with. However, it was not possible for CBO leaders in these first workshops to present here clear plans of what they would do, since these had to be developed through discussion with their members, and consequently ideas for indicators and participatory monitoring, which are an integral part of the adaptive learning process, could also not be discussed or coordinated at the initial stage.

Examples of good practices identified for further testing by CBOs at this stage included: fish sanctuaries and improvements to them, reintroduction

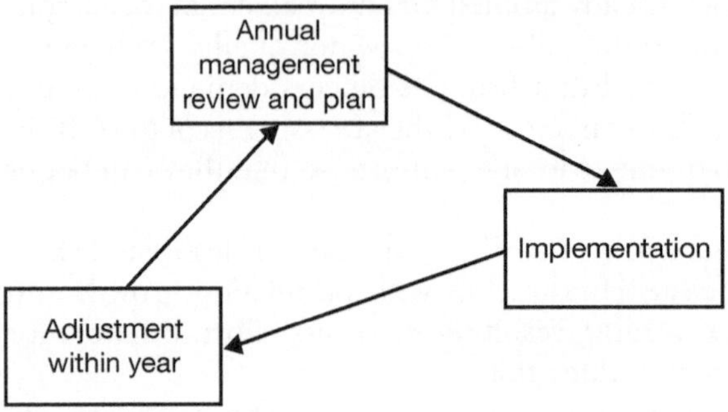

Figure 5.1a Adaptive management by CBOs prior to FHRC support.

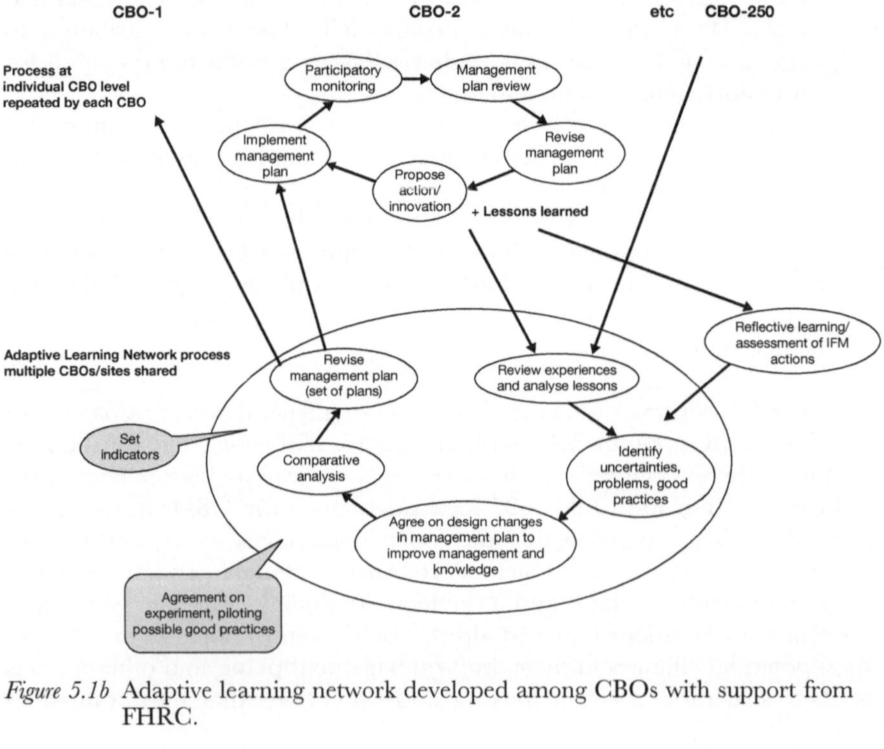

Figure 5.1b Adaptive learning network developed among CBOs with support from FHRC.

of native fish species, improvements in stocking practices in closed beels (see Box 5.1). Some of the potential lessons that addressed common issues experienced by the CBOs came from experience of a limited number of CBOs that had worked previously with the research team on integrated floodplain management, such as adoption of alternative dry season crops with lower irrigation water demand, and alternative ways of processing jute to reduce water pollution. To this extent over the subsequent years the research team encouraged this IFM approach, but it also arose naturally through exchanges between CBOs that had earlier focused on different aspects of floodplain natural resource management.

Box 5.1 Some of the main lessons and successes identified by CBOs at the outset

- Leadership: Leader selection through showing hand is good for the committee, transparency in selection comes if leaders are elected through ballot paper, but it is bad for management of the organization to change leader frequently. Decisions are better when there are educated/appropriately qualified leaders in the executive committee, and as a result management is peaceful. Including local elites/influential persons in the CBO or advising it is often helpful for resource management.
- Closed beels: Releasing larger fingerlings allows multiple harvest. Application of lime before fingerling release can decrease the incidence of fish ulcers. Application of fertilizer, cow dung and oilcake increases production.
- Fish sanctuaries increase fish catches: One big sanctuary shows a better conservation impact than two small sanctuaries. The deepest place in the waterbody is most suitable for sanctuary establishment. Sanctuaries enhance populations of rarer species of fish.
- Other fishery management initiatives: Observing a closed season during Baishakh-Asher (mid-April–June) is necessary to increase wild fish populations. Gear restriction helps to increase fish production (exactly which gears should not be used when is open to further testing and assessment).
- Integrating floodplain resource management: Construction of sluice gates and/or re-excavation of canals have increased production of crops and cultured fish. Water can be used for other aquatic resources (such as aquatic plants). Siltation is reduced by planting trees beside streams, and the trees attract insectivorous birds and as a result use of pesticides is reduced which makes the water better for fish.

Examples of gaps in understanding and constraints that the CBOs wanted to address and generate additions to knowledge and best practices included:

- appropriate areas and conditions to create fish sanctuaries (although this was an increasingly widespread management intervention, CBOs in floodplain systems have little to guide their planning other than local knowledge);
- institutional development and annual planning processes to bring different stakeholders into the process; and
- integrating multiple floodplain uses so that the total expected returns from the floodplain are maximized/increased (the overall returns from the resource base and from scarce dry season water can be higher and less flood prone when less-water-demanding crops are used, sluice operation modified, and surface water is protected to maintain natural fish stocks; Sultana *et al.* 2005).

The CBO leaders then went back to their own communities where they discussed the shared learning in the workshop. They called executive body meetings to decide whether they wanted to change their previous management plans or not, and discussed the issues with their general bodies. After obtaining general body responses, the CBOs revised their management plans finding room to change their decisions on the basis of not only their own experiences but also experiences of the other CBOs as shared in the workshops. As a result CBOs developed specific plans for testing improvements in their resource management that were informed by the lessons and ideas of the other CBOs in the network and were endorsed by their general membership. Box 5.2 gives two examples of how CBOs adapted their practices in response to lessons and experiences shared through the network.

Originally it had been thought that annual workshops among CBOs, complemented by a round of exchange visits between CBOs and a newsletter, would be enough for the adaptive learning process to function. However, the research team decided in the first year that more frequent workshops were needed to enable this process. Workshops were preferred as they provide a forum where a larger critical mass of CBOs can meet face-to-face together (through their leaders) to share experiences and lessons, and they make efficient use of both CBO leader and researcher time. Workshops have taken place at approximately six month intervals, and for this to be more cost-effective and logistically easier a switch was made to holding these on a regional basis, with the initial 150 CBOs divided into three regions (south-west, north-west and centre-east/north-centre).

It was found that there were more similarities than differences in experiences, gaps and lessons between CBOs from different environmental contexts, so holding workshops within a region reduced travel time and costs. But wider sharing was still covered by organizing exchange visits between regions. It was also more effective to ensure that all CBOs spoke

Box 5.2 Examples of adaptive learning network influence on CBOs

- Peer pressure: When Borobila CBO joined the adaptive learning network in 2007 there were severe conflicts between members over resource use, a failed sanctuary, and few poor executive members. By late 2009 this had been turned around. Mr. Arzun, the current chairperson, in explaining this and referring to the influence of learning from other CBOs said, 'We were ashamed of our own management and discussed the issues frequently with all members and we cited examples from other CBOs. The members decided that changes were needed for their sustainability, conservation of the waterbody and good governance.'
- Resource management: Mr. Mizanur Rahman, CBO representative from Saster Baor (which had been stocking carps since the early 1990s after initial support from a fisheries project) said,

> We always thought we have to keep the *baor* clean [clear of vegetation] so that the cultured fish can grow properly. Usually we were not bothered about the native fish. When I came to the adaptive learning workshops and went to the north-west region for the exchange visit, I saw CBOs had established sanctuaries in their waterbodies. When I asked them about the use of it they showed me their harvest record. I never knew that establishing sanctuaries can increase income so much. We CBOs compared the prices of cultured fish with native fish and found out that the price of native fish is three to four times more than for cultured fish. When I came back I talked to our committee, but some of the general members were not convinced. However, on a majority decision we established a one hectare sanctuary in 2008 in the *baor*. In 2007 native fish contributed about 7 per cent of the total income, but this has increased to 26 per cent by the beginning of 2010. . . . We did not spend any extra to achieve this. All the members are now very happy and worked voluntarily to keep the sanctuary in good shape.

in workshops. This revealed which CBOs had more or less experience and encouraged direct sharing and support between CBOs. In the first round of these regional workshops in November 2007 to January 2008 the CBOs presented their planned improvements in resource management – many as 'schemes' for which they sought small grants or other support – to all the other CBOs of that region as well as the research team. Potential indicators for success and participatory monitoring methods were also discussed. This enabled peer review of plans by the other CBOs, and a coordinated approach to testing resource management improvements. In the networking workshops CBOs proposed indicators for the different initiatives. Because workshops were held in different regions there were differences in CBO views on indicators; moreover it was difficult for CBOs to make operational in quantitative terms some of the general indicators they proposed, and in some cases CBOs were optimistic in the amount of data that they proposed themselves collecting compared with the time demands on their members. Consequently the research team helped to consolidate the CBO proposals into a set of common indicators and recording formats for each initiative they wanted to try. This process has allowed the networked CBOs to understand better impacts and to undertake monitoring by themselves.

Thus these plans were coordinated by the networked CBOs so that alternative views and improvements in their management practices could be compared between locations, to generate knowledge more systematically of what does or does not work, so that appropriate innovation and improvements based on experience can spread more quickly. Because CBOs are not in direct competition, they have been keen to compare results between sites, although this has depended on the team facilitating reflective learning and discussions among CBOs. The research team made follow up visits to assess the feasibility of proposed actions and to trouble shoot any problems faced by the CBOs in their record keeping, etc., and to conduct assessments of CBO performance to relate with the adaptive learning process.

The project team provided some limited capacity support to CBOs through trainings and visits and piloting to address knowledge gaps, but unlike most projects the emphasis was on the CBOs themselves taking initiatives and gaining and sharing knowledge. Where the piloting proposed by specific CBOs and agreed in the network involved new activities and some risks, modest funds were disbursed for demonstration purposes. Requests for funds were submitted to the project team, the plans were reviewed in the field by members of the research team and then by a project steering committee (comprising representatives from the project partners and from relevant government agencies) for approval. Before disbursing any fund or support in kind a general meeting was held in each CBO and the general members were informed about the amount to be disbursed. The funds were disbursed as support in kind or into the bank accounts of the CBOs in the presence of CBO members. In each CBO a three-member Project Implementation Committee (PIC) was formed with members from outside the executive committee for implementing and supervising the initiative. When the scheme had been implemented the top three executive committee members endorsed and certified that the initiative had been implemented according to schedule. The main challenges faced related to the timing of cash flows and seasonality of initiatives – for example, providing quality seeds in time for the start of the dry season, and the reluctance of a lead organization that was not agriculture-oriented to advance funds in time. In addition a few CBO leaders took advantage of the system for personal gain in community initiatives such as fish sanctuaries, but the combination of PIC oversight and reporting back to peers in the adaptive learning workshops limited this.

This process has continued with every six months a round of regional workshops involving the leaders of all participating CBOs. Figure 5.1b represents the adaptive learning network process that has emerged. The large circle in the bottom represents these regional workshops among CBO leaders held roughly every six months, in the top is the modified process of adaptive management followed by each of the participating individual CBOs (rather than the earlier independent cycle shown in Figure 5.1a), and in the mid-lower right is an additional step of reflective learning on specific interventions that was added as the initiative progressed.

Participatory monitoring by the CBOs has been supported with advice and standardized simple recording formats by the research team. Analysis has combined reflective learning by the CBOs, based on their data and qualitative assessments and interpretations, with data analysis by the research team. Lesson sharing among the CBOs has led to further adaptive improvements and wider uptake among CBOs. This has been complemented by annual assessments of CBO performance made by the research team, and CBOs' own assessments of their strengths and weaknesses, which have been shared in workshops, and have encouraged some CBOs to address gaps and also directed capacity building by the project partners. The main themes for adaptive learning have been the expected ones integrating fisheries and agriculture; good governance including a stronger role for women in CBOs; and learning. However, CBOs have also brought an interest in alternative natural resource based enterprises (e.g. bee keeping, duck rearing) and in saving and revolving fund mechanisms, which have spread through the learning network.

Incorporation of past research outputs into adaptive learning

Most research related to floodplain natural resource management has, like development assistance, been divided on sectoral lines. This was also reflected in the activities adopted by the CBOs, which were usually limited to a specific aspect of fisheries management or of water management for rice. However, the CBOs involved in the adaptive learning network did not limit themselves to their original 'comfort zone', and this was one of the expectations of the research team. We had been involved in initial research in the early 2000s in Bangladesh supported by the Land-Water Interface component and Natural Resources Systems Programme of DFID's Renewable Natural Resources Research Strategy (RNRRS). This identified the potential gains from system-based integrated floodplain management (IFM) – trying to optimize the productivity of floodplain systems by considering the tradeoffs and returns from dry season water used not only for irrigating rice but also to conserve fish stocks that could repopulate the floodplain in the monsoon, thereby maximizing floodplain productivity and returns in ways that would benefit the poor who depend more on floodplain commons (Shankar 2002; Shankar *et al.* 2004). This identified key IFM components: dry season refuges for fish, closed fishing seasons in the early monsoon, adoption of shorter duration and less-water-hungry alternate crops to conserve drier season water, and opening sluice gates earlier in the pre-monsoon so as to allow fish to migrate and spawn.

The IFM recommendations were piloted with two of the participating CBOs with some success (CNRS *et al.* 2005), and this formed a framework for natural resource management that could be adopted by CBOs through adaptive learning. Apart from the Bangladesh Environmental Lawyers Association (BELA) all of the partners in the present initiative were involved in this piloting, which worked with two existing CBOs and had highlighted how a multi-disciplinary team from NGOs and researchers could work together supporting CBOs. One of these sites is a 350 ha seasonal floodplain in southwest Bangladesh where for IFM the team brought together, in a central committee, existing separate institutions involving fishers, farmers and sluice operators. Then in 2003–2005 farmers gradually replaced 20 per cent of dry season irrigated rice with pulses, potatoes and garlic. These crops were profitable and left more surface water in the dry season, which the community protected as a fish sanctuary. This enhanced fish survival and reproduction, resulting in higher fish catches. Other related innovations that these communities adopted through collective action included: an alternative process for jute retting to reduce the problems of poor surface water quality in floodplains; using water from 'drainage congestion' for increasing fish and crop production; integrated pest management; rehabilitation of locally rare/extinct fish species; and excavation of canals for restoring water for fish and crops.

Coupled with this was the earlier research, also under Natural Resources Systems Programme (NRSP), on IFM institutions that had identified lessons on the factors that are important to local institutional performance in floodplain management, along with process documentation approaches (Lewins 2004; Sultana and Thompson 2008). These informed the design of IFM and the approach of working with a range of types of CBOs. In addition, we recognized that the starting point of the CBOs involved in the project varied considerably; in particular, some communities and associated CBOs had not gone through any well-structured local participatory planning process to consider the needs and opportunities of their areas, and did not fairly represent the range of local stakeholder views, particularly the poor. Through the RIU support the opportunity was taken to support 25 communities and their CBOs through Participatory Action Plan Development, an approach developed in Bangladesh through previous RNRRS action research (Sultana and Thompson 2004) that has been shown to be effective in bringing together a range of local stakeholders using a single floodplain to share perspectives and identify common needs and interventions for which there is a consensus (Sultana and Abeyasekera 2008). Hence this was a potential component of adaptive learning within a particular location or CBO command area.

An important adjustment took place in 2008 when an expansion of the network was possible through a linked project under the Research Into Use (RIU) programme. An additional 100 CBOs were integrated into the process from January 2009 onwards, resulting in more CBOs participating in the original three regions and the addition of CBOs in the north-east region (the totals are shown in Table 5.1). Already through sharing of experience and issues between CBOs with different backgrounds CBOs had started to consider the overall productivity of their floodplain resource systems. RIU support enabled additional support including training to be provided to CBOs that based on adaptive learning wanted to take up natural resource management options consistent with the concept of IFM.

Further evolution of the adaptive learning network

By late 2008 it was a demand of the CBOs trying the same types of initiatives that they should meet and discuss impacts and why and how their initiative worked or did not work. The research team also recognized that the larger regional workshops gave limited scope for CBOs to assess the performance and detailed lessons from the various initiatives, which had become quite diverse. Consequently the team proposed and introduced the additional step of smaller reflective learning meetings among CBOs taking up common initiatives within a region, as shown in Figure 5.1b. These have been held once a year but have depended on the interest of the relevant partner from the research team – so they have not taken place in all regions covered by the CBO network. One issue that emerged was that returns from the same crop innovation could differ greatly between farmers (mainly from differences in the weather), leading to those who had poor returns wanting the project to arrange compensation.

Culturally in Bangladesh for the Muslim majority rural women are very rarely involved in economic activities outside the homestead – few cultivate land and very few women catch fish. However, there are some CBOs where women play an active role, particularly in floodplains where women use natural resources. Women from the floodplain beels participated in the workshops along with men, and contributed to the lessons and the proposed adaptive learning activities. To address this gap among some of the CBOs, women members were oriented and trained to be resource persons in IFM, and they then passed on what they had learned to about 15 other women in their community. This process takes time to achieve changes in practices, but women are particularly interested in innovative non-rice crops, such as garlic, that they can cultivate in small plots close to their home, and already about a quarter of these women have taken up such crops. Moreover, in all three regional committees of CBOs one woman representative was selected from the CBO leaders to become member of the Federation of CBOs.

Within the adaptive learning approach the CBOs were sharing lessons and experiences, and could take up a more integrated approach to floodplain management. But the CBOs and research team saw another potential, which was a particular concern of the CBOs – that they could have strength in numbers around shared concerns and issues, particularly those identified at the outset of the process (earlier bullet points). The network of CBOs could form a lobby for continued access to resources (addressing attempts by local elites to capture waterbodies), and to provide a forum where CBOs could interact with government agencies. This could strengthen accessing services, notably from the Department of Fisheries, Local Government Engineering Department, and Department of Agricultural Extension, but could also create a voice in policy debate representing thousands of poor floodplain resource users through the member CBOs.

In the first year of the project the CBOs agreed to establish a more formal network based on the three regional committees (north-centre, south-west and north-west). In January 2008 all 13 members of each regional networking committee (39 CBO leaders in total) met and agreed to register a federation of CBOs and formed an ad hoc central committee to convene this and work on registering the organization. This central committee finalized a constitution, and applied and succeeded in obtaining registration as a legal entity under the Society Registration Act, 1860 on 29 April 2008. This federation is officially named the Society for Water Resources Management (SWRM). In late 2008 its membership expanded to just over 250 CBOs through support of the RIU project Integrated Floodplain Management. A key development during 2009 was the inclusion of CBO representatives from SWRM in a combined steering committee for the two projects (IDRC and RIU supported). This has provided a forum where CBO representatives can interact with higher level government officials from relevant agencies to raise their concerns and create pressure to resolve their problems, and thereby has added an element of learning between the CBOs and government agencies.

Furthermore, in March 2010 with support from the RIU project the first convention and annual general meeting of the SWRM was held where leaders of all of the 250 participating CBOs attended. It was opened by the chair of the Parliamentary Standing Committee on Land, which raised the profile of the federation and gave the CBOs access to policy levels on issues over the leasing of and access rights to waterbodies. The activities of the federation were reported back to members, elections for the leadership of all committees were held, and the north-east region was incorporated as an expansion of the network. While the individual CBOs are already independent and self-sustaining, at this stage organizing a convention (especially for such a dispersed network of CBOs) required assistance from the research team. Although the CBOs are optimistic about the future of SWRM, there are many challenges for this to become a self-sustaining federation. The CBO leaders are all volunteers and several of the more capable and respected leaders who coordinate the federation in its committees are increasingly busy with other demands on their time, but are reluctant to step down as they have status among the CBOs. The modest subscriptions that CBOs can afford to pay for SWRM membership cannot cover the costs of interactions and events needed for adaptive learning, and the member CBOs lack the capacity to write proposals for potential funding agencies. In addition while the CBO leaders have shown they are capable of advising one another and addressing common problems at local level, they still depend on having advice from the research-facilitation team, especially to organize and coordinate events such as workshops. To move forward from the present arrangement and capacity might for example need support for SWRM to employ an organizer-facilitator-fundraiser.

Outcome and impacts

Governance and knowledge

The adaptive learning process has helped CBOs gain confidence to contact local officials and extension workers to access services. Also those officials have found that the CBOs are a more efficient means for extending good practices than simply targeting individual farmers. CBOs have also advised one another, and the federation has taken initiatives to overcome internal conflicts and external threats faced by member CBOs. The CBO association has also raised concerns with senior government officials over policy changes, continued security of access to waterbodies, and pollution.

Assessments of the performance of 153 participating CBOs that have been active since 2007 (averaging about 200 members each, and in total representing just over 1,000 villages inhabited by over 300,000 people and managing just under 50,000 ha of floodplains) shows that CBO performance and governance have improved, with significant enhancement of natural resource management and empowerment of the poor.

Table 5.3 shows significant gains in a scoring system for seven thematic indicators and Figure 5.2 shows the numbers of CBOs improving their performance. These changes can be attributed to lessons shared between CBOs during adaptive learning. Natural resource related rule breaking and conflict was already relatively low and has fallen during the adaptive learning process. There are also cases where CBOs have worked together to overcome either internal conflicts within a member CBO or external threats.

Women have also now on average a greater role in decision making in the CBOs, and the adaptive learning process has been the reason – male CBO leaders heard from women involved in integrated approaches to

Table 5.3 Changes in mean performance of CBOs by seven themes (% of maximum possible score)

Theme	*Closed beel (26 CBOs)*			*Open beel (86 CBOs)*			*River (36 CBOs)*		
	2007	*2009*		*2007*	*2009*		*2007*	*2009*	
Resource management	64.6	75.3	*	64.7	73.1	*	64.1	75.7	*
Pro-poor	74.1	82.7	*	63.2	69.9	*	69.4	80.7	*
Women's role	5.0	12.7		37.0	50.8	*	20.0	28.1	*
Organization	72.0	80.8	*	72.3	73.3		71.4	72.4	
Governance	64.4	69.9		70.0	73.3		73.8	72.2	
Financial	50.5	52.5		57.9	61.1		47.6	56.2	*
Linkages	49.0	40.8		52.5	46.4		56.3	53.1	
Overall	54.3	59.2	*	59.6	64.0	*	57.5	62.6	*

Note: * significant difference $p<0.05$ in paired t-test.

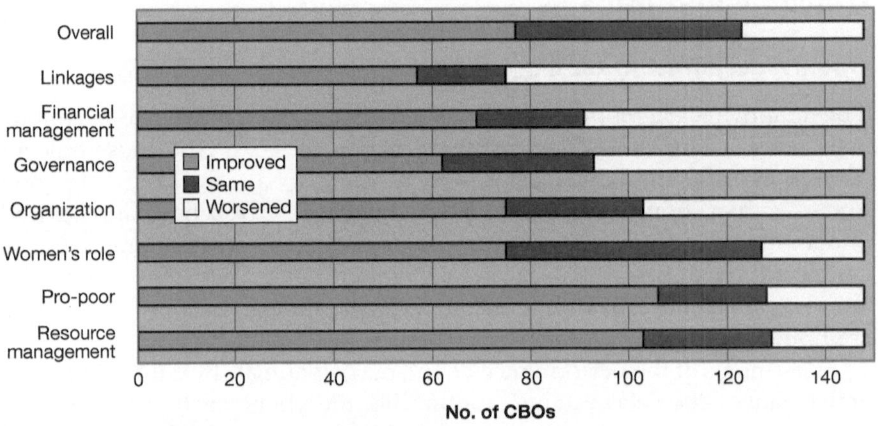

Figure 5.2 Changes in CBO indicators 2007–2009, Bangladesh.

floodplain management (as documented in Sultana and Thompson 2008) and then acted to consult with women in their own areas and involve some in taking up activities. However, there are few women members in some types of CBO – particularly those in closed beels managed by male fishers.

More transparent governance practices have been reinforced and spread among CBOs, such as revising management plans annually, holding AGMs, electing office bearers, internal audits, and savings and revolving funds (Figure 5.3).

Some changes are dependent on the context and purpose of the CBO. For example, CBO leaders identified advisory committees of locally respected non-members as helpful in overcoming local disputes. More than 20 per cent of open beel and river CBOs added advisors, but very few closed beels did because they would then face problems of requests and influence over their sharing of income, whereas a majority of river CBOs (also largely fisher members) had advisory committees as they face problems of outsiders making brushpiles, and do not control significant financial assets.

Without peer pressure through the adaptive learning network these CBOs would at best have continued as before. Even so, some CBOs have not strengthened their performance; this depends on the interest of their leaders, or membership, and in some cases their remit limited their scope to adopt good practices. It would appear that the closed beel and river CBOs benefited more in resource management and being pro-poor (in part by reducing conflicts), while floodplain CBOs gained more in governance practices.

Table 5.4 Returns from low water demand crops taken up by CBOs compared with irrigated rice in 2008–2009 dry season

Crop	*Water demand (mm)*[1]	*No. of farms*	*Net return as % of HYV rice*	*Benefit–cost ratio*	*Net return (US$/100 mm water/ha)*
Grasspea	90–100	10	168	4.6	485
Garlic	150	45	323	3.4	625
Mustard	160	38	122	2.9	221
Wheat	200	5	58	2.2	85
Maize	240	29	62	2.7	74
Sunflower	300	29	249	3.7	240
Potato	350	13	558	2.4	462
HYV rice	835–1000	9	100	2.1	8

Notes:

1 Biswas and Mandal (1993) quoted in Shankar et al. (2004).

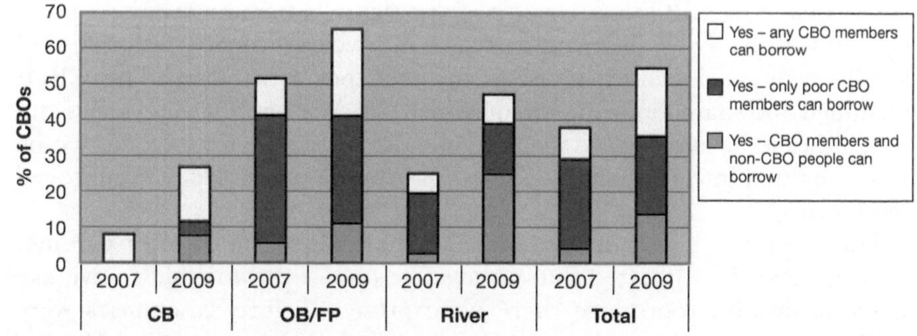

Figure 5.3 CBOs operating revolving funds, Bangladesh.

Sustainability and productivity

Up to early 2010 over 300 farmers from 72 CBOs tried different low water demand alternative crops (garlic, potato, maize, sunflower, wheat and mustard) which were not previously grown in their areas. On average all of these crops proved profitable according to the farmers. Uptake within each CBO area is slow but farmers who tried the crops have expanded their areas. After successful CBO demonstrations of maize and sunflower in the south-west region, during December 2009–January 2010, 40 CBOs adopted sunflower and made linkages to sell their harvest. Most of these crops are short duration ones and need very little investment, which is attractive to the poor sharecroppers, marginal farmers and part-time fishers. Although the areas cultivated are modest, dry season crop diversification has been shown to reduce irrigation water demand, and maintain more surface water. Table 5.4 shows the reduction in irrigation water use from these crops compared with rice and the favourable returns for participating farmers.

Through the CBOs farmers also tested a new rice variety (Berke-3004) that yields a good quality rice grain and had been developed by earlier participatory research projects with farmers. Over 300 farmers linked with 68 CBOs tested the new variety. The CBO network enabled farmers spread throughout much of the country to test a variety in the field and to compare the results. Interestingly the farmers in the south-west generally reported good results, and were saving seeds to grow it again, whereas in the north-west where the variety was originally developed it did not yield well in drought conditions.

Research evidence and CBO learning up to mid-2010 found that fishery conservation (sanctuaries, closed seasons, etc.) has restored fish species diversity and catches, even in closed waterbodies (110 CBOs established new (51) and/or improved fish sanctuaries (59), Figure 5.4). Sanctuaries are constructed by adapting traditional fish aggregating devices made from brushpiles (branches of trees) that were previously used to catch fish. This provides a shelter from which adult fish can repopulate the floodplain system during the monsoon. Fish sanctuaries in part of a waterbody (the average for participating CBOs is 6 per cent of dry season water area) were reported by CBOs to be associated with increases in species diversity, restoration of locally rare species, and increases in catch per unit effort. The CBOs identified and shared information on which trees are best for sanctuaries and which should be avoided. As many households catch fish for food or income this translated into increased incomes for fishing households and improved food security.

For example, CBOs in closed beels that added sanctuaries through adaptive learning reported that on average only seven to eight native fish species that can reproduce there were present before sanctuaries were established, but with sanctuaries they found 14 such species, and likewise fish catches of these species rose substantially because fish after over-wintering

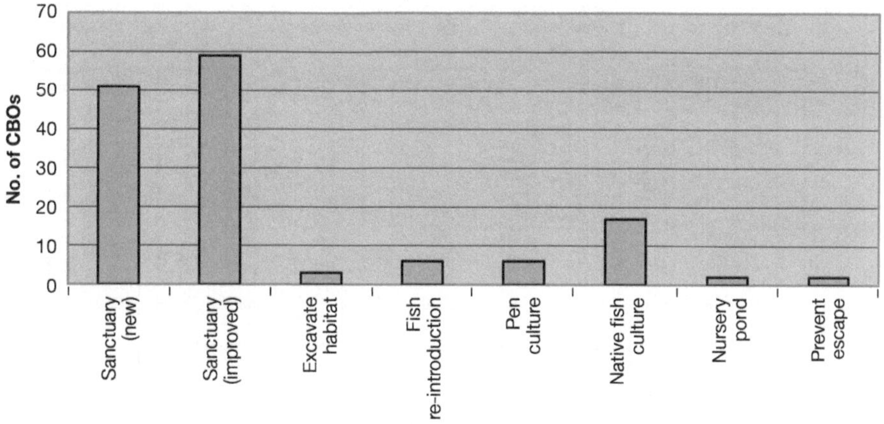

Figure 5.4 Number of CBOs taking fishery management actions up to 2010, Bangladesh.

in the sanctuaries spawned in the pre-monsoon and populated the rest of the baors. Thus the addition of small sanctuaries in these closed waterbodies brought an additional benefit at minimal cost to the CBOs.

Equitable outcomes

Surveys of households in a sample of the participating CBOs indicate that incomes from all sources and for all household types increased more than inflation between 2007 and 2009. There has also been a more general improvement in food security for all categories of household (Figure 5.5). This directly reflects the improvements in fishery management based on conservation achieved by many of the CBOs.

This finding is also supported by a case study which indicated that while changes in access rules have enabled professional fishers to earn more than under pre-CBO conditions, the CBOs have ensured that all poor people can fish for food. CBO level data also confirms that CBOs have enhanced access of the poor to natural resources (Figure 5.6). Considering the negative trend elsewhere in capture fisheries productivity, this indicates that the poor have benefited from CBO activities in managing common pool resources.

Discussion – lessons and issues

On pages 145–150 we reviewed the factors associated with the change towards adaptive co-management of floodplains (water and fisheries) in Bangladesh. This has led to the establishment of an increasing number of CBOs, each with rights and responsibilities recognized by government, and a link with a sponsoring government agency.

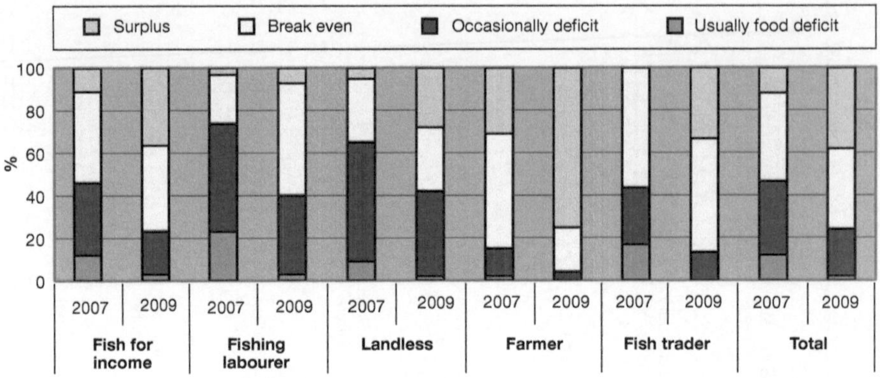

Figure 5.5 Changes in food security of surveyed households.

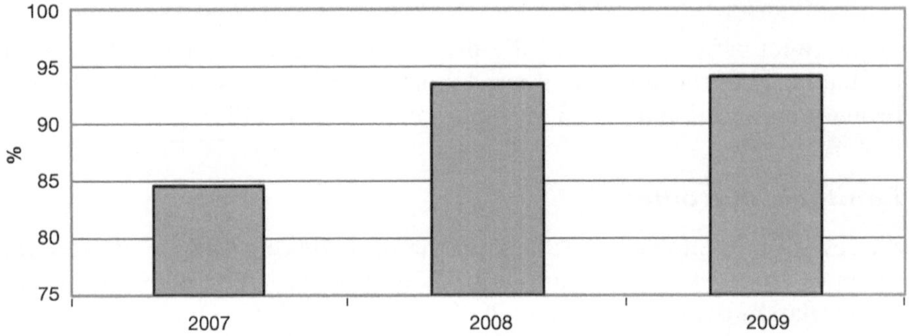

Figure 5.6 Percentage of sites where access of poor to natural resources improved under CBO management.

The primary precondition for an adaptive learning network has been the existence of CBOs holding rights over and actively managing floodplain natural resources. This was also necessary if communities were to take a more integrated or holistic view of the productivity of floodplain natural resources. These CBOs have continued after their founding projects ended, and their enthusiasm to work together and take an adaptive approach has been the main driving factor. This has in part been because knowledge has been enhanced and there are economic benefits from the combination of adaptive learning and IFM components. For example, a farmer in the south-west said: 'I have had plots of land lying fallow for years and sometimes I tried to grow something there depending on rain but most of the time it was unsuccessful. Now that I learned how to maximize cultivation with very little water, I will continue to try different crops. Other farmers are also trying to adopt the practices I am doing.' Overall when the CBO federation leaders reviewed their experience they concluded that there are material benefits for the community members of CBOs from IFM activities and networking. New ideas/innovations are being shared and added all the time, and in reflective learning sessions the CBOs can see which are the best/most effective strategies and innovations. They also said that previously the CBOs individually worked in a limited sphere but each CBO is now expanding its influence and recognition within its wider community. The CBO leaders also value face-to-face meetings and the knowledge that they are not alone in facing similar problems, and aspire to collectively strengthen their access to public resources.

The project team was already familiar with many of the CBOs, having been involved in the projects that formed those CBOs, which facilitated the early stages of the process. The project team had also been involved in research, development and action research and already had an orientation and experience in participatory action research. Hence the team had some understanding of their role in supporting participants in planning and design of actions and associated assessment of impacts through a series of iterations, rather than a more top-down or rigid approach. The project team comprises NGOs and research bodies, and a diversity of partners with different experiences relevant to their responsibilities in the project, for example BELA with legal expertise, Banchte Shekha with experience working with women, CNRS with experience in the north-eastern *haors*, and the Flood Hazard Research Centre (FHRC) with research expertise and experience with a wide range of CBOs. While this offered a range of skills and experience, having a partner responsible for each region meant that not all of the CBOs were supported in adaptive learning by a partner with experience of working in natural resource management and there were variations in understanding of the adaptive learning approach and of IFM. More capacity building among the team might not have resolved this as the ultimate constraint has been the level of motivation between partners.

Attempts to ensure that indicators are set with the CBO network and are generated from participatory monitoring have been limited by the interest of CBOs in trying innovations and the time and skill of CBOs and their members to record experiences. Reflective learning which uses a mix of qualitative and semi-quantitative findings has been found to be more effective for CBO adoption and learning, but does not generate traditional experimental research findings. In addition, as our initiative has limited links with traditional research sectors (government research institutes and universities), so far there is no sign of those bodies participating in adaptive collaborative management among a network of CBOs. In this regard the experience differs from and is more bottom-up than many cases of adaptive management in developed countries where there is an emphasis on research institutes designing experiments and piloting with management bodies, with or without substantive participation at the grass-roots level.

While there have been obvious advantages and benefits from this approach, CBOs are widely dispersed so it is hard for them to make the network function and take up issues at higher levels on their own. The SWRM central committee has said that considering overall the experience gathered from the federation and its meetings and workshops this is so substantial that it exceeds what they can do or gain alone. Nevertheless considering the limited financial resources of the CBOs, the transaction costs and financial costs of the network also limit the number of larger workshops and activities. Yet the significant amounts of time given to this activity by a considerable number of CBO leaders – not only joining meetings but also writing articles to disseminate their lessons, and taking initiatives to advise and assist other CBOs – indicate that it has a value.

Participation in workshops has been necessarily limited to one representative from each CBO, and the same applies to exchange visits; there is a risk that a CBO's chosen representative is less interested than other executive committee members in the potential to learn and improve floodplain management. At each stage the team asked the CBO leaders to feed back their experience to other members of their CBO, and for any proposed action of the CBO originating from the adaptive learning process to be endorsed by that CBO's executive committee. But further ways need to be found of ensuring that more members and leaders from each CBO are involved.

Close coordination needs frequent interactions for which face-to-face meetings, workshops and visits are more effective. When CBOs are scattered this requires more funds than the CBOs can contribute. We consider that the improvements in natural resource productivity, CBO governance, and participation of the poor that are linked with the adaptive learning network justify modest public investments in subsidizing continued adaptive learning. Moreover to function effectively the federation should have an annual convention. CBOs within a region can more easily share lessons and cooperate for conflict resolution. Although they are scattered, CBOs do spend time and resources communicating with each other and with regional committees when facing problems. The rapid expansion and falling costs of mobile phone ownership in rural Bangladesh has facilitated networking and communications.

CBOs have a certain idea of their objectives from their parent projects. Taking up new ideas takes time to understand and trust other CBOs' lessons and to then change management plans. The annual cycle of seasons means that several years are needed for testing of new actions in IFM to show results and be shared between CBOs, and lessons adapted more widely.

Some women from the floodplain *beels* participated in the workshops along with men, and contributed to the lessons and the proposed adaptive learning activities. The aim should be to ensure that priorities of women as well as men are addressed in trials. However, tradition limits the involvement of women in some floodplain natural resources, and also means that there are few women office bearers in CBOs. While we have attempted to address this, for example by developing the capacity of women active in some CBOs to orient women (and men) in other CBOs, substantive change will take time.

The project partners being outside of government agencies enabled a flexible approach to be adopted based on empowering local poor floodplain users – the floodplain communities themselves decide on interventions to take up within a range of options and issues covered by the system view of natural resources. As relevant government agencies have not been directly involved in the adaptive learning network, this has not been incorporated in their normal programmes. However, this may not be undesirable, as government officials tend to dominate over local people, links to service providers and policy tiers are vital, but for the CBOs to forge such links from a position of greater strength and confidence requires that they first invest in links among themselves such as this adaptive learning network. Consequently there is a gap between the aspirations of the CBO network to raise their problems and views on government policies and practices, and their ability to access policy makers and organize suitable events. The project bridged this gap to some extent by developing a steering committee where CBO regional committee members sit with government officials and the project team tried to play the role of intermediaries in a move towards collaboration between CBOs and government. With hindsight some of the potential of the network to advocate change at higher levels and to follow up on planned actions and indicators could have been strengthened and given greater emphasis.

This approach can also strengthen links with government and access to knowledge. Networks of CBOs are a cost effective way for government agencies to link with communities with mutual benefits. Agencies can deliver services and messages to many people through CBOs, and should cover the costs of communication and networking among CBOs. Networks of CBOs can also help communities to access science and new technologies (such as alternate jute retting methods to avoid water pollution and to restore sufficient oxygen level in water to support fish). However, the story is different when CBOs advocate for maintaining their rights, for example access to waterbodies. One of the gaps in achievement so far has been the lack of recognition of CBO network representatives in existing forums between government and civil society, for example the relevant District Jalmohal Committees (fisheries committees) that allocate waterbody access.

The network established through this project has been formalized by the CBOs, but its short term sustainability will depend on continued access to outside resources and facilitation. For example, member CBOs need to have access to a pool fund for small one-time grants or subsidies to initiate innovations. In the long term there are threats to the continued access of CBOs to floodplain waterbodies, and networking could enable CBOs to collectively play a greater role in influencing policy and practice of government in favour of their sustained access. Unfortunately the framework for engagement in debate over waterbody access is set by administrators that deal with individual cases and waterbodies, and the land administration has avoided interacting with the network or federation. At this stage where alternative forums have to be found, and considering the scattered nature of the CBOs, this still requires external facilitation and resources to help build the capacity of the CBO federation and its members.

Considering the above discussion, we identify several challenges that need to be addressed while promoting an adaptive learning network in community based natural resource management:

- If application by one community of a lesson generated by another community does not work, even if this is due to a natural calamity, then there is a loss of faith and the community is less likely again to take up lessons.
- Incentives or subsidies such as small grants were found to be necessary for communities to test innovations identified through an adaptive learning process.
- If one or more members of a community do not physically see an innovation then that community is unwilling to adopt lessons and innovations generated elsewhere, hence exchange visits are an important component of adaptive learning.
- CBOs differ greatly in their interest in generating data on impacts through participatory monitoring for comparisons and lesson learning, so compiling comprehensive or consistent data sets is difficult.
- A wide geographic scattering of CBOs results in relatively high costs for horizontal linkages and for coordinating efforts to influence policy among a set of CBOs. CBOs are only willing to make modest cash contributions to this process, although there are considerable benefits to members.
- Institutionalization of learning between CBOs has to some extent taken place spontaneously through competition between CBOs and mobile phone communication between CBOs.
- Incorporating adaptive learning within government agencies is a challenge. NGOs and projects are more willing to adopt this approach, but government is unwilling to change its interactions with other stakeholders.

Conclusions

We have found that an adaptive learning network among relatively diverse CBOs that share some common aims and concerns is possible, effective and worthwhile. In this case it has contributed to the adoption of an integrated systems-view of floodplain resource management that brings increases in overall productivity that benefit the poor and better off. It has also through peer pressure, lesson sharing and regular interaction among CBOs resulted in improvements in their organizational functioning and governance. The CBO network has seen some strengthening of links with government service providers for individual CBOs, but is still challenged when trying to raise its views and issues with higher policy levels of government. In the short term the adaptive learning process has enhanced the sustainability of the individual CBOs and the productivity of their resources, but developing this approach takes time and resources. The sustainability of an adaptive learning network and federation of CBOs is dependent on some level of facilitation and financial support for the social services it provides.

Overall adaptive learning among networks of CBOs has been demonstrated to bring benefits from adaptation in integrated floodplain management to the communities served by CBOs. These benefits from collaboration are greater than those achieved by CBOs managing resources in isolation. This approach has potential to be adapted to other types of CBO, and to other countries where common pool resources such as wetlands and forests are managed through collective action particularly when it is formalized through CBOs.

Acknowledgements

This chapter is an output of research funded by the Canadian International Development Research Centre (IDRC) and the Research Into Use programme, managed by NR International, a project funded by the UK Department for International Development (DFID) for the benefit of developing countries. The views expressed are not necessarily those of IDRC, DFID or NR International. We thank Hemant Ojha for valuable comments and suggestions for improvement of the first draft. We thank the leaders and members of the more than 250 participating community based organizations for their interest and enthusiastic cooperation in adaptive learning. We also thank for their contributions to the actions and research our partner team members from Bangladesh Environmental Lawyers Association, Banchte Shekha and Flood Hazard Research Centre in both projects, and Center for Natural Resources Studies and MRAG in the DFID supported project.

References

Adger, W.N., Brown, K. and Tompkins, E.L. (2005) The political economy of cross-scale networks in resource co-management. *Ecology and Society* 10(2): 9. [online] URL: http://www.ecologyandsociety.org/vol10/iss2/art9/.

Agrawal, A. (2001) Common property institutions and sustainable governance of resources. *World Development* 29(10): 1649–1672.

Ahmed, M., Capistrano, A.D. and Hossain, M. (1997) Co-management of Bangladesh fisheries. *Fisheries Management and Ecology* 4(3): 233–248.

Ali, M.Y. (1997) *Fish, Water and People*. University Press Ltd., Dhaka, Bangladesh.

Argyris, C. (1996) *Organizational Learning II: Theory, Method, and Practice*. Addison-Wesley Longman, Reading, MA.

Armitage, D., Marschke, M. and Doubleday, N. (2008) Adaptive co-management and the paradox of learning. *Global Environmental Change* 18(1): 86–98.

Armitage, D.R, Plummer, R., Berkes, F., Arthur, R.I., Charles, A.T., Davidson-Hunt, I.J., Diduck, A.P., Doubleday, N.C., Johnson, D.S., Marschke, M., McConney, P., Pinkerton, E.W. and Wollenberg, E.K. (2009) Adaptive co-management for social–ecological complexity. *Frontiers in Ecology and Environment* 7(2): 95–102.

Arthur, R.I. and Garaway, C.J. (2005) Learning in action: a case from small waterbody fisheries in Lao PDR. pp. 191–198. *In* Gonsalves, J., Becker, T., Braun, A., Campilan, D., de Chavex, H., Fajber, E., Kapiriri, M., Rivaca-Caminade, J. and Vernooy, R. (eds.) *Participatory Research and Development for Sustainable Agriculture and Natural Resource Management: A Sourcebook.* Vol. 3 Doing participatory research and development. International Potato Center-Users' Perspectives With Agricultural Research and Development, Laguna Philippines and International Development Research Centre, Ottawa, Canada.

Beck, T. and Fajber, E. (2006) Exclusive, moi? Natural resource management, poverty, inequality and gender in Asia. pp. 297–320. *In* Tyler, S. (ed.) *Communities, Livelihoods and Natural Resources: Action Research and Policy Change in Asia.* Intermediate Technology Development Group and International Development Research Centre, Reading, UK and Ottawa, Canada.

Berkes, F. (1994) Property rights and coastal fisheries. pp 51–62 *In* Pomeroy, R.S. (ed.) *Community Management and Common Property of Coastal Fisheries in Asia and the Pacific: Concepts, Methods and Experiences.* International Center for Living Aquatic Resources Management, Manila.

Berkes, F., Colding, J. and Folke, C. (eds.) (2003) *Navigating Social-Ecological Systems.* Cambridge University Press, Cambridge, UK.

Biswas, M.R. and Mandal, M.A.S. (eds.) (1993) *Irrigation Management for Crop Diversification in Bangladesh.* University Press Ltd., Dhaka.

Borrini-Feyerabend, G., Taghi Farvar, M., Nguinguiri, J.C. and Ndangang, V. (2000) *Co-management of Natural Resources: Organizing, Negotiating and Learning-by-doing.* GTZ and IUCN, Kasparek Verlag, Heidelberg, Germany.

Brammer, H. (2000) Controversies surrounding the Bangladesh Flood Action Plan. pp. 302–315. *In* Parker, D.J. (ed.) *Floods* (vol. 1). Routledge, London.

CNRS, WorldFish, ITAD and MRAG. (2005) Better Options for IFM: Uptake Promotion. NRSP Project R8306 Final Report. Center for Natural Resource Studies, Dhaka.

Davidson-Hunt, I.J. (2006) Adaptive learning networks: developing resource management knowledge through social learning forums. *Human Ecology* 34(4): 593–614.

DOF. (2006) *National Fisheries Strategy.* Department of Fisheries, Dhaka.

Fabricius, C., Folke, C., Cundill, G. and Schultz, L. (2007) Powerless spectators, coping actors, and adaptive co-managers: a synthesis of the role of communities in ecosystem management. *Ecology and Society* 12(1): 29. [online] URL: http://www.ecologyandsociety.org/vol12/iss1/art29/.

FAO. (1988) *Land Resources Appraisal of Bangladesh for Agricultural Development.* 7 vols. Food and Agriculture Organization, Rome.

Freeman, M.M.R. (1989) Graphs and gaffs: a cautionary tale in the common-property resources debate. pp 92–109. *In* Berkes, F. (ed.) *Common Property Resources.* Belhaven Press, London.

Garaway, C.J. and Arthur, R. (2004) *Adaptive Learning: A Practical Framework for the Implementation of Adaptive Co-Management – Lessons From Selected Experiences in South and Southeast Asia.* MRAG Ltd., London.

Halls, A.S. (1998) Impact of flood control schemes on river fish migrations and assemblages in Bangladesh. *Journal of Fish Biology* 53: 358–580.

Hanchett, S. (1997) Participation and policy development: the case of the Bangladesh Flood Action Plan. *Development Policy Review* 15(3): 277–295.

Holling, C.S., Berkes, F. and Folke, C. (1998) Science, sustainability and resource management. pp 342–362. *In* Berkes, F. and Folke, C. (eds.) *Linking Social and Ecological Systems*. Cambridge University Press, Cambridge, UK.

Huda, A.T.M.S. (2003) Fishing in muddy waters: policy process for inland fisheries in Bangladesh. Community Based Fisheries Management Project (CBFM-2) Working Paper 3. WorldFish Center, Dhaka.

Jentoft, S. (2007) Limits of governability: institutional implications for fisheries and coastal governance. *Marine Policy* 31(4): 360–370.

Lewins, R. (2004) *Integrated Floodplain Management: Institutional Environments and Participatory Methods, Final Technical Report*. ITAD Ltd., Hove, UK.

Margoluis, R. and Salafsky, N. (1998) *Measures of Success: Designing, Managing and Monitoring Conservation and Development Projects*. Island Press, Washington, DC.

Minkin, S.F., Rahman, M.M. and Halder, S. (1997) Fish biodiversity, human nutrition and environmental restoration in Bangladesh. pp 183–198. *In* Tsai, C. and Ali, M.Y. (eds.) *Openwater Fisheries of Bangladesh*. The University Press Limited, Dhaka.

Muir, J. (ed.) (2003) *Fisheries Sector Review and Future Development: Theme Study: Economic Performance*. World Bank, Danida, USAID, FAO and DFID, Dhaka.

MWR. (1999) *National Water Policy*. Government of People's Republic of Bangladesh, Ministry of Water Resources, Dhaka.

MWR. (2001) *Guidelines for Participatory Water Management*. Ministry of Water Resources, Government of the People's Republic of Bangladesh, Dhaka.

Oakerson, R.J. (1992) Analysing the commons: A framework. *In* Bromley, D.W. (ed.) *Making the Commons Work: Theory, Practice and Policy*. ICS Press, San Francisco, CA.

Olsson, P., Gunderson, L., Carpenter, S., Ryan, P., Lebel, L., Folke, C. and Holling, C.S. (2006) Shooting the rapids: navigating transitions to adaptive governance of social–ecological systems. *Ecology and Society* 11(1): 18. [online] URL: http://www.ecologyandsociety.org/vol11/iss1/art18/.

Ostrom, E. (2005) *Understanding Institutional Diversity*. Princeton University Press, Princeton, NJ.

Senge, P. (1990) *The Fifth Discipline: The Art and Practice of the Learning Organization*. Doubleday, New York.

Shankar, B. (2002) *Final Report: Maximisation of Joint Benefits From Multiple Resource Use in Bangladesh Floodplains*. Project R 7868, Natural Resources System Programme, Department for International Development, UK.

Shankar, B., Halls, A. and Barr, J. (2004) Rice versus fish revisited: on the integrated management of floodplain resources in Bangladesh. *Natural Resources Forum* 28 (2004): 91–101.

Stern, P.C., Dietz, T., Dolsak, N., Ostrom, E. and Stonich, S. (2002) Knowledge and questions after 15 years of research. pp 445–486. *In* Ostrom, E., Dietz, T., Dolsak, N., Stern, S. and Weber, E.U. (eds.) *The Drama of the Commons*. National Academy Press, Washington DC.

Sultana, P. (2010) *Improving Floodplain Management: Implications of Floodplain Aquaculture Enclosure*. Policy Brief 4. Flood Hazard Research Centre, Bangladesh, Dhaka.

Sultana, P. and S. Abeyasekera. (2008) Effectiveness of participatory planning for community management of fisheries in Bangladesh. *Journal of Environmental Management* 86(1): 201–213.

Sultana, P., Johnson, C. and Thompson, P. (2008) The impact of major floods on flood policy evolution: insights from Bangladesh. *International Journal of River Basin Management* 6(special issue): 1–10.

Sultana, P. and Thompson, P. (2004) Methods of consensus building for community based fisheries management in Bangladesh and the Mekong Delta. *Agricultural Systems* 82(3): 327–353.

Sultana, P., and Thompson, P.M. (2008) Gender and local floodplain management institutions: a case study from Bangladesh. *Journal of International Development* 20: 53–68.

Sultana, P. and Thompson, P.M. (2010) Local institutions for floodplain management in Bangladesh and the influence of the Flood Action Plan. *Environmental Hazards* 9(1): 26–42.

Sultana, P., Thompson, P., Ahmed, H. and Hossain, A. (2005) *Better Options for Integrated Floodplain Management in Bangladesh: Uptake Promotion Piloting of IFM Options*: Narail site. Centre for Natural Resource Studies and WorldFish Center, Dhaka.

Thompson, P.M., Sultana, P., Islam, M.N., Kabir, M.M., Hossain M.M. and Kabir, M.S. (1999) An assessment of co-management arrangements developed by the Community Based Fisheries Management Project in Bangladesh. Paper presented at the International Workshop on Fisheries Co-management, 23–28 August 1999, Penang, Malaysia.

Thompson, P.M., Sultana, P. and Islam, N. (2003) Lessons from community based management of floodplain fisheries in Bangladesh. *Journal of Environmental Management* 69(3): 307–321.

Toufique, K.A. and Gregory, R. (2008) Common waters and private lands: distributional impacts of floodplain aquaculture in Bangladesh. *Food Policy* 33(6): 587–594.

Tyler, S.R. (2008) Adaptive learning in natural resource management: three approaches to research. Working Paper 22. Rural Poverty and Environment Working Paper Series. International Development Research Centre, Ottawa, Canada.

World Bank. (1990) *Flood Control in Bangladesh: A Plan for Action*. Washington: Asia Region Technical Department, World Bank.

World Bank. (1996) *The World Bank Participation Sourcebook*. World Bank, Washington.

WorldFish Center. (2003) *Community Based Fisheries Management Phase 2 (CBFM-2) Annual Report September 2001–December 2002*. WorldFish Center, Department of Fisheries, Banchte Shekha, BELA, BRAC, Caritas, CNRS, CRED, FemCom, Proshika, Dhaka.

6 Learning in contested landscapes

Applying adaptive collaborative management in forested landscapes of Zimbabwe

Tendayi Mutimukuru-Maravanyika and Frank Matose

Introduction

This chapter reviews the experience of the adaptive collaborative management (ACM) approach and strategies as they were applied in practice in Mafungautsi State Forest in Zimbabwe. In doing this, we draw key lessons on what worked or did not work, why, and how ACM could be applied in a contested resource governance situation. The ACM approach was initiated by the Center for International Forestry Research (CIFOR) in 1999 with the aim of exploring whether it would simultaneously result in positive changes in people's well being and enhance the sustainable management of resources. The new approach was implemented in several countries, including in Mafungautsi State Forest in Zimbabwe, the focus of this chapter. The approach has its roots in ideas from several disciplines relating to complex system behaviour, adaptive management, social learning theory, and ideas by sociologists (and others) about the roots of human cooperation and competition. Because of the complex nature of natural resource management, the ACM approach aims to facilitate collaboration among stakeholders in consciously learning together about the impacts of their management actions and adapting their practices accordingly.

In this research, we make use of reflexivity as a methodology to help us attain objectivity when looking back at the experiences of the ACM project in Mafungautsi Forest from 1999 up to 2007. There are several definitions of reflexivity in literature (Aamodt, 1991; Guillemin and Gillam, 2004; Mason, 1996; Hertz, 1997; Côté & Levine, 2002; Huberman and Miles, 1994), but they come all down to the same critical reflection of the researcher's own position and views, and how this influences his/her interpretation of the research findings (Burr, 1995). Some definitions go further and relate reflexivity to the influence of the researcher on the researched situation and the participation of others, non-researchers in knowledge generation (Giddens, 1990; Bourdieu, 1992). Reflexivity consists of two processes: self-reflection and identification of the researcher's feelings and pre-conceptions; and the ability to put aside these feelings and pre-conditions (Ahern, 1999). Huberman and Miles (1994) argue that sharing experiences and insights fully makes researchers accountable to readers. Reflexivity therefore helps to make qualitative research findings credible and objective by accounting for researcher's values, beliefs, knowledge and biases (Cutcliffe, 2003; Bourdieu, 1992).

In the research, we play the role of retelling the ACM project story as we were part of the implementing team in Zimbabwe. The first author (Tendayi) was one of the field researchers responsible for implementing the ACM approach in the field. Tendayi joined the CIFOR's ACM project as a participatory action research (PAR) researcher in January 2001 and worked for the project until 2003 when it ended. Her terms of reference as an ACM team member included: facilitating the implementation the ACM approach in Mafungautsi State Forest; contributing to the further development of the ACM concept; conducting field work related to participatory resource management using PAR as the main methodology; writing field reports; analyzing findings from the field work and preparing written outputs (such as scientific papers and manuals) for different audiences; and finally acting as a resource person in training workshops to equip forest extension officers with the ACM approach. Tendayi, as part of her Ph.D. research work, also conducted post-project fieldwork in Mafungautsi – her final field work was done in 2007 when she was no longer working for CIFOR and was a full-time Ph.D. student. The second author (Frank), who was on secondment from the Forestry Commission (FC), CIFOR's key partner, was the team leader responsible for directing the whole research process. He worked for the project from 2000–2003. His responsibilities included taking lead on the conceptual issues for the research as well as giving the research direction and dealing with the policy aspects of it.

Using the reflexivity methodology, we critically review our experiences on the ACM project and its outcomes. We start by making explicit the assumptions we (the authors as well as the other members of the team) had before initiating the ACM research as these influenced our methodologies in implementing the ACM approach in Mafungautsi. We later put these initial assumptions aside by adopting a new analytical framework (consisting of new institutionalism and political ecology) to critically analyze the project and its outcomes.

The chapter begins by presenting the analytical framework, followed by the general context for the study, and then a description of the pre-ACM situation in the study area. The paper later describes how the ACM approach was implemented in Mafungautsi, and several challenges we encountered in the process. In the conclusion section we return to the broader questions we raised on the efficacy of the adaptive and collaborative management approach. Experiences from Mafungautsi provide some concrete insights into the potential and limitations of ACM approach in complex and contested resource management situations. We also provide evidence on the types of outcomes ACM can generate, and analyze factors and conditions needed to sustain them over time. From this analysis, we point to the need for users of the ACM approach not to only focus on enhancing social learning processes, but to also ensure that other crucial elements (like addressing root causes of natural resource management problems and putting in place resource management institutions and means for enforcing them) are also addressed.

Analytical framework

Following the failure of top-down resource management initiatives, there has been a shift in the past two decades and an increase in pressure for developing countries to implement decentralized,[1] community-based natural resource management (CBNRM) approaches (Wilshusen *et al.*, 2002; Borinni-Feyerabend, 1996; Dzingirai and Breen, 2005; Leach, 2002). CBNRM projects have taken various forms, and a diversity of names including joint forest management, co-management, collaborative management, and shared forest management. This shift fits well with calls by proponents of participation to incorporate the views of local communities, whose livelihoods depend on the resources in question. Proponents of participatory approaches then seek to make local self-interest compatible with defined conservation objectives. This shift towards participation by local communities was incorporated into the World Conservation Strategy (WCS)[2] in 1980, and then in various other international conventions including the Convention on Biological Diversity (CBD). In regard to Africa, this shift towards CBNRM aimed to integrate rural development goals with conservation objectives, and to ensure the participation of local people (IUCN, 1980). Advocates for participatory approaches argue that taking local values and indigenous knowledge into account is the only way to achieve sustainable resource management.

Participatory approaches however have also produced disappointing results (Kiss, 1990; Wells *et al.*, 1992; Franks and Worah, 2003; Emerton, 2001; Oates, 1995; Gibson and Marks, 1995; Barrette and Arcese, 1995; Alpert, 1996; Hannah, 1992), resulting in people wondering which way to take in order to ensure sustainable management of resources. An emphasis on participation often misses important aspects of learning and collaboration (Folke *et al.*, 2002; Jiggins and Rolling, 2000). This has sparked new thinking about strengthening participatory and community-based resource management practices, and CIFOR's ACM initiative was one among them. In the context of increasing criticisms to produce positive results on sustainability and poverty reduction from participatory and CBNRM approaches, the Center for International Forestry Research (CIFOR) initiated a multi-country, multi-site adaptive collaborative management (ACM) research project – to try out and further develop the ACM approach. The idea was to take a value-adding approach in terms of enhancing collaboration among stakeholders and learning, rather than looking for a replacement of participatory and CBNRM approaches.

Unlike earlier participatory approaches, the ACM approach recognizes the complexity and unpredictability of natural resource management systems and recommends resource management activities to be experimental and based on learning. As we briefly mentioned in the introductory section, ACM draws on a number of theoretical strands including complex systems theory, adaptive management theory, social learning theory, ideas about equity in development, notions of cooperation and competition, models of human interaction and pragmatic processes of facilitating change (CIFOR, 2000; Diaw *et al.*, 2009). The approach embodies two main orientations: (a) collaborative management,[3] as applied in various fields of natural resource management; and (b) adaptive management,[4] seen as a way to address the complexity and uncertainty inherent in natural resource situations.

Colfer (this volume) outlines how the ACM concept evolved within CIFOR – through the dynamic and often contentious engagement among social and biophysical scientists. Our team in Zimbabwe was also involved in a series of meetings and reflections to discuss and develop the concept and approach of ACM. Our work with CIFOR in Zimbabwe was guided by the understanding that ACM was a value-adding approach whereby individuals or groups who use or manage a forest agree on the problems they face, jointly come up with solutions, implement them, monitor the outcomes and learn from the impact of these actions and use the lessons to improve. Though we were not sure of how the processes would unfold in practice as none of us has been involved in such open learning and collaborative approach before, we were excited to try it out as a team. In our discussions with the local stakeholders, we further agreed that any resource management plans were to be implemented with the recognition that they might not fulfil their intended objectives. What was crucial however was to learn from both failures and successes. It was therefore important for various stakeholders involved in managing the forest to learn collaboratively from the implementation, so as to make joint decisions about the next steps. At the beginning of the process, we were a bit hesitant and unsure and struggled to convince various stakeholders to embark on such a learning journey whose outcomes were uncertain. We, however, gained confidence as time passed on, seeing the excitement that various stakeholders had in the processes that were being facilitated.

General context

Mafungautsi State Forest is 82,000 hectares in size and lies in Gokwe South District, Midlands Province, north-western Zimbabwe (Figure 6.1). The greater part of Gokwe South District lies in the agro-ecological region III,[5] receiving 600–800mm rainfall per annum mainly concentrated in the season from November to March. This is higher than rainfall received by the rest of Gokwe District mainly in agro-ecological regions IV and V. Gokwe South District therefore has a higher potential for agriculture. Although local communities practise mixed farming (Vermeulen, 2000), the most viable agricultural activity for Gokwe District is animal production (Katerere *et al.*, 1993). Local people in Gokwe rely mainly on agriculture to make a living, and they produce both subsistence and cash crops, including maize, cotton, groundnuts, and cattle (Vermeulen *et al.*, 1996). The communities around Mafungautsi Forest are dependent in part on the forest for subsistence and a substantial number of community members benefit economically from the sale of forest products such as honey, thatch grass, broom grass and wild fruits. The forest provides a wide range of goods and services, including: timber, thatch grass (*Hyparrhenia femitina*), broom grass (*Aristida junciformis*), honey, mushrooms, poles, herbs, firewood, wild animals (only a few wild animals are found in the forest, and these include, zebra, kudu, bush pig, warthog, reedbuck, bushbuck, buffalo, hare, hyena and duiker (Maturure *et al.*, 1994)), medicinal plants such as *intolwane* (*Elephantorriza goetzei*), wild vegetables, construction and roofing materials, natural honey collected from trees, bush meat, pasture, mopane worms (a source of protein and considered a delicacy by many Zimbabweans) and wild tea leaves (common

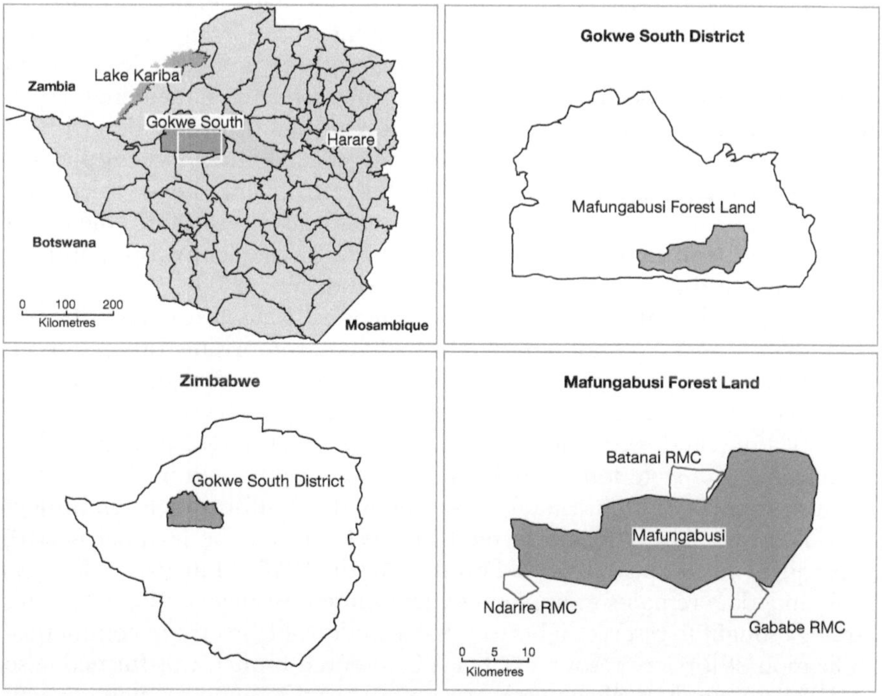

Figure 6.1 Map showing the location of Mafungautsi State Forest and the study sites, Zimbabwe.

name in Eastern Highlands of Zimbabwe – Makoni tea from leaves of *Fadogia ancylantha*) (Matose, 2002).

The forest derives its name from the Mafungautsi[6] Plateau on which it stands. This plateau lies at an attitude of 1,100–1,390m (Matose, 2002) and is flat, except where dissected by four rivers – Sengwa, Ngondoma, Mbumbusi and Lutope. The rivers drain into the Zambezi. The dam on the Zambezi, Lake Kariba, is an important source of hydroelectricity power for Zimbabwe. Mafungautsi Forest has over time been a landscape of contested values and meanings. To the government forestry department and other agencies involved in environmental management issues, it is a remnant of an original vegetation type associated with the Mafungautsi Plateau, and a water catchment area for the important rivers in the country – key reasons why the forest must be preserved. For the local community members, however, the forest represents a reservoir of future arable land because of its numerous wetlands and is also an ancestral home, especially for the early settlers who used to stay in the forest before it was claimed by the state.

The dominant soils in the forest are derived from Kalahari sands[7] up to 75m deep. These sandy soils have low clay and silt fractions (3 per cent clay, 2 per cent silt) and therefore cannot stabilize organic matter significantly (Zingore *et al.*, 2005). They are therefore low in organic matter content, a crucial element for sustainable crop production by farmers oriented towards household food self-sufficiency, who sell little and have limited cash to buy fertilizers, and so mostly rely on nutrients mineralized from organic matter (Zingore *et al.*, 2005). Zingore *et al.* (2005) found that soil organic matter in the Kalahari sands declines rapidly after cultivation and no meaningful crop production can be achieved after five to ten years of crop production without inputs of manure or fertilizer. The river valleys in Mafungautsi cut through the Kalahari sands into Karoo Sandstones, thereby giving rise to varied soils. For example, soils found in valleys include colluvial deposits from the upper slopes; these tend to be deep, moderately well-drained loamy soils, and are highly favoured for agricultural purposes (Mudekwe, 2003).

Mafungautsi is the third largest indigenous[8] state forest after Gwaai (144,265ha) and Ngamo Forests (102,900ha), comprising dry miombo[9] forests. State forests were created to protect them from destruction and are managed for various ends, including protection of watersheds, biodiversity conservation, and extraction of timber and other forest products. The management of state forests in Zimbabwe has been vested in the Forestry Commission (FC), a parastatal body under the Ministry of Environment and Tourism. The FC was formed in 1954 (during the colonial period) subsequent to the passing of a Forestry Act in 1948 (Mapedza, 2007). Its main mandate remains to implement government strategy on forestry. This strategy sought to reserve rights to private landowners so ensure commercial utilization of forestry resources. The FC research branch was formed (also in 1954) to conduct silvicultural research geared to support management and exploitation of forest resources by the private landowners. Other roles of the FC include controlling and managing all state forests in the country, protecting private forests, controlling the cutting of timber resources in all areas, promoting forestation programmes in the country, regulating trade in forest produce, and finally creating an institutional framework to ensure these tasks were effectively carried out.

The forest is dominated by *Brachystegia spiciformis* and *Julbernadia globiflora*. The forest also has a high proportion (about 25 per cent of all stems) of the Zimbabwe teak (*Baikiaea plurijuga*), a leguminous tree valuable for timber (Vermeulen, 1994). *Baikiaea plurijuga* is well adapted to the very deep, nutrient-poor Kalahari sands due to its very deep rooting system. Most of the large specimens of *Baikiaea plurijuga* were removed by selective logging between 1989 and 1992. A few other commercial timber species, such as *Pterocarpus angolensis*, are also found in the forest. Grasslands in the forest occur only in river valleys where high water tables during the rains exclude the trees that are generally sensitive to water logging. Grass cover is low in the forests on the plateau where the deep-rooted trees out-compete grasses in accessing soil water.

Stakeholders in the management of Mafungautsi State Forest can be grouped in two – district level and local level stakeholders. District level stakeholders include the FC, Rural District Council (RDC), the District Administrator (DA), the Ministry of Youth, Gender and Employment Creation (MYGEC), the Department of Natural Resources (DNR), the Department of Agricultural Research and Extension Services (AGRITEX), and district level political party structures. Except for AGRITEX and the FC, none of the other district level stakeholders is actively involved in facilitating processes in communities. Local level stakeholders include local leaders (chiefs, headmen, and village heads), members of local government structures (Village Development Committees (VIDCOs), Ward Development Committees (WADCOs), local councillors, political party structures,[10] Resource Management Committees (RMC) and agricultural extension agents). The responsibilities of the various stakeholders in Mafungautsi are shown in Table 6.1.

Table 6.1 Stakeholders and their roles in Mafungautsi (adapted from Matose, 2002)

Level	*Stakeholder/ organization*	*Role in the management of the forest*
District level stakeholders	FC	A state body specifically mandated to provide advice on and control, management and exploitation of forest resources. The FC has regulatory roles as well as extension roles.
	AGRITEX	Government department that carries out agricultural extension in communal areas. It is responsible for providing technical extension services to forest users and training, e.g. on beekeeping.
	DNR	Secretariat to the Natural Resources Board. Nationally its role is advisory, regulatory and supervisory in relation to all natural resources.

continued . . .

Table 6.1 continued . . .

Level	*Stakeholder/ organization*	*Role in the management of the forest*
	MYGEC	Ministry that coordinates government activities on rural development activities. It is responsible for providing training to communities and RMCs on various issues including proposal writing. It is also responsible for monitoring community participation in development projects.
	RDC	This is the local government authority. The RDC has a Rural District Development Committee (RDDC) that coordinates all council activities. It is responsible for formulating and implementing by-laws, and consolidating ward development plans into the district five-year plan.
	DA	The DA chairs the RDC RDDC and is an employee and representative of the Ministry of Local Government, Rural and Urban Development.
Local level stakeholders	Traditional leaders	They play a supportive role to the RMCs and also help in policing issues.
	Councillors	These are elected people who chair the WADCOs and are responsible for forwarding the ward development plans to the Rural District Council.
	VIDCOS and WADCOs	VIDCOS were created by the 1984 Prime Ministerial Directive to give a democratic orientation to the process of planning for local development. The VIDCO is the lowest unit of government administration, whose role is to identify needs of people in their villages and articulate them in the form of village development plans. These plans are presented to the WADCO. The WADCO draws its memberships from the chairpersons of the VIDCOS in the ward and is chaired by an elected councillor. It is responsible for consolidating the various VIDCO plans into a WARD development plan to be presented to the Rural District Development Committee.
	RMCs	RMCs were formed on FC initiative to represent their communities and act as a link between the communities and the FC. Their main roles include issuing permits to resource harvesters, controlling and monitoring resource harvesting and keeping the community bank account.
	Resource users	These are local people who utilize forest resources as well as help in the management of the forest. Their participation is mainly through the RMC.

Politics (and the political party one belongs to) is a dangerous issue in Batanai and Ndarire areas. Most people in these two sites support the ruling Zimbabwe African National Union Patriotic Front (ZANU-PF) party. In Gababe the majority support the government opposition party, the

Movement for Democratic Change (MDC), probably because they have not benefited much from the ZANU-PF government, and were left out of development processes. Politicians were more active in Batanai, which complicated participatory forest work in the area. For instance, during the initial stages of the project research process, meetings were delayed by people having first to introduce themselves by chanting political slogans, and then by stating their positions in the ZANU-PF party.

Forest governance systems and practices: the pre-ACM situation

People living on the edges of Mafungautsi Forest have a history of struggles over land. Before the 1940s, some of the early settlers in Mafungautsi North lived at Raji (now Chomagora small-scale commercial farms), located on the south eastern edge of Mafungautsi State Forest (see Figure 6.2). These people were forced to move by the colonial government when the area was converted into a commercial farm, and they settled into what is now Mafungautsi Forest land. The displaced families joined others and settled at Bandakamwe area around the head waters of Lutope River in the eastern part of the forest. In 1954, the forest was gazetted as a state forest. However, up to 1963, the takeover of the forest by the FC had little effect on the lives of the forest dwellers, who continued to perform religious rituals and ceremonies in the forest, to ensure plentiful harvests as well as continued flow of water in rivers.

In 1963, the inhabitants of Bandakamwe were asked to move to the western part of the forest, to an area that became known as the Zanda

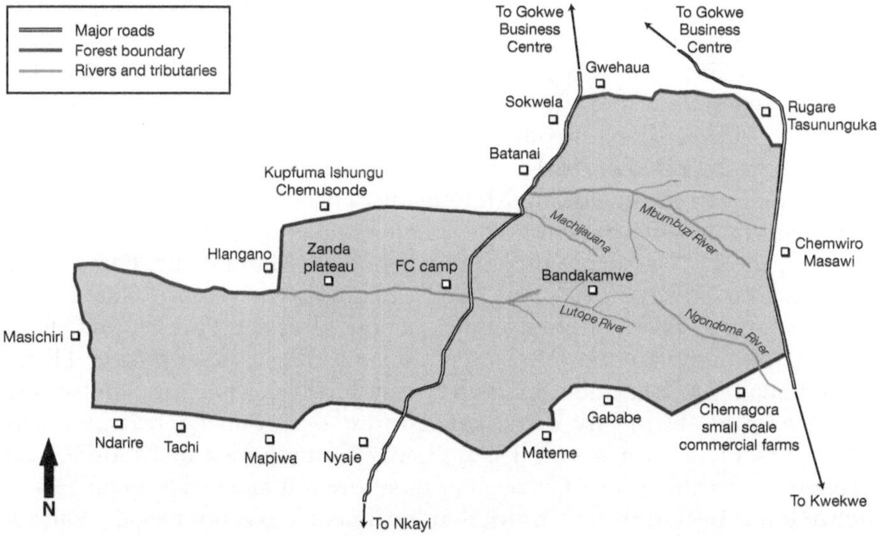

Figure 6.2 A map of Mafungautsi State Forest drawn by local communities, Zimbabwe.

Plateau. The FC recommended this area as suitable for settlement. The relationship between the FC and the forest residents was cordial, and residents helped the FC to put out forest fires without any payment. However, in the 1970s the FC began to issue rules and regulations to be followed by the residents. For example, large game hunting was prohibited. The FC officials also began to issue permits for people to harvest forest products. During the late 1970s, at the peak of the liberation war struggle, the FC employees stopped their forest service. Some families took this opportunity to move back to Bandakamwe, while others moved into the forest to settle. Movement into the forest and reclaiming of land by local people was based on the assumption that the war was being fought to regain access to land lost to the colonial government (Matose, 2002).

At independence, FC employees returned to work, and their first assignment was to evict settlers who had moved into the forest in 1981, and those who had moved back to Bandakamwe. The relationship between the FC and the forest residents remained cordial, even after the evictions and the reintroduction of the FC permit system for harvesting resources. However, things changed by 1983, due to the rise of dissident activities in the forest. In 1985, dissidents in the forest set the FC camp and its equipment on fire (Matose, 2002). The burning of the FC camp, and rumours that forest residents were harbouring the dissidents, provoked the government to issue a three-month eviction notice to the residents of the Zanda area in 1986. But before the eviction notice period had elapsed, the army came and set structures belonging to forest residents on fire. These residents then moved out of the forest to be accommodated in villages on the forest margins. When hostilities between the FC and local communities ended in 1987 most displaced residents were anxious to return to Zanda, but they were never invited back. The residents were now afraid to use their own initiative, as some of them had been evicted several times before. As a result, they continued to live on the fringes of their forest homeland, hoping that the FC would one day allow them access to forest land for cultivation and settlement (Matose, 2002). The introduction of the Fast Track Land Reform Programme (FTLRP) in 2000 was, therefore, seen as an opportunity by some local people to reclaim land in Mafungautsi Forest.

When Mafungautsi State Forest was converted into a protected state forest in 1954, the FC was made the custodians of the Mafungautsi Forest. The FC, due to the Forestry Act, was given all authority to manage state forests in the country. The first point under section 41 of the Forestry Act states that, unless authorized in terms of sub-sections (3) or (4) of section 44, no person shall cut, fell, injure or destroy any forest produce or remove any forest produce from any demarcated forest or protected private forest (Government of Zimbabwe, 1996). Local communities in Mafungautsi, because it was also a state forest, were therefore not allowed to manage and utilize forest resources. To ensure that local people did not illegally harvest forest resources in Mafungautsi Forest, members from the FC Forestry

Protection Unit (FPU), the policing arm of the FC, patrolled the forest area arresting community members who did not abide by the set rules. For instance, if one was found with a dog in the forest, it would be shot. However, because the FPU members were few and could only be at certain parts of the forest at a time, community members continued to access resources from the forest illegally. This resulted in many conflicts between the local communities and the FC and the forest continued to be degraded. Upon this realization, the FC decided to try out new methods of managing the forest.

Now, following the success of the Communal Area Management Programme for Indigenous Resources (CAMPFIRE) that was launched for the management of wildlife resources in the country, the FC decided to initiate a pilot Resource Sharing Project (RSP) in Mafungautsi in 1994. In the RSP, local people were invited to participate in the management of the state forest as well as benefit from minor non-timber forest products (such as grasses). Major products such as timber were however excluded from the RSP and these continued to be sources of conflict with forest authorities as local communities continued to access timber illegally. Under the RSP, communities around the forest were divided into 14 Resource Management Committee (RMC) areas with each area given a certain portion of the forest to manage. RMCs were later put in place to manage and control the resource use. Their main roles involved administering permits for resource users to harvest the minor forest products, monitoring the harvesting process, opening and keeping a community bank account where the moneys raised through the permit system were to be kept and finally advising the community on how the funds could be spent.

The way the RMCs were formed, however, led to many problems. The selection procedure for members was one difficulty. In one case the FC officer sent word that seven people from a specific community should come and see her, and those who came were selected as the RMC committee (Mapedza and Mandondo, 2002). People who went to see the officer were those who could afford bus fares to travel from their areas to Gokwe Business Centre, where the FC office was located. In other cases, the officer asked the traditional leadership authority to make nominations. Because of the way the RMCs were put together communities were unclear about their roles, and saw them as an extended arm of the FC. This generated a lot of tension. The RMCs themselves were accountable only to the FC, and they paid scant regard to the existing traditional leadership, for instance. This, in turn, resulted in unaddressed conflicts between traditional leadership and RMC members. For instance, some village heads and their community members failed to attend meetings organized by the Batanai RMC as they regarded the committee as having no authority to make such arrangements.

Upon election RMCs were handed the RMC constitution, developed by the FC in collaboration with the Ministry of Youth, Gender and Employment

Creation. Regardless of the low literacy rate in Gokwe area,[11] the RMC constitution was in English, and still had not been translated into vernacular languages up to the time the ACM project was initiated. The constitution gave the RMCs authority to exercise all powers necessary to achieve community objectives. It also gave the RMCs powers to make decisions, but these decisions were subject to approval by the FC officer, who had power to reject them. For instance, RMCs were asked by the FC to open bank accounts in order to deposit money raised through issuing permits to resource harvesters. The FC officer, however, was a signatory to the account and the communities could not withdraw the money without approval from him/her. Although the money was said to belong to communities, and that they could use it for their own needs, the FC only approved use of the funds on forestry related projects and activities.

However, up to 1999, there was little progress in terms of collaboration between the FC and communities living around the forest. Although on paper they were jointly managing the forest, in practice the FC continued with the use of force to stop the harvesting of illegal resources such as timber. Ordinary community members still continued to be sidelined in the management process and they continued to harvest resources that were even outside of the resource sharing agreement. Things remained business as usual even after the introduction of the RSP and the forest continued to be degraded. Several conflicts still remained among stakeholders. These conflicts were between the two major stakeholders, the FC and local communities as well as among the local level stakeholders in the communities. Conflicts among the local level stakeholders in the communities included the following:

- Resource boundaries: Each RMC was allocated an area by the FC where they could harvest resources such as broom and thatch grass. In most cases, however, the boundaries were not clear and this was a big problem in all RMC areas.
- Resource theft within and between RMCs: Some community members were said to steal resources from the forest to avoid paying the permit prices. It was also alleged that those who stole the resources used poor harvesting practices such as digging, resulting in depletion of the valuable resource. In addition, it was alleged that members from other RMCs came and stole resources in areas outside their RMCs, and this resulted conflicts.
- Fire management: Some community members were accused of starting forest fires, especially those involved in honey harvesting. It was alleged that community members without livestock (especially cattle) were responsible for starting fires, probably during hunting expeditions, since they lost nothing when pastures were burnt.
- Lack of transparency by RMCs: Community members accused the RMCs of lack of transparency. In some cases RMCs were said to issue permits, especially to outsiders, before the grass was ripe. In other cases,

RMCs were said to misuse money raised in their areas and communities missed out.

- RMC members and traditional leadership: There were conflicts between traditional leaders and RMC members. It was said that traditional leaders were not supportive of RMC activities. For instance, in Batanai area, some village heads did not notify community members of meetings organized by their RMCs, making it difficult for the RMC to operate.
- Incentives for the RMC members: RMC members wanted to be paid for their work in controlling resource harvesting and raising money for the communities. However, most communities did not want to pay them, saying, 'you work for the Forestry Commission, and it must pay you for your work'. In the Chemwiro Masawi RMC area, when the RMC appealed for payment from the community, some community members said '*kana zvakukona rega, tinotsvaka vamwe*' – 'if you are fed up with your work you can leave, we can always find someone to replace you'. The FC, however, also refused to pay the RMCs, saying, 'your communities must pay you as you are working for them'.

Conflicts between the communities and the FC included the following:

- Boundaries: There were conflicts related to the forest boundary ever since the establishment of the forest as a conservation area. At one time, for instance in Nyaje RMC area, the FC, using their map, came and evicted some people they said were living in the forest. The people of Nyaje were very bitter about this because some lost their fields whilst others were forced to move out of the forest areas. According to the people, the FC had encroached on their village in an attempt to increase the forest area. Similar stories were also narrated by community members in Gababe area.
- Fire management: The FC accused communities of causing forest fires. Community members believed, however, that the FC started these fires to punish them by destroying pastures for their animals.
- Resources communities were allowed to harvest: The FC allowed the villagers to harvest non-timber forest products (NTFP) such as grasses, mushrooms and herbs. This made villagers bitter as they wanted to harvest high-value commercial timber found in the area. According to communities, there were many mature trees that could be harvested in the forest. Some RMC areas such as Ndarire had no grasses in their area and, for them, harvesting timber was the only possible income-generating activity from the forest.
- Perceptions on the role of the forest: According to the FC, Mafungautsi Forest was a water catchment area for four of the major rivers in Zimbabwe that drain into the Zambezi river, the major source of hydroelectricity in Zimbabwe. According to the FC, the forest was predominantly growing on very fragile sandy soil and hence needed to

be protected in order to prevent siltation of the major rivers. For some community members, the forest was arable land, and they wanted to settle there.

Because the situation in the RSP remained 'business-as-usual', there was a lack of collaboration and learning among the various stakeholders in Mafungautsi. Also, despite the fact that this was a pilot project for the FC, no deliberate learning processes were facilitated to generate important lessons for influencing forestry policy in the country – the forestry policy that was in place in 1948 still remained in place and local communities were legally not allowed to manage state forests.

Although on paper stakeholders in Mafungautsi were jointly managing the forest in the RSP, in practice the situation did not change mainly because of the following key reasons. First, no capacity building was done to prepare the forest extension officers on the new participatory approach – they therefore continued with their perceptions that they had full knowledge and that communities had no knowledge. Community members also continued to believe that their knowledge was useless and that the extension workers knew everything. Second, nothing was done to deal with the conflicts that existed among stakeholders and this also hindered collaboration processes. The newly developed ACM approach therefore offered hope for the situation in Mafungautsi as it values collaboration and learning processes among stakeholders. When using the ACM approach, one has to identify the key problems that hinder collaboration among stakeholders and deal with them. The approach also emphasizes learning at various levels and therefore was ideal for influencing the forest policy.

Implementing ACM in practice

To implement adaptive collaborative management in Mafungautsi, the ACM team followed a number of steps and these are described below. To guide the process, the team came up with a series of steps to follow. The team was however flexible and willing to add new steps depending on how the process was unfolding in the field. For example, after conducting the context studies and discovering that the community members were passive, the team came up with interventions to break this passiveness. What happened at each step, key findings and lessons learned as well as decisions made after the step are described below:

Introducing ACM to the local communities

To introduce the project to local community members in Mafungautsi, introductory meetings were organized with communities in the study sites. The researchers presented the aims and objectives of the project and also the implementation approach. Because of the several conflicts that existed between the community members and the FC, community members were suspicious of the project. Some of the issues they raised during the discussions include the following:

- How do we know that you are not the FC that has come under cover?
- Our relationship with the FC has no free play, the environment is not conducive enough for us and FC to discuss and meet both our needs.
- We and the FC are always running away from each other; when we saw your vehicle, some of us ran away.
- We held a meeting sometime ago with the FC, and nothing came out of it, how is this meeting different from the one we had?
- Are benefits from Mafungautsi Forest going to accrue to us, or people from elsewhere?
- Is your organization related to the FC?
- Is the CIFOR research focused on the people only, or the people and FC?

An explanation of the project by the CIFOR researchers generated more questions and suspicions among community members. They particularly wondered about the idea that they would have to undertake their own research, and would be equal partners with the ACM team, and the FC. Some were quick to point out that they could not undertake their own research, since they lacked knowledge to do so. Other community members wanted to know examples of where the ACM approach had been used and resulted in tangible benefits accruing to the people. This particular question caught the ACM researchers by surprise; ACM was still a relatively new concept and there were still no examples to refer to. Others asked how it would be possible for them to be equal partners with the FC when they had so little power over the forest, relative to the FC – the FC had legal power to manage the forest whilst local communities did not have such power. It took several meetings and more explanations by the researchers for communities to understand what the project was trying to do as well as their own role.

The introductory experiences by the ACM team clearly show that a number of issues played in the background, unknown to the ACM team. Suspicion towards FC officials, poor relations among stakeholders, power issues at the community level, and previous experiences with other projects played a role in these first encounters. In general, the ACM team learned that community members were often suspicious. In the research communities, the lack of confidence in the FC was evident as a result of earlier unfulfilled promises. As a consequence, the ACM team had to come up with ways to build trust as well as improve relations among stakeholders.

Context studies

After introducing the ACM project, the ACM team carried out contextual studies. These studies aimed to understand key aspects of the local level situation. They were also an important opportunity for both researchers and communities to share information, and develop their views on current resource issues. The contextual studies had both ecological and socio-economic components. The ecological survey aimed to make an inventory of the resources and the quantities present in the forest (of both grass and trees). The survey also assessed the rate of tree cutting for either timber or harvesting of honey. The socio-economic survey identified organizations working in the study sites, the type and location of resource extraction activity, and the problems faced by communities in collecting and utilizing forest resources.

Important findings from the context studies included: the realization that communities in Mafungautsi were not composed of homogenous groups of people, but contained several stakeholder groups with different interests, perceptions and views concerning the role and use of Mafungautsi Forest resources (some community members saw the forest as land for cultivation and settlement whilst the FC viewed the forest as a fragile water catchment area that needed to be protected from human activity); several conflicts existed among stakeholders, especially between the local communities and the FC; local community members were passive when it came to resource management activities, and they felt they were poorly represented in the RSP. From the ecological surveys, reports of vegetation species in areas managed by the research sites, their location and abundance were identified. The ecological surveys showed that a large number of trees were being cut for honey harvesting and other purposes, and that the incidence of fire in the forest was also high (Mudavanhu, undated).

Following the context studies, report-back sessions were organized with community members in the research sites. The main objectives of these sessions were to provide opportunities for (a) the researchers to present their findings, get feedback from the communities and validate their findings and (b) discussion of the implications of the findings. Not only did the reflection sessions help the ACM team clear up some misinterpretations of findings, but they also benefited local resource users. When reporting the findings from the context studies the researchers avoided the land issue because of its high sensitivity. To their surprise, however, community members brought the issue into the dialogue themselves, as they felt that this was a priority issue. The ACM researchers, however, being scared for their lives in a politically dangerous environment like Gokwe South District as well as feeling that the land issue was beyond the scope of CIFOR work, decided to work with community members who were interested in managing resources in the forest only. This, as we realize now, was a big mistake. When opportunity came, people who were interested in the land moved into the forest and settled, disturbing resource management activities.

As already mentioned above, during the initial stages of the ACM work in Mafungautsi, it became clear that: several conflicts existed among stakeholders (including those, for instance, between the traditional leadership structures and the newly formed resource management committees; and between the FC and local communities); not all community members were interested in all the forest resources; and local communities felt disempowered to participate in the management of forest resources. Participatory action research (PAR) was a key element of the ACM approach and required all stakeholders to actively participate in the process. The ACM team, therefore, developed and implemented several interventions to manage conflicts and build stakeholder capacities before the PAR process could progress. These interventions included empowerment training, conflict resolution, training on leadership skills and finally the formation of resource user groups. All these interventions, which are described separately below, were done to prepare the stage for PAR.

Empowerment training

Because of the way local communities were engaged in resource management in the past, they felt disempowered and lacked confidence to participate in the joint resource management project. To build their confidence and also encourage them to effectively participate, the ACM team decided to conduct Training for Transformation (T-for-T), an empowerment training. The team got the idea of the T-for-T workshop from its use by the Intermediate Technology Group (ITDG) in Chivi Ward 21 in Zimbabwe. In Chivi, ITDG discovered that T-for-T resulted in farmers demanding changes in the approach and attitudes of extension workers. It also stimulated farmers to carry out their own experiments as a means to solve their own problems. It also led to greater democratization of leadership and more transparent decision-making (ITDG, 1997). T-for-T was first developed in Kenya in 1974, and was later adapted to Zimbabwean conditions by Hope and Timmel (1995). T-for-T draws to some extent on Paulo Freire's *Pedagogy of the Oppressed*. Freire (1970) advocates that change should be driven by the people themselves rather than by outsiders. This idea provided the motivation to stimulate and build confidence among marginalized groups, and to encourage them to initiate a process of change (Freire, 1970). T-for-T emphasizes experiential learning, empowerment of local people to control their own destiny through active participation in their development, and the sharing of ideas and knowledge. It strengthens people's confidence through mottos such as 'nobody knows everything and nobody knows nothing'. The ACM team saw T-for-T as a suitable intervention approach to overcome inertia among forest users.

The ACM team then organized a one-week T-for-T workshop facilitated by hired outside T-for-T experts. The workshop took place at Shingai Training Centre, at Gokwe Centre. A total of 43 participants attended the workshop from all the research sites: 7 people from Ndarire (all men), 15 from Batanai area (11 men and 4 women), 14 from Gababe (9 men and 5 women), the 2 workshop facilitators (the T-for-T experts), 3 CIFOR researchers (1 woman and 2 men), the RSP Coordinator and the District Administrator for Gokwe South District. The workshop facilitators developed a programme to challenge community members to be critical about their situation. They aimed to make the community members think of possibilities they had for improving their situation, instead of not doing anything about it. In the workshop facilitators made extensive use of visualization techniques, games, stories and short plays (sketches) to help participants reflect on and challenge their 'mental models' of their contexts and behaviours.

Conflict resolution

After the T-for-T training, the ACM team organized several workshops and took advantage of other organized meetings (by the traditional leadership, the FC, or the RMCs in the research sites) to build the capacity of stakeholders for conflict resolution. In terms of process, the conflict resolution workshops started with presentations from resource persons, mostly hired conflict resolution experts. They made presentations on the importance of conflict resolution and the various methods that could be used to resolve conflicts. Participants were later asked to recall conflicts they had faced in the past, individually or in the RSP, and critically look at what mechanism they had used to deal with the conflicts and how effective they were. Later on, participants identified their current conflicts. Then, in the workshop, they tried to resolve these conflicts through facilitated negotiation: participants first had to agree on the status of the conflict, and the available alternatives, and then come up with jointly agreed solutions. The conflict resolution processes were facilitated by the researchers and experts during the various types of workshops and meetings organized by FC, RMC and traditional leaders. In all cases, facilitation was crucial to cope with suspicions and help stakeholders open-up to others, to bring their conflicts out into the open, and to discuss, negotiate and resolve them. These discussions between stakeholders provided great learning opportunities.

Training on leadership skills

The ACM team also organized trainings that especially focused on competencies for leadership. One of these workshops was organized immediately after the T-for-T and conflict resolution workshops. Others were organized when the ACM researchers felt the need. In addition to these trainings, ACM researchers made use of other community meetings and organized platforms. The ACM team felt that leadership capacities were important for effective functioning of key institutions, namely, the RMC and the traditional leadership. Leadership training was based on the assumption that the sustainability of projects depends on the quality of both leaders and their followers. During the leadership training sessions, participants discussed the various roles of office bearers, how to organize meetings, the importance of coming up with an agenda before each organized meeting, issues of representation, accountability and transparency, the need to hold democratic elections, and good leadership qualities.

Formation of resource user groups

From the context studies the ACM team already had learnt that not all community members were interested in forest issues. The ACM team therefore decided to form groups of people who shared an interest in a particular resource. These resource user groups were to ensure that specific resources were managed sustainably. In Batanai and Gababe, three different resource user groups were formed, focusing on thatch grass, broom grass and honey. These were the main resources for which harvesting was permitted under the resource sharing agreement. The ACM team invited community members in the study sites to join the resource user groups and made explicit that this was on a voluntary basis. One could join in or drop out when one wished. Resource users were also free to join any other group and could thus belong to more than one group, especially if the group meetings were not taking place at the same time. Members of the resource user groups had opportunities to share their experiences and learn from each other in order to avoid making mistakes that others already made. The ACM team was responsible, during the early stages, of facilitating the process for resource users to address their problems. Initially, a variety of people came to the resource user group meetings and participated in the early stages of the PAR process, maybe out of curiosity. Many dropped out as time went on and only a few persons participated in all the stages of the research. Asking people to voluntarily join in or drop out as and when they wished, though it helped the ACM team to work with people with a genuine interest tackling problems associated with certain resources, was, on the other hand, a key reason why the learning processes were slowed down. When new people joined the learning groups, they needed to be taken through the processes that others had done to date.

It was only after the above mentioned interventions, that participatory action research (PAR) processes could be initiated with various resource user groups. The PAR process is described in the next section. Stakeholders at a range of scales (including resource user groups, resource management committees, and FC officers and researchers), were included in processes to develop visions, develop action plans to realize their visions, and implement action plans, monitor and evaluate the outcomes and reflect and learn together about the impact of their actions. An example to illustrate the PAR process that the resource users went through is given below.

The PAR process with resource users

The earlier top-down management approach to the forest, and unfulfilled promises in the RSP, disillusioned many local community members. They had lost hope of getting anything positive from the project. Because of this, the ACM team decided to use scenarios and visioning – tools that stimulate creative ways of thinking and help stakeholders break out of established patterns of assessing situations (Wollenberg *et al.*, 2000). This, as we shall see later on, proved to be a powerful method in giving local communities hope for positive change as well as giving them energy to participate in the project.

To come up with their visions, the ACM team asked resource users to describe how they wanted things to look in two years' time. In the description, ACM team asked them to include the status of forest resources, the number of people harvesting resources, harvesting methods, availability of markets, prices of the resources (both permits and selling prices), revenue from resources, and finally, the rules and regulations for governing use of resources.

After developing and presenting vision statements resource users were asked by the ACM team to describe the current situation with regard to their resources. This showed how different the current status was, compared to the desired outcomes. Group members were thereafter asked to identify what they thought might hinder transition from the current situation to the envisioned situation. The resource user groups later came up with action plans to deal with possible hindrances they had identified. The groups then implemented these actions, and opportunities were created for them to reflect on lessons learned from the impact of these interventions. The lessons were later used in decision-making processes. An example of one of the many PAR processes facilitated with broom grass resource users at one of the research sites, Batanai, is presented in the next section.

Several visioning meetings were organized for different resource user groups in each RMC in the period of 2001 to early 2002. The majority of these meetings were facilitated by CIFOR community partners on behalf of CIFOR. From the numerous meetings, it is clear that community members during the 2001–2002 period spent a lot of time in resource management meetings and workshops, and this raises a recurrent but unresolved question in participatory development – the opportunity cost of the time thus spent in 'workshopping', and how and when to try to reduce it to a minimum.

PAR with broom grass resource users at Batanai RMC area

This group consisted of collectors of broom grass and was dominated by women. Because the group, like all community members in the research, was disillusioned and had no hope of anything good coming out of the joint resource management project, the team decided to use visioning as an action planning tool to generate interest and hope. The team organized several meetings with the broom grass resource user group members to assist them to develop their vision, assess the current situation for their resource and come up with action plans on how they would move from the current to the desired future situation. A description of the current and desired future situation is presented in Table 6.2.

After the visioning exercise, the broom grass resource users identified the possible hindrances for them to attain their vision. The ACM team, in asking

Table 6.2 Current and future scenarios developed by the broom grass group at Batanai

Indicator	*Current situation (2001)*	*Vision of the broom grass resource (in 2003)*
Number of people involved in broom grass	A small number are involved. For instance in Mafavillage, only 15 out of 194 households are involved in harvesting broom grass. Those near the road are the ones involved. Mostly women involved. Few men. Children also involved.	At least 50–80% of the total households in the RMC area. Men and women should all be equally involved. With more people involved, we can all help each other.
Harvesting methods	Most people cut using sickles. There are some who dig, though fewer.	Wish that all of us use the same harvesting method. All people harvesting by cutting. No one should be digging.
Markets	Sell at Gokwe, Bulawayo, Kadoma, Nemangwe, Sayi, Manoti, and Gweru and also to our neighbours.	Get a huge market close by like at Gokwe Business Centre.
Rules and regulations	Digging is not allowed. We pay Z$30.00/day (USD0.54/day)[1] for a permit. Permit prices go up every year, and it is difficult to cut all the broom you need in only one day.	Hope that permit price will match the number of days the permit will be valid for. We would like it to be Z$80.00/ 4 days (USD1.43/4 days). All RMCs to have same rule on digging or cutting.
Incomes	Selling brooms is a reliable way of raising income. Currently get Z$15,000.00/ annum (USD268.34/annum), from brooms.	Hope to get as much as Z$30,000.00/annum (USD536.67/annum).

Note:
1 1USD = Zim$55.90.

Source: Mutimukuru-Maravanyika, 2010.

them to do this, was already part of the action planning process. Resource users would later come up with actions to offset the hindrances so that they could move from the current state of things to their desired future situation. During a meeting organized and facilitated by the ACM team, broom grass resource users at Batanai identified the following hindrances:

- lack of cooperation among resource users in the use of the cutting method of harvesting;
- lack of cooperation across the RMC;
- failure to establish a local market for buyers of brooms from our area.

During this meeting, resource users narrated how community members in their area had conducted an experiment together with the FC, to learn together the most sustainable method of harvesting the broom grass (Box 6.1). This was done because of the problem they had faced on the depletion of the broom grass in their area. The experiment, to the ACM team, was an important opportunity for joint learning and decision-making by the stakeholders in Mafungautsi. After realizing that digging the broom grass was leading to the depletion of the grass, stakeholders in Mafungautsi made a joint decision to cut instead of dig the grass. In this case, joint learning by stakeholders helped them to make a decision to utilize the broom grass resource in a sustainable manner.

Box 6.1 The Machije experiment

At the inception of the resource-sharing project in 1994, in an effort to enhance joint learning about sustainable methods of harvesting broom grass, community members in the Batanai RMC, on the initiative of the Forestry Commission, decided to conduct an experiment in Machije wetland (the area where Batanai RMC harvests broom grass). The experiment was conducted in two small plots staked out jointly by the broom grass resource users and the FC. In one of the plots, resource users harvested grass by digging. In the other plot they harvested the grass by cutting, using sickles. The stakeholders then monitored to see how the grass would grow in each of the plots. In the seasons that followed no new broom grass germinated in the plot where grass was harvested by digging. Instead, a new grass variety which could not be used for making brooms emerged. Stakeholders concluded that the best method for harvesting their grass without depleting it was cutting. For two years after the experiment, no one dug broom grass in Batanai RMC. However, after the two years, some people resumed digging despite the fact that they knew its adverse impact on the resource.

After hearing the story, the team asked several questions to find out if any other such joint learning and decision-making opportunities had been created in the project. The team realized that no other joint learning opportunities had been created. The ACM team realized that after this experiment, there had been no planned opportunities for the participants to come together and discuss the undesirable change in the harvesting practices by resource users and no other joint learning opportunities had been created, and yet the ACM approach encourages resource managers to learn from the outcomes of their management activities. For the ACM approach, learning is a continuous process especially when managing complex systems and not just a one-off thing like the Machije experiment. Because of this, during the action planning meetings, the team encouraged resource users to discuss why people had reverted back to unsustainable harvesting methods. During these meetings, broom grass resource users identified several factors that had contributed to the sudden change in harvesting methods. One of these factors was the continued market demand for 'dug brooms', i.e. brooms made from dug grass. In most places where people were selling their brooms, the customers wanted 'dug brooms' and these were selling faster than 'cut brooms'. It was alleged that customers wanted dug brooms as these lasted longer than cut brooms – the roots of the grass was said to make the grass stick to each other. This resulted in many of the Batanai residents returning to the practice of digging the grass even though they knew the adverse effects of such practices. When reflecting upon this process now, we now realize that we should have brought in other important issues also in the discussion – for instance, the issue of not just dealing with the supply side of the broom grass but also the demand side. The Machije experiment only focused on the suppliers of the grass and convinced them that digging was an unsustainable harvesting method. However, no effort was made to also convince the buyers (including key stakeholders like consumer associations and environmental groups) to buy cut brooms. It is therefore the duty of facilitators of ACM processes to bring in issues left out in discussions, probably because the stakeholders do not have knowledge on them.

After identifying the reasons why resource users had reverted to unsustainable harvesting methods, the broom-grass collectors, through facilitation of the ACM team, used the lessons learned to come up with various ways to ensure that their resource was used in a more sustainable manner. These include the following:

- Everyone, including those who were not present at this meeting, should tell the RMC members whenever they see someone digging broom grass. It was agreed that RMC members cannot do their duty single-handed, as they had other commitments. Moreover, according to the community members, the RMC members are not paid like the FC Forest Protection Unit officers, hence could not devote most of their time to arresting people who transgressed. The RMC members had to work in their own fields as well, in order to survive.

- Instead of giving a permit before harvesting grass, it was suggested that it would be better if people would pay after harvesting. This would enable the RMC members to inspect and check if the grass was harvested by cutting or digging. The Batanai resource users, however, thought that there would be a risk that people might harvest and disappear before paying for their permits.
- In order to deal with the problem of the market preference for 'dug brooms', resource users suggested four options: (1) All broom grass harvesters (within and outside the Batanai RMC) to cooperate and only provide cut brooms. This would force consumers to buy cut brooms since these would be the only ones available on the market. (2) RMCs to negotiate with the Gokwe Rural District Council for a law to prohibit the sale of dug brooms. This would then force all broom grass collectors to cut instead of dig the grass. (3) Broom grass harvesters to come up with new bundling methods that could make the cut brooms last longer. This would make customers prefer buying the cut brooms. (4) A suggestion was made for resource users to advertise their brooms so that customers would come to Batanai instead of the Batanai sellers taking the brooms elsewhere. This would give more opportunity for the RMC to inspect and check if all the sold brooms are cut brooms and not dug brooms.
- To deal with the problem of other RMCs around Mafungautsi harvesting by digging instead of cutting, the participants suggested that a 'look and learn' workshop would be organized in which they would share their experiences from the other RMCs to come and learn from the Machije experiment. They could visit the plots in the forest and see for themselves. One of the Batanai community members would explain how the experiment was conducted and what the findings were on the digging vs. the cutting of broom grass.

In April 2002 the broom grass resource user group organized a 'learning' workshop, together with the FC officer, with an aim to learn about new ways of bundling cut brooms. The local women who were knowledgeable on the alternative bundling method showed others how they bundled their brooms and the resource users and the FC officer agreed to jointly market the newly bundled brooms. The FC officer took the brooms to the annual Provincial Agricultural Show in August 2002. Eight out of the 32 decoratively bundled brooms were sold at a price of Zim$50/broom (USD0.88/broom). This was Zim$42 (USD0.74) more per broom than undecorated brooms. The remaining 24 brooms were sold at the FC office for the same higher price. The production of the decorated brooms is much more time-consuming, compared to the conventional bundling method: it takes 20–30 minutes more to make a decorated broom. People therefore felt that the decorated brooms were being sold at a fair price. Undecorated cut brooms (which were bundled using the conventional method) were selling at Zim$8/broom (USD0.14/broom) at Gokwe Centre.

In addition, the Batanai women started teaming up to go and sell their brooms in faraway places. Some of these women would not have managed to do this alone as their husbands did not allow them to travel to such far places unaccompanied. For instance, in an interview on 17 July 2002 with Mrs Siwela, a broom grass resource user from Batanai area, Mrs Siwela revealed that, together with her neighbour, she had begun to travel to faraway places to sell their brooms.

Collaborative monitoring

The ACM team, like other researchers (Frost and Mandondo, 1999), believed that adaptive management required a functioning monitoring and resource information system to enable resource managers to assess the condition of their environment, the status of key resources and the effect of their management processes. The team believed that such a monitoring system in multi-stakeholder situations had to be collaborative in nature for it to enhance joint learning processes. When several stakeholders participated in the monitoring process, it was termed collaborative monitoring (CM). CM therefore played a central role in the ACM approach, and helped stakeholders to generate information required to check whether they were moving towards attaining their goals (Guijt, 2007).

The team followed a number of steps in initiating the CM process and these included the following: conducting a survey to understand the existing monitoring system, generating a common understanding of what CM was and discussions with key informants on how CM could be conducted in Mafungautsi, wider consultations with the public to present the outcomes from the discussions with the key informants, wider stakeholder consultations to finalize the terms of reference for the monitoring sub-committee that was given the role of leading the CM process (Mutimukuru *et al.*, 2007). The project, however, ended before the developed CM system was implemented.

Linking the Mafungautsi outcomes with national policies

As already mentioned above, although the RSP was meant to be a pilot project, it failed to generate lessons to influence forest policy. This was mainly because no one made effort to critically reflect on the project and its outcomes. With the ACM approach, cross-scale linkages are crucial and important as they help to create a conducive environment for joint management. To ensure that this component was not left out in the project, the second author of this paper – the ACM team leader – was tasked to take the lead in ensuring that the results from the field would influence forestry policy. To do this, the ACM team implemented a number of interventions including writing and presenting policy briefs to policy makers (including the FC), organizing 'look and learn' tours for policy makers to the ACM research sites, and facilitating policy makers to attend international conferences to enhance their learning on which policies work. For example, the Permanent Secretary in the Ministry of Environment and Tourism attended a workshop in Indonesia where she met with other policy makers and had the opportunity to share experiences.

The ACM researchers also organized policy round table discussions for policy makers on emerging findings from the field, and invited them to participate in ACM training workshops. For example, the FC top management was invited to a half-day training workshop in June 2006. In addition, the ACM team organized regular update meetings with the FC top management where they briefed them about the ACM activities and findings. The strategy was to use results from the field to influence the national forestry policy formulation process. It took quite a while before these positive outcomes were seen in Mafungautsi, and little had been done to influence national policies when the project ended. However, some funds were secured later on for an up-scaling phase of the ACM approach, and part of the money was used to continue working on influencing policies. With time, small positive outcomes began to surface with regard to the national forestry policy making process. The FC management decided to up-scale the ACM research to seven other provinces in the country with one district participating as an ACM Learning Centre – officers from these districts also participated in the various ACM training workshops. This was made possible by funding from the Canadian International Development Agency (CIDA.) This up-scaling phase gave birth to the *ACMZim News* newsletter (the first author of this chapter was the chief editor of this newsletter), which acted as a platform for the various officers in the different districts to learn from each other and share experiences. Experiences from the ACM centres were supposed to generate concrete lessons to influence forestry policy in Zimbabwe.

Outcomes of the ACM processes and their sustainability

The ACM project in Mafungautsi Forest was meant to add value to the Resource Sharing Project (RSP) which had so far not been successful. Towards the end of the ACM project in 2003 the situation in the project had improved (Mutimukuru-Maravanyika, 2010). First, the participation of local communities in decision-making and management of the forest had improved. Previously, it was only the RMCs that actively participated in this. Key factors that contributed to this positive change include: the improvement in relations among stakeholders that came about after conflict resolution processes, the gaining of confidence by the local communities through capacity building such as the Training for Transformation workshop, as well as their contribution being taken seriously by outside partners such as the Forestry Commission. This resulted in the emergence of genuine partnerships, with both stakeholders jointly working together to solve their problems.

Second, as already mentioned above, at the beginning of the ACM project there were serious conflicts between the RMCs and the community members mainly because of the way the RMCs functioned. As time went on in the project, relations between these two stakeholders improved mainly because of the positive changes in the operations of the RMCs. The RMCs became transparent and downwardly accountable to their communities. Key factors that led to this include, among others, the capacity building that was done for all stakeholders on good leadership qualities and the encouragement that the FC gave the RMC to be accountable to their communities. Third, improvements in incomes through value addition to the forest resources. For instance, the decorated brooms that came about through joint learning processes sold at a much higher price than the undecorated ones. Also, the increased sale of products through sourcing of outside markets and value addition also helped resource users to generate much needed income. Finally, stakeholders adopted and promoted use of sustainable practices generated through joint learning processes.

All the above positive outcomes gave the ACM team and the Mafungautsi stakeholders confidence that the ACM approach had indeed added value to the RSP. Although they were aware that this was a research project to test as well as contribute to the further development of the ACM concept, the ACM team members were anxious and under pressure to produce positive results to influence forestry policy in the country. The team, to some extent, was worried that failure of the approach to produce positive results would mean that they themselves had failed to implement ACM in practice.

Developments after the ACM project had ended

During the ACM project period (2001–2003), the ACM team members took lead in facilitating the ACM approach in Mafungautsi. The FC officer participated in most of the organized meetings and workshops, and the CIFOR researchers shared with the officers highlights and insights from the meetings they did not attend. This helped the officers to learn by observing the ACM team facilitating the various processes. When the project ended in 2003, the CIFOR researchers handed over the overall facilitation of the process to the FC officer.

The team did not pull out of the area drastically, but gradually, offering considerable back-up support initially and gradually reducing it as time went by, and as the FC officer became more confident in organizing and facilitating the learning processes on his own. As part of the exit strategy from Mafungautsi the ACM team organized five training workshops which the FC officer (and several personnel at different levels, including the FC top management and personnel from the Forestry College of Zimbabwe) attended. The workshops aimed at equipping participants with facilitation skills and knowledge of the ACM approach, and the ACM team acted as resource persons during the workshops. The FC officer remained actively involved in the processes after the ACM team had pulled out of the area to become – eventually – observers of the process (Mutimukuru and Kozanayi, 2005).

Joint learning processes at the resource user group level stopped after the ACM project ended in all sites, partly because the community partners, who used to receive an allowance from CIFOR, stopped facilitating them. Because of continued economic decline and increasing hunger and poverty in the country some community partners left for neighbouring countries in search of work. The FC officer had limited resources and chose to focus on the learning processes at a much higher level. These included 'look and learn' tours, pre-grass cutting workshops and discussion meetings (Mutimukuru-Maravanyika, 2010). The FC officer continued to receive financial back-up from CIFOR because, due to the declining economic conditions in the country, he had no access to other financial resources. In September 2003, for instance, the officer faced problems of transport and fuel, and these made it difficult for him to visit communities frequently. The FC officer had further problems when his vehicle was involved in an accident, and he could no longer visit communities. When ACM researchers asked him to join their trip to Gababe, he refused, saying 'communities will not take me seriously because last time I failed to meet my promise to go to the village and sort out some issues we agreed upon'. The FC officer died in October 2005.

A follow-up study that was conducted by the first author of this paper four years after the project had ended to check on the learning and collaboration processes revealed:

- Following the developments in the country concerning the land issue, several people had moved into the forest and the number had continued to increase. The new settlers had cleared forest areas both for settlement and for agricultural fields. Resource management activities by RMCs operating in the areas where the new settlers were located had stopped. The new settlers privatized resources in areas where they settled by staking-out and fencing-off their plots, and members from communities outside of the forest area could no longer access forest resources freely. The new settlers set fees for extraction of resources like broom grass from the staked-out fields. In the forest area that the ACM research sites were managing, there was increased cutting down of live trees and poaching of wild game.
- The remaining joint learning processes at the RMC level that the FC officer continued to facilitate after the ACM project ended had stopped when the FC officer passed away. The person who took over as the RSP coordinator had a new focus and only received a brief training on the ACM approach (a half day ACM training workshop in June 2006) even though he had not been exposed to similar approaches before.
- The functioning of RMCs, which had improved over time, had deteriorated partly due to lack of input from the FC officer. In Gababe area

RMC performance initially continued to improve after the ACM project ended, but RMC members stopped being accountable to communities on their activities after the death of the FC officer. This resulted in communities becoming suspicious that RMC members were embezzling community funds. Several conflicts among the stakeholders broke out, and these were left unaddressed when the resource management activities and joint learning processes stopped.

Reflection and discussion

The Mafungautsi ACM experience as well as the subsequent study reveal a number of issues linked to the future and prospect of adaptive collaborative approaches to natural resource management.

First, the importance of building trust and confidence for stakeholders to work together, especially when there is a long history of conflict. During the initial stages of the Mafungautsi work, it became clear that: several conflicts existed among stakeholders (both internal and external conflicts); the FC officers, though implementing the joint forest management project, lacked knowledge on what this entailed; and local communities felt disempowered to participate in the resource management activities. For us, effective participation of stakeholders in resource management was extremely crucial for the success of the joint management initiative. We therefore developed and implemented several interventions (empowerment training workshops, conflict resolution workshops, training on leadership skills and finally the formation of resource user groups) to manage conflicts and build stakeholder capacities before the joint learning and experimentation could be initiated.

Besides identifying existing conflicts and dealing with them, we came up with interventions to boost the confidence of especially local community members who, because of their previous engagement with the FC, saw themselves as powerless and looked up to outsiders to solve their problems. In addition, there was need to build capacity of the FC officers to change their perceptions on their own role and that of community members in participatory management of the forest. Unfortunately, after the heavy investment in building the capacity of the FC officer, he passed away in 2005. The new officer who was appointed to take over the coordinator's role had no knowledge of the approach, resulting in all the gains of the project being reversed. This raises questions on the type of arrangements that need to be in place to ensure that institutionary memory is not lost when those trained in new approaches are transferred or die.

Second, the struggle to address the issue of power and politics that underlie resource management practices. Although we became aware of the complexities of issues of power and politics associated with Mafungautsi during the early stages of the project, we did little to respond to this understanding, apart from engaging in some empowerment training with marginalized groups. At the beginning of the project we were quite aware of the highly unequal distribution of power between the local communities and the FC, for instance, but we failed to think through the implications of this inequality. The FC held all the cards. It has legal authority to manage the forest derived from the existing Forest Act. Although local communities were allowed to participate in resource management under the RSP, they had no legal authority to manage the forest reserve. Although some effort was put in trying to influence the forest policy, it was frustrating for us that the project ended when little was done in this area.

This unequal distribution of power resulted in conflicts among stakeholders. Like Scott (1985) clearly points out, the weak and powerless have their forms of resistance to encounter the powerful people and advance their own agendas. This resistance is regarded as the 'weapons of the weak' (Scott, 1985). Powerless people in Mafungautsi had their own ways of advancing their agendas through, for instance, poaching of timber (as it was not included in the resource sharing project) and the invasion and settling into the forest by local communities to reclaim their ancestral land. Although contextual studies revealed historical struggles over land between local communities and the FC, we saw this as being beyond the scope of our work at that time. We however had a strong belief that through implementing the participatory action research (PAR) process, local stakeholders in Mafungautsi would eventually gain the confidence to tackle the bigger land issue in the area on their own. This, from our own view now, was putting too much faith in participatory and joint learning processes. The belief in the power of participation to resolve the land issue seems in retrospect only wishful thinking. Several researchers (Logan and Moseley, 2002; Chauveau and Richards, 2008; Kaimowitz and Shell, 2007) have suggested that resource management initiatives will not succeed even in their most limited conservation aims if they shy away from analysis and resolution of fundamental societal conflicts.

Reflecting on our experiences now, it is clear that initiating the process for reform of the Forest Act to also give communities legal authority to participate in the management of state forest ought perhaps to have been a precondition for the implementation of the project. As Mapedza and Mandondo (2002) point out, without meaningful legal power over the resources they are supposed to manage, it is difficult to get the commitment of local communities. Before implementing ACM in Mafungautsi, we should have put all effort to try and influence policy through various means other than waiting for the ACM outcomes. One option would have been to start with thorough desk studies of successful participatory initiatives, and present these to the forest policy makers in Zimbabwe so as to challenge them to rethink the forest policy in the country. Failure to engage with issues of power and politics has been one of the strongest criticisms of participatory approaches (Cooke and Kothari, 2001) and successful participatory development initiatives have been found to be those that explicitly aimed to focus on issues of power and politics and challenge existing power relations, rather than simply work around them for service delivery (Hickey and Mohan, 2005).

Third, working with donor and public institutions which are not appreciative of the extra efforts involved in learning based and collaborative approaches. The ACM project in Zimbabwe (as well as in other countries where CIFOR's ACM projects were implemented) had a project time frame of three years. Within this period, we were expected to learn about implementing the new approach in practice, help to refine the ACM concept, as well as contribute to the improvement of both human wellbeing and the status of the forest.

With the benefit of hindsight, we now realize that this was an over estimation of what could be achieved within such a short project period. A huge amount of time was spent in setting the stage for joint learning processes to take place. The project however came to an end when resource users were just beginning to implement their action plans, before a solid foundation was put in place on the benefits of joint learning in resource management. When the project ended, the road ahead was still unclear, especially for the resource users. Without being convinced on the benefits of joint learning in resource management, it is not surprising that stakeholders stopped facilitating joint learning processes when the FC officer passed away, and the positive outcomes from the project were not sustained.

Fourth, ensuring that ACM initiatives bring significant contribution to the local community's livelihood. When we look back now, we realize that like the Resource Sharing Project (RSP), the ACM project also did not make a significant contribution towards the local community's livelihood. The project also continued to focus on low value resources like broom and thatch grasses. Even with value added, such resources still sold at low prices, thereby not generating significant income for the resource users. It would have been more useful if for instance we had also focused on improving resource users' thatching skills to encourage them not only to sell grass but also to work as skilled thatchers. A similar approach might have been taken with the beekeepers – e.g. training them to generate more income through value addition and productions of products such as candles and body oil. Other skills (for instance, marketing skills) might have been equally important. Lobbying for high value timber to be included in the resource sharing agreement should have been a priority action for the project. Because of the short project time frame, we did not have an opportunity to even think about these options which could have been implemented to deal with the challenges that resource users were facing. Value addition and the inclusion of high value resources such as timber in the RSP agreement could have significantly contributed to poverty alleviation by giving local communities a better stake in the timber trade.

Fifth, the need to engage with best practice guidelines for natural resource management institutions. In Mafungautsi, although we concentrated our effort on influencing the RMCs to become transparent and downwardly accountable institutions, we still neglected best practice guidelines in the area of institutions. The issue of rules and regulations has been clearly identified by a growing body of literature (Ostrom, 1990; Bromley and Cernea, 1989; Cleaver, 2000; Kayambazintu *et al.*, 2003) as important in determining the success or failure of participatory resource management initiatives. Although resource users were clear that it was important for effective rules to be in place for managing various resources, we did not do much to enable them to come up with clear rules and means for enforcing them. Without these, again it is not surprising that the positive outcomes from the ACM project were not sustained.

Sixth, the need for huge initial financial and human resource investments, especially during the initial stages of the ACM process. From our own experiences in facilitating ACM in Mafungautsi, the initial stages of the ACM project required huge human and financial investments. As time went by, after capacity building for the FC officer and the community partners we gradually handed over the facilitation of the ACM processes to them. When following up events after the project ended we however realized that the FC officer, who was supposed to up-scale the ACM initiate to other areas, lacked basic resources for him to do his work, such as transport and fuel.

Despite our own knowledge of the amount of human and financial investments required especially during the early stages, we now realize that leaving the future ACM work in the hands of the FC was indeed an overestimation of what the FC officer could possibly achieve on his own resources once the project ended. We failed to properly assess how an under-staffed and under-funded organization, with personnel earning meagre salaries hardly sufficient for survival, would manage alone.

Seventh, the nightmare of organizing learning and deliberative processes in an unstable socio-economic and political climate. Our project was implemented in a rapidly changing socio-economic and political climate in the country. The collapse of the Zimbabwean economy (1 billion per cent inflation and lack of monetary denominations to undertake any transactions) resulted in the out migration of most economically active persons from the countryside in general but more specifically from the ACM sites to earn a living. Economically active individuals sought other forms of livelihood outside of the forest environments, such as the lucrative illegal gold panning in nearby centres of Kadoma and Kwekwe or further abroad in Botswana and South Africa. Some of the people who migrated were key people in the joint learning processes. This greatly slowed down learning processes as new people with no clue of what others had done before came to the resource user group meetings.

Concluding remarks

Implementing the ACM approach in Mafungautsi was easier said than done. In addition to conducting research on the ACM approach and its outcomes, we had a daunting task of convincing our local stakeholders to embark on this learning by doing journey, whose outcomes we were not sure of. We however gained confidence when some positive outcomes began to surface in the project. As we had envisioned, the ACM project indeed produced a number of positive outcomes. These include: an emergence of genuine partnerships with former adversaries working together to solve their problems; improvements in the functioning of resource management institutions; increased incomes through value addition and sourcing of alternative markets; and adoption and promotion of sustainable resource harvesting practices. However, a follow-up study four years after the project had ended revealed that the positives were not sustained. This was partly because of our failure to address some of the challenges that we encountered during implementation of the ACM approach.

We now realize that these challenges are important and should be dealt with if future ACM initiatives are to produce lasting positive changes. These include: building trust and confidence for stakeholders to work together, especially when there is a long history of conflict; addressing issues of power and politics that underlie resource management practices; working with donor and public institutions which are not appreciative of the extra efforts involved in learning based and collaborative approaches; ensuring that ACM initiatives bring significant contribution to a local community's livelihood; ensuring that natural resource management institutions are put in place as well as their means of enforcement; the huge initial financial and human resource investments, especially during the initial stages of the ACM process; and how to organize learning and deliberative processes in an unstable socio-economic and political climate.

We draw a number of lessons from our ACM experiences. First, from our experiences, though difficult to implement due to its numerous embedded concepts, ACM is doable in practice and has potential to produce positive changes to both the status of resources and the well-being of resource based communities. Sustainability of these outcomes is however dependent on how the ACM facilitators deal with challenges encountered in implementation, for instance: (a) putting in place effective institutions to support management efforts. Institutions in this case refers to both the rules for resources use and organizations (these must be downwardly accountable and transparent) responsible for resource management. Such rules need to be recognized by outside law enforcing agents and must have means of enforcement. And (b) dealing with more fundamental and underlying causes of natural resource management problems. This includes radically challenging unequal distribution of power and empowering the marginalized. In stable socio-economic and political climates where there is political will by the national government to support resource management efforts, it is possible, for instance, for ACM facilitators to facilitate dialogue among stakeholders on the value of contested resources. In Cameroon, when the ACM team realized that land shortage was the underlying forest management problem, the team facilitated platforms for stakeholders to discuss and resolve the problem. After the discussions and dialogue, stakeholders finally agreed to come up with a new land use plan for the area. This included the creation of a buffer zone within the forest to provide additional farm land to local communities (Jum, 2003; Jum *et al.*, 2009). Agro-forestry activities were also allowed within the zone and the forestry department provided the new agro-forestry management skills required. Institutional structures were also put in place to support the implementation of agreements, and those who did not abide by the rules were reported and arrested (Jum *et al.*, 2009).

Second, we also learnt that project time frame is a crucial factor in the success and failure of learning and collaboration based approaches. Such projects need to be given longer time frames, (probably from five to ten years) than other projects so that stakeholders have adequate time to implement their actions, monitor them and learn from their impacts. In situations where various groups of stakeholders are managing different parts of the same resource (like in Mafungautsi with different user groups, with each focusing on addressing problems related to their resource), more time is required to enable stakeholders to share their activities and outcomes with each other as well as monitor their combined impact to the resource in question.

Third, as the ACM approach embodies several key concepts, there is risk that some crucial elements (such as dealing with issues of power and politics and setting up effective resource management institutions and their means of enforcement) are forgotten as facilitators focus on other issues that seem important to them. There is therefore a need for building the capacity of ACM facilitators in the different ACM concepts so that they are aware of them and able to engage them in the process. Having a multi-disciplinary ACM implementing team is therefore a plus in ensuring that the various concepts are considered in the process.

Notes

1 Decentralization has been recommended as a way to reduce inefficiency (Ostrom *et al.*, 1993). It is a process by which entrustments (including regulatory and decision making powers, responsibility for planning and implementation and administrative capacity) are transferred to local groups (Nemarundwe, 2003; Ostrom *et al.*, 1993; Mandondo, 2000; Agrawal and Gibson, 1999). There are two forms of decentralization: (a) devolution – when entrustments are completely transferred to local groups; and (b) deconcentration – in which entrustments are transferred to local groups that remain accountable to the central government.

2 The World Conservation Strategy (WCS) was commissioned by the United Nations Environment Programme (UNEP) which, together with the World Wildlife Fund (WWF), provided the financial support for its preparation and contributed to the evolution of its basic themes and structure.

3 The collaborative management approach has made headway as a result of a growing consensus that without some basic agreements among affected parties about resource use, sustainability would never be achieved and degradation would continue.

4 The adaptive management approach is an iterative process providing regular analysis of progress towards established objectives and adjustment to changing circumstances.
5 Zimbabwe has been divided into five natural regions (NR)/agro-ecological zones: NRI receives 1050mm+ rainfall per annum with some rain in all months of the year; NRII receives 700–1,050mm rainfall per annum with rainfall only in summer; NRIII receives 800–700mm rainfall per annum and is characterized by high temperatures, infrequent heavy rains and seasonal droughts; NRIV receives 450–600mm rainfall per annum and is characterized by high temperatures and seasonal droughts; and NRV receives less than 500mm rainfall per annum and this rain is erratic (Vincent and Thomas, 1962).
6 Mafungautsi means 'a place that smokes', referring to the clouds that hang over the plateau due to its altitude in relation to the surrounding areas (Matose, 2002).
7 This is the most northerly extension of the Kalahari Sands geological surface in Zimbabwe (Vermeulen, 1994).
8 The indigenous forests of Zimbabwe fall under three tenurial categories: forests in communal areas; forests in protected areas and state land; and forests in large scale commercial farming areas (McNamara, 1993)
9 Miombo forest is the most extensive tropical seasonal forest and dry forest formation, covering substantial regions of south and central Africa (Campbell *et al.*, 1996; Frost, 1996). The regions where Miombo forest is found receive more than 700mm of rainfall and are often characterized by nutrient poor soils.
10 In Zimbabwe there are two major political parties, the ruling ZANU-PF party and the Movement for Democratic Change (MDC) opposition party. These two parties have similar structures at district and communal levels. In communal areas, members of political parties hold the following positions at both the ward and village: ward/village chair; vice chair; ward secretary; vice secretary; treasurer; and committee members.
11 The average literacy rate for Gokwe was low, with only 39 per cent of the people having received formal education in 1992 (Central Statistics Office, 1992).

References

Aamodt, A. M. (1991). Ethnography and epistemology: generating nursing knowledge. In J. M. Morse (ed). *Qualitative Nursing Research: a contemporary dialogue*, pp. 40–53. Newbury Park, CA: Sage

Agrawal, A. and Gibson, C. C. (1999). Enchantment and disenchantment: the role of 'community' in natural resource conservation. *World Development*, 27 (4), 629–649.

Ahern, K. A. (1999). Ten tips for reflective bracketing. *Qualitative Heath Research*, 9 (3), 407–411

Alpert, P. (1996). Integrated conservation and development projects. *Bio-Science*, 46, 845–855

Barrett, C. B. and Arcese, P. (1995). Are integrated conservation and development programmes sustainable? On the conservation of large mammals in sub-Saharan Africa? *World Development*, 23, 1073–1085.

Borrini-Feyerabend, G. (1996). *Collaborative Management of Protected Areas: tailoring the approach to the context*. Gland, Switzerland: IUCN

Bourdieu, P. (1992). *Invitation to a Reflexive Sociology*. Chicago: University of Chicago Press.

Bromley, D. W. and Cernea, M. M. (1989). The management of common property natural resources: some conceptual and operational fallacies. *World Bank Discussion Papers* 57.

Burr, V. (1995). *An Introduction to Social Constructionism.* London: Routledge.

Campbell, B. M., Frost, P. and Byron, N. (1996). Miombo woodlands and their use: overview and key issues. In B. Campbell (ed). *The Miombo in Transition: woodlands and welfare in Africa*, pp. 11–57. Cifor, Bogor, Indonesia, 266p.

Central Statistics Office (1992). *Zimbabwe National Census Report.* Harare: Government Printers.

Chauveau, J. P. and Richards, P. (2008). West African insurgencies in agrarian perspectives: Cote d'Ivoire and Sierra Leone compared. Wageningen University and Research Center Publications.

CIFOR (2000). The ACM Researcher's Handbook: a work in progress. Draft 2, 5 October.

Cleaver, F. (2000). Moral ecological rationality: institutions and the management of common property resources. *Development and Change*, 31, 361–383.

Cooke, W. and Kothari, U. (eds) (2001). *Participation: the new tyranny*? London: Zed Books.

Côté, J. and Levine, C. (2002). *Identity Formation, Agency and Culture: a social psychological synthesis.* Mahwah/New Jersey/London: Lawrence Erlbaum, 2002.

Cutcliffe, J. R. (2003). Reconsidering reflexivity: introducing the case for intellectual entrepreneurship. *Qualitative Health Research*, 13, 136.

Diaw, M. C., Aseh, T. and Prabhu, R. (eds) (2009). *In Search of Common Ground: adaptive collaborative management in Cameroon.* Center for International Forestry Research, Bogor, Indonesia.

Dzingirai, V. and Breen, C. (eds) (2005). *Confronting the Crisis in Community Conservation: case studies from Sothern Africa.* Center for Environment, Agriculture and Development, University of KwaZulu-Natal, South Africa.

Emerton, L. (2001). The nature of benefits and the benefits of nature: why wildlife conservation has not economically benefited communities in Africa. In D. Hulme and M. Murphree (eds). *African Wildlife and Livelihoods: the promise and performance of community conservation*, pp. 208–226. Oxford: James Currey.

Folke, C., Carpenter, S., Elmqvist, T., Gunderson, L., Holling, C. S. and Walker, B. (2002). Resilience and sustainable development: building adaptive capacity in a world of transformations. *Ambio*, 31, 437–440, Royal Swedish Academy of Science.

Franks, P. and Worah, S. (2003). New directions for integrated conservation and development: a CARE/WWF perspective. Gland: WWF International Improving Conservation and Development in Eco-regions Programme: *Issues in Natural Resource Management*, 1, 2–3.

Freire, P. (1970). *Pedagogy of the Oppressed.* New York: Herder and Herder.

Frost, P. G. H. (1996). The ecology of miombo woodlands. In B. Campbell (ed). *The Miombo in Transition: woodlands and welfare in Africa*, pp. 11–57. Cifor, Bogor, Indonesia, 266pp.

Frost, P. G. H. and Mandondo A. (1999). Improving rural livelihoods in semi-arid regions through management of micro catchments. IES working paper No. 12.

Gibson, C. C. and Marks, S. A. (1995). Transforming rural hunters into conservationists: an assessment of community-based wildlife management programmed in Africa. *World Development*, 23, 941–957.

Giddens, A. (1990). *The Consequences of Modernity.* Cambridge: Polity Press.

Government of Zimbabwe (1996). *Forest Act.* Chapter 19:05. Revised Edition. Government printer, Harare.

Guijt, I. (2007). *Negotiated Learning: collaborative monitoring in resource management.* Washington DC: Resources for the Future (RFF).

Guillemin, M. and Gillam, L. (2004). Ethics, reflexivity and ethically important moments in research. *Qualitative Inquiry*, 10, 261.

Hannah, L. (1992). *African People, African Parks: an evaluation of development initiatives as a means of improving protected Area Conservation in Africa.* Washington, DC: Biodiversity Support Program.

Hertz, R. (1997). Introduction: reflexivity and voice. In R. Hertz (ed). *Reflexivity and Voice*, pp. vi–xviii. Thousand Oaks, CA: Sage.

Hickey, S. and Mohan, G. (2005). Relocating participation within a radical politics of development. *Development and Change*, 36 (2), 237–262.

Hope, A. and Timmel, S. (1995). *Training for Transformation: a handbook for community workers* (rev. ed.), Books I–III. Gweru, Zimbabwe: Mambo Press.

Huberman, A. M. and Miles, M. B. (1994). Data management and analysis methods. In N. K. Denzin and Y. S. Lincoln (eds). *Handbook of Qualitative Research*, pp. 428–445. Thousand Oaks CA: Sage.

ITDG (1997). *Chivi Food Security Project: a process approach.* Harare, Zimbabwe: ITDG publication.

IUCN (1980). *World Conservation Strategy.* Gland, Switzerland: International Union for the Conservation of Nature and Natural Resources.

Jiggins, J. and Rolling, N. (2000). Potential and limitations for ecological governance. *International Journal of Agricultural Resources, Governance and Ecology*, 1, 28–42

Jum, C. (2003). New thinking: local actors in a buffer zone management enterprise in Cameroon. ACM news 4 (1), Cifor, Bogor, Indonesia.

Jum, C., Abega, M. and Bengono, F. (2009). Action research as a strategy for collaborative management in Ottotomo. In M. C. Diaw, T. Aseh and R. Prabhu (eds). *In Search of Common Ground: adaptive collaborative management in Cameroon.* Bogor, Indonesia: Center for International Forestry Research.

Kaimowitz, D. and Shell, D. (2007). Conserving what and for whom? Why conservation should help meet basic human needs in the tropics. *Biotropica*, 39 (5), 567–574.

Katerere, Y., Moyo, S. and Mujakachi, L. (1993). The national context: land, agriculture and structural adjustment, and the Forestry Commission. In P. N. Bradley and K. McNamara (eds). *Living with Trees: policies for forest management in Zimbabwe*, World Bank Technical Paper No. 210. Washington DC: The World Bank.

Kayambazintu, D., Matose, F., Kajembe, G. and Nemarundwe, N. (2003). Institutional arrangements governing natural resource management of the Miombo woodland. In G. Kowero, B. M. Campbell and U. Rashid Sumaila. *Policies and Governance Structures in Woodlands of Southern Africa.* Bogor, Indonesia: CIFOR Publication.

Kiss, A. (1990). *Living with Wildlife: wildlife resource management with local participation in Africa.* Washington DC: World Bank. Technical Paper No. 130.

Leach, M. (2002). Plural perspectives and institutional dynamics: challenges for community forestry. In J. A. E. Oglethorpe (ed). *Adaptive Management: from theory to practice.* Gland, Switzerland and Cambridge, UK: UUCN. VI and 166pp.

Logan, B. I. and Moseley, W. G. (2002). The political ecology of poverty alleviation in Zimbabwe's Communal Areas Management Programme for Indigenous Resources (CAMPFIRE). *Geoforum*, 33, 1–14.

Mandondo, A. (2000). *Forging Democratic Resource Governance Systems from the Relic of Zimbabwe's Colonial Past.* Berkeley, CA: Institute of International Studies.

Mapedza, E. (2007). Forestry Policy in colonial and postcolonial Zimbabwe: continuity and change. *Journal of Historical Geography*, 33, 833–851.

Mapedza, E. and Mandondo, A. (2002) *Co-management in the Mafungautsi State Forest Area of Zimbabwe: what stake for local communities?* Working Paper no. 5, Working Paper Series on Environmental Governance in Africa. Washington DC: World Resources Institute.

Mason, J. (1996). *Qualitative Researching.* London: Sage

Matose, F. M. (2002). Local people and reserved forests in Zimbabwe: what prospects for co-management? Ph.D. thesis, Development Studies, University of Sussex.

Maturure, P. N., Mazambani, D., Mujakachi, L. R., Phiri, C. and Sibanda, S. (1994). *Community-based Natural Resource Management Project: Pilot Forest Grazing and Wildlife Management Program in Mafungautsi Forest – preliminary results of the feasibility study.* Harare: Forestry Extension Services, Forestry Commission internal document, 44pp.

McNamara, K. (1993). Key policy issues. In P. N. Bradley and K. McNamara (eds). *Living with Trees: policies for forestry management in Zimbabwe.* World Bank Technical Paper 210. Washington DC: World Bank.

Mudavanhu H., Undated. Report of the ecological survey for Mafungautsi State Forest, Zimbabwe.

Mudekwe, J. (2003). Adaptive co-management project: Mafungautsi Forest Reserve. Baseline study of the forest resources and collaborative monitoring mechanisms guidelines. Unpublished.

Mutimukuru, T. and Kozanayi, W. (eds) (2005). Keeping the momentum: sustaining projects/processes when outside support has been withdrawn. *ACM News Zimbabwe*, August 2005, 2 (2). A newsletter published by Mirror Publications for CIFOR Zimbabwe.

Mutimukuru, T., Kozanayi, W. and Nyirenda, R. (2007). Initiating a dynamic process for monitoring in Mafungautsi State Forest, Zimbabwe. In I. Guijt, *Negotiated Learning: collaborative monitoring in resource management.* Washington DC: Resources for the Future (RFF).

Mutimukuru-Maravanyika, T. (2010). Can we learn our way to sustainable forest management? Adaptive collaborative management in Mafungautsi State Forest, Zimbabwe. Ph.D. thesis, Wageningen University and Research Centre, the Netherlands.

Nemarundwe, N. (2003). Negotiating resource access: institutional arrangements for woodlands and water use in Southern Zimbabwe. Ph.D. thesis, Swedish University of Agricultural Sciences, Agraria 408, Uppsala.

Oates, J. F. (1995). The dangers of conservation by rural development: a case study from the forests of Nigeria. *Oryx*, 29: 115–122.

Ostrom, E. (1990). *Governing the Commons: the evolution of institutions for collective action.* Cambridge: Cambridge University Press.

Ostrom, E., Schroeder, L. and Wynne, S. (1993). *Institutional Incentives and Sustainable Development: infrastructure policies in perspective.* Boulder, CO: Westview Press.

Scott J. C. (1985). *Weapons of the Weak: everyday forms of peasant resistance.* New Haven: Yale University Press.

Vermeulen, S. J. (1994). Consumption, harvesting and abundance of wood along the boundary between Mafungautsi and Gokwe Communal Area, Zimbabwe. M.Sc. thesis, University of Zimbabwe, Harare.

Vermeulen, S. J. (2000). *Setting the Context of the Mafungautsi (Zimbabwe) Project in a Criteria and Indicators Framework.* Harare: CIFOR ACM Programme, Typescript 53pp.

Vermeulen, S. J., Campbell, B. M. and Matzke, G. E. (1996). The consumption of wood by rural households in Gokwe Communal Area, Zimbabwe. *Human Ecology*, 24 (4).

Vincent, V. and Thomas, R. G. (1962). *An Agricultural Survey of Southern Rhodesia.* Part I Agro-ecological survey. Harare: Government Printers.

Wells, M., Brandon, K. and Hannah, L. (1992). *People and Parks: linking protected area management with local communities.* Washington, DC: World Bank.

Wilshusen, P. R., Brechin, S. R., Fortwrangler, C. L., and West, P. C. (2002). Reinventing the square wheel: critique of a resurgent 'protection paradigm' in international biodiversity conservation. *Journal of Society and Natural Resources*, 15, 17–40.

Wollenberg, E., Edmunds, D. and Anderson, J. (2000). *Anticipating Change: scenarios as a tool for adaptive forest management: a guide.* Bogor: Centre for International Forestry Research.

Zingore, S., Manyame, C., Nyamugafata, P. and Giller, K. E. (2005). Long-term changes in organic matter of woodland soils cleared for arable cropping in Zimbabwe. *European Journal of Soil Science*, 56, 727–736.

7 Learning to improve livelihoods

Applying an adaptive collaborative approach to forest governance in Nepal

Mani Ram Banjade

Introduction

This chapter outlines my experiences in catalyzing an adaptive collaborative approach (ACA) in Nepal's forestry sector from 2000 to 2011. During this period, I was part of a series of action and research projects as field researcher, team leader for the ForestAction component, and team coordinator for another project.[1] These projects aimed at improving community forestry policies and practices for forest conservation, livelihoods improvement, contribution to poverty reduction, and promotion of inclusion and equity. Although these were the separate projects, they all shared a concern for social learning and collaborative governance. In this chapter, I explain how different strategies and tools of inclusion, deliberation and learning were applied to improve collaborative action and learning among forest stakeholders from local community to national levels, and identify where we succeeded, and where we failed and why.

The ACA resonated with the notions of 'action learning', 'deliberative processes' (Chambers 1997), 'deliberative planning' (Forester 1999), adaptive collaborative management (Fisher *et al.* 2007), and others, who advocate collaboration and learning in resource management and development processes. We went beyond an emphasis on participation and local control, and actively catalyzed interactions among multiple sources of knowledge, as well as facilitated reflexive learning among actors at different levels of governance. Our strategy was guided by the concern for empowering the disadvantaged actors in the management process.

During the period of 2000–2011, I was involved in a series of four action and research projects that comprised three phases of ACA. The 'deliberative planning and self-monitoring' (2000–2002)[2] and 'piloting self-monitoring'

(2003–2004) together formed the formative stage of the approach, which had a focus on transforming Community Forest User Group (CFUG) level decision-making processes to enhance inclusion and equity. A CFUG is a legally recognized local institution, which is established in relation to management of a patch of forestland called community forest (CF) with members presumably comprising the traditional users of the CF. The 'piloting self-monitoring' project was largely an application of the lessons of an earlier project in a different project area, so it is not discussed in this chapter separately. The second phase, 'ACA at multiple levels' (2004–2007), was instrumental in linking different levels of governance, i.e. CFUG, meso (district and/or sub-district level actors and institutions), and national levels. The third and, in this chapter the final, phase – 'innovation systems' (2008–2011) – focused on applying earlier research results into effective use by forging linkages across wider sectors and actors for efficient and equitable resource governance and innovation. My experiences presented here were primarily with ForestAction Nepal (2003–2011) and with the Center for International Forestry Research (CIFOR) (2000–2001), which was an international partner during 2004–2007.

The principal thrust of this chapter is to discuss how ACA was conceptualized as an idealized mix of different traditions to facilitate inclusive, deliberative and co-learning processes, and how these idealized notions were distorted in practice because of ongoing power relations, struggles and movements. By reviewing three phases of ACA-based initiatives, I identify several challenges and issues that we confronted in catalyzing ACA processes, and identify how we dealt with the issues and challenges and, if any, with what level of success.

Unlike a conventional best practice story, this chapter details my experience throughout the trajectory of ACA, presents empirical evidence of discrepancies between commonly held assumptions and actual practice, and reflects on the ways in which ACA can be put into effective use in the context of community-based natural resource management. With the benefit of hindsight, what I find critical is that ACA facilitators have to encounter enormous resistance and challenges in multiple institutional domains, and should be prepared to muddle through vested interests and power dynamics at different levels and institutional milieus, while also managing frustrations and excitements within the ACA team.

The rest of the chapter is organized in four sections. The section that follows outlines the underlying assumptions and expectations of adaptive collaborative approach that we applied in the field, over three stages or phases, followed by a phase-wise recounting of issues and challenges confronted. The section preceding to conclusion and the way forward analyzes and discusses the issues and challenges of applying ACA into the context of community forestry and explores new frontiers of ACA in the context of natural resource management.

Assumptions and expectations in applying adaptive collaborative approaches in a complex world

Policies and strategies of forest ecosystem management have moved away from top-down, linearly conceived views towards more collaborative approaches. One turning point for this shift in international forestry discourse was the adoption of 'forestry for people' as the theme of the Eighth World Forestry Congress in 1978 (Leslie 1987: xi; also see Westoby 1987); this focus appeared again as the theme of the International Year of Forests 2011 (UN 2011). Coinciding with the Eighth World Forestry Congress was the landmark publication of *Forestry for Local Community Development* by FAO (1978), which played a key role in re-establishing the stakes of local people in forest management. Similarly, since then forestry has moved beyond utilitarian-based traditional forestry and economics (Bengston 1994) or sustained timber yield (Kant 2007), and recognized its diverse environmental and socio-cultural values and services. With the expanded scope of forestry, a number of new actors emerged in the sector, namely non-governmental organizations (NGOs), community based organizations (CBOs), networks and markets.

The addition of social aspects in traditional forestry involved the integration of additional complexity, as management has to take into account the complex interaction of ecological, environmental and social factors. These complex and dynamic relationships pose uncertainty and unpredictability about the outcomes of an intervention (McDougall and Braun 2003). In this context of uncertainty and dynamism, Holling (1995) and Lee (1999) suggested adaptive approaches to ecosystem management that placed a greater focus on dynamic learning and adjustment. These approaches are only partially useful in integrating social and biophysical systems as they failed to embrace adequately the social and political contexts of the intervention. The approaches that they suggested had little role for the local dwellers of the ecosystem. Fals-Borda and Rahman (1991) recognized that local people have knowledge of their environment accumulated through generations. They argued that participatory action research (PAR) could break the monopoly of the external experts-driven top-down approach of knowledge generation by engaging and empowering local people in the overall development process. Several scholars have demonstrated a need to recognize the local community's knowledge and their role in development and knowledge generation processes (Brokensha *et al.* 1980; Berkes 1993; Thapa *et al.* 1995; Walker *et al.* 1995; Blaikie *et al.* 1997; Escobar 1998; Messerschmidt and Hammett 1998; Shivaramakrishnan 2000; Finger and Schuler 2003; Mahesh and Hemlata 2004; Timsina *et al.* 2004). Justification of community forestry in Nepal was largely based on the recognition of indigenous knowledge and management practices on forestry and its intricate links with rural livelihoods (Gilmour and Fisher 1991).

Despite successful attempts for securing rights of local people in forest governance, and the deliberative frontiers that have increased as a result, many argue that the process of integrating traditional wisdom, such as that of local communities in protecting their forests, into professional forestry- and science-driven knowledge, often undermines local knowledge and institutions, and local ways of knowing (Agrawal 1995; Bryant 1998; Escobar 1998; Nightingale 2005; Giri and Ojha 2010). For example, Ojha (2006: 142) points out the deliberative gap between citizens and forest bureaucrats, and shows that the paucity of deliberation constrains democratic forest governance.

Nepal's Community Forestry (CF) policy is widely recognized as institutionally unique and progressive in the world arenas of community-based natural resource management (Kumar 2002; McDougall *et al.* 2007). In the post-1990 Nepal with democracy and pluralism, a growing number of government institutions, NGOs, private sector and civil society groups became active in CF in Nepal (Banjade *et al.* 2006a) (see Table 7.1 for the key categories of actors and institutions operating at different levels of CF governance). The stakeholder landscape in CF has also expanded due to increased demand for services such as those linked to value chain promotion.

Table 7.1 Actors and institutions at different CF governance levels

Governance level	*Key actors and institutions*
Local (CFUG)	Autonomous CFUG Community-based organizations (women's groups, saving credit groups, interest groups) Forest-based entrepreneurs
Meso (district or sub-district)	District Forest Office (DFO) and its subsidiary wings (Area Forest Office, Range Post) Local governments (District Development Committee (DDC) and Village Development Committee (VDC)) and other line agencies Networks such as Federation of Community Forest Users, Nepal (FECOFUN) NGOs, INGOs Donor projects Forest-based entrepreneurs
National	Ministry of Forest and Soil Conservation, Department of Forest Other government line agencies – Ministries and Departments FECOFUN, NGOs Donors, bilateral projects, INGOs Private sector
International	Global Alliance of Community Forestry (GACF) International non-governmental organizations (INGOs) International treaties, agreements Donors Multinational companies

Over the past three decades of its history, CF has made notable contributions to forest conservation and local community development. In the recent years, however, there have been increasing concerns over limited contributions to livelihoods, equity and poverty reduction (Malla 2000; Malla 2001; Timsina 2003; Khadka and Schmidt-Vogt 2008; McDougall *et al.* 2008). In this context, one of our major learning questions in all the research projects was: how and to what extent can adaptive collaborative approaches (ACA) overcome fundamental relations of power that sustain exclusion and inequity in natural resource management?

Our motivation to engage actively in ACA-oriented initiatives was also reinforced by the continuation of techno-bureaucratic control of CF practices despite successful cases of devolution and community empowerment. Studies have demonstrated that, while CFUGs are legally recognized autonomous institutions representing the interests of local communities, in practice they are forced to retain patron-client relationships with the state officials (Pokharel 1997; Malla 2001; Ojha 2006; Paudel *et al.* 2008). For this reason, local communities are forced to seek prior approval from the government forestry authority on forest decisions for which communities are legally empowered. However, there are also some examples whereby CFUGs have successfully challenged such techno-bureaucratic hegemony, especially where suitable facilitative support is available (Ojha *et al.* 2010). Moreover, we also found the possibility for collaboration between different CF actors at district and sub-district levels, which can create conducive environment for more effective interaction between government and civil society in democratic forest governance (Banjade *et al.* 2007).

In this context, our hope was, and still is, that ACA-based support to CFUGs could empower them to interact effectively with external stakeholders, as well as help create more inclusive and deliberative institutional procedures internally within the CFUGs. Such approach, we hoped, could enable the CFUG to fulfil mandatory legal provisions as well as draw much needed resources, knowledge and skills from external organizations.

Our approach sought to integrate active reflection and learning in the lifeworld of CF in Nepal, which involved diverse groups of forest dependent people as well as other stakeholders in the complex and dynamic socio-ecological system of the post-1990 Nepal of multi-party democracy and pluralism. ACA integrates this and other concepts that are discussed below (see Figure 7.1).

Similarly, we also hold that when various actors of CF are engaged in a conscious learning process, i.e. social learning, they can be more effective in understanding the existing social and ecological systems. In such a process, the actors would also engage in action and reflection, to learn from their engagement, be responsive to the complex and dynamic social and ecological systems, and improve both practices and processes. One basic assumption is that diverse actors, holding diverse knowledge or worldviews, need to collaborate for effective CF governance and management. Community

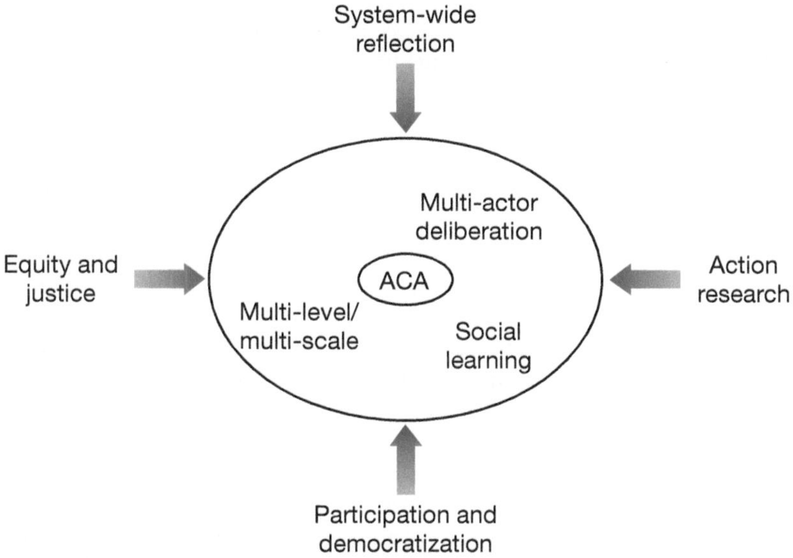

Figure 7.1 Conceptual framework of ACA.

forestry in Nepal has offered incentives for collaborative co-learning processes mainly because: i) they can utilize the knowledge from other actors to increase their efficiency and effectiveness, beyond those that could be realized while working alone; ii) they can advance their own ideas and interests, and legitimize them through the endorsement of others; and iii) they can strongly legitimize their existence or identity by their participation in, and recognition by other actors involved in CF governance.

Over time our approach placed an emphasis on multi-scalar issues in addition to inclusion, deliberation and learning in community forestry of Nepal. We found that the success of CF depended upon the governance at community, meso and national levels as well as deliberative links between these levels (McDougall *et al.* 2007). Multi-stakeholder deliberation formed the basis for these linkages, as well as for framing the governance processes at each level. In theory, an ideal deliberative process gives room for the empowerment of those at the margins because reasoned and domination-free communication is at the heart of deliberation (Habermas 1987; Dryzek 2000; Ojha 2008). In an idealized deliberative governance situation, participants could present their reasons for making certain choices against other alternatives, to help arrive at more rational decision outcomes (Habermas 1996), and interaction and exchanges induce 'reflection on preferences, values and interests in a non-coercive fashion' (Dryzek 2000: 76). In practice, however, the CF governance process encounters the prevailing power asymmetries among various actors resulting into contestable exclusion and inequity. As ACA recognizes this fact, inclusion – i.e. participation and voice – forms an inherent guiding principle of ACA in practice, which should foster equity and justice.

Other normative principles endogenous to ACA include action research and system-wide reflection, i.e. all the interventions in the CF should have a component of action and learning – reflection on the implications to various components of a system because of an intervention – and these should form a part of planning, implementation and (self-) monitoring. In the process, empowerment and external facilitation are generally aimed at releasing different actors from their prior norms, i.e. disturbing existing imbalances between actors so that what Hendriks (2009) describes as coercive and generative power can have the positive space in the process of deliberation. While coercive power of the marginal groups helps increase their space in the deliberative frontiers, generative power provides reflexivity among the actors to promote deliberation.

Informed by these theoretical developments and methodological innovations in policies and practices of community-based natural resource management, and their limitations, we applied an adaptive collaborative approach (ACA) in the context of Nepal's community forestry (CF). The following section provides the details of the trajectory of applying ACA in the community forestry context in Nepal.

Overview of adaptive collaborative projects

Nepal's CF sets an appropriate context for applying ACA. The legal framework provides space for various actors and institutions an opportunity to collaborate across a range of issues including effective forest management, livelihood improvement, and promoting social inclusion and equity.

We began by considering CFUGs as the key actors to partner with in facilitating adaptive and collaborative approaches since they were the managers of, and directly dependent on, the forests. A CFUG is a legal entity to manage a designated area of forest. Once a CFUG obtains approval from the government, it often becomes a strong and vibrant local institution. It aims to pursue forest sustainability and enhance livelihoods of local people (Pokharel *et al.* 2007). However, in practical arena, a CFUG has to confront a range of challenges such as effective management of forest resources, equitable distribution of benefits within the CFUG, engaging with market forces, and complying with regulatory standards and government directives. Various other institutions at different levels, such as Federation of Community Forest Users Nepal (a CFUG network), NGOs and donors also support CFUG efforts.

The Forest Act (1993) and Forest Regulations (1995) have defined a CFUG as a legal entity with institutional autonomy and perpetual succession. In practice, however, the autonomy of the CFUG is curtailed during the negotiation with the government forestry authority during preparation of constitution and forest management plans (Paudel *et al.* 2008; Ojha *et al.* 2009a).

According to the existing regulatory provisions, a CFUG is formed after identification of traditional users of a forest area and their willingness to assume forest management responsibility. Then, household members of the CFUG elect an executive committee (EC), and other thematic subcommittees as appropriate, to oversee the day-to-day activities of the CFUG, as well as to implement and monitor CF and community development activities. General assembly is the most powerful body of a CFUG, which can elect and reshuffle the EC, endorse budgets and plans presented by the EC, and respond to unforeseen circumstances. A constitution and an operational plan for forest and community development activities are the basis for the functioning of a CFUG. Thus any new initiatives should be taken within the provisions made in these documents.

Within this context, the first phase of the ACA (2000–2004) principally aimed to address the issues of exclusion of marginalized people, especially women and 'low' castes (or *dalits*)[3] and/or classes, from internal planning and decision-making processes within CFUGs, and the limited livelihood benefits from CF due to its protectionist orientation. We catalyzed improved internal planning and decision-making processes within CFUGs, focusing on more active and equitable use of forest resources. In this phase, the key assumption for applying ACA was that human well-being and resource conservation could be achieved by facilitating inclusive, deliberative and learning-based processes within the CFUGs.

Four CFUGs from two hill districts of Nepal were identified as case studies for ACA processes (see Figure 7.2 for districts and number of CFUGs involved in each district in different phases of the research). The activities were facilitated by external researchers, and were oriented towards recognizing heterogeneity, collective visioning and goal-setting, indicator-based self-monitoring, collective learning, and ensuring equitable political space and sharing of CF benefits. During ACA facilitation in the CFUGs, nodes of learning cycles such as inclusive planning, implementation and reflection formed the part of the CFUG planning, decision-making, implementation and evaluation processes.

The second phase (2004–2007) of ACA was built on some of the lessons from the earlier phase. Despite the ACA's success in promoting inclusive governance at CFUG level, with some visible improvements on equity and democratization, we confronted a challenge of scaling up the lessons and the innovation processes. A number of issues were identified which needed further study and innovation, such as developing effective linkages of communities with higher governance levels, developing local capacity, and broadening the geographical coverage (to cover areas from Terai which has different resource and socio-cultural contexts than in the hills of Nepal where our work was concentrated in the first phase).

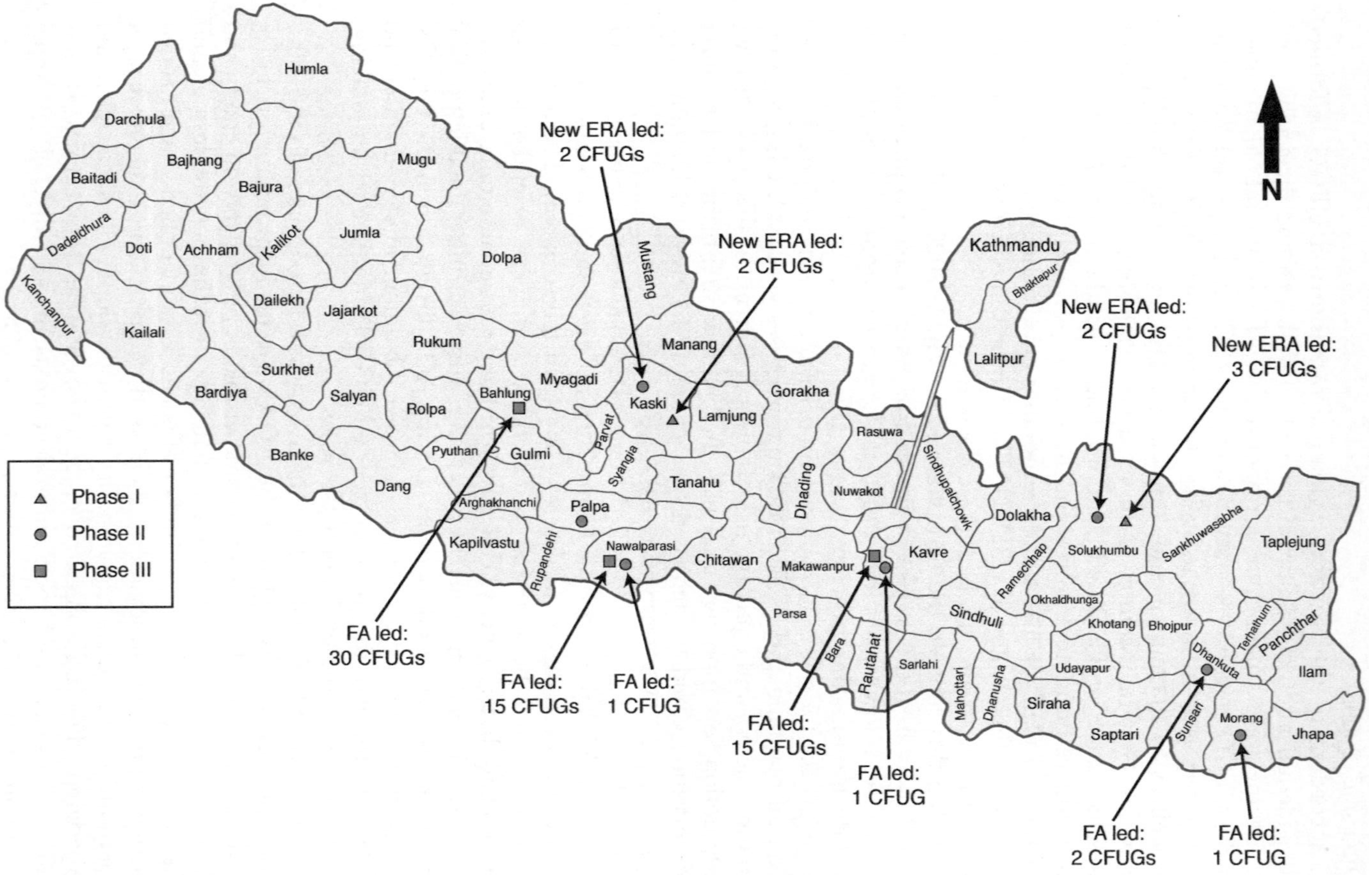

Figure 7.2 Districts and number of CFUGs involved in each district in different phases of the action research.

To address these issues, the strategy in this phase was that local volunteers were selected as ACA facilitators rather than involving external researchers as in the first phase. They were trained and mentored to facilitate ACA processes at community and meso levels. Similarly, the need of broadening the geographic coverage of ACA application was addressed by including several other districts and CFUGs from each of Nepal's geographical regions, i.e. Tarai, middle-hills and mountain (7 districts and 11 CFUGs as compared to 2 districts and 4 CFUGs in the previous phase). Likewise, to forge effective deliberative linkages and policy learning, this phase worked at community, meso and national levels simultaneously, and at the interface of these levels with different stakeholders.

The two phases were also different in terms of institutional engagement and partnerships. For example, while Phase 1 was designed by the Center for International Forestry Research (CIFOR), and three Nepali partners (ForestAction, New ERA and Kaski Team) were involved later for project implementation, Phase 2 was designed jointly by CIFOR and two Nepali partners, namely ForestAction and New ERA. Similarly, in the first phase there was a limited opportunity to forge linkages of CFUGs with actors and institutions at other levels, i.e. meso and national. Thus this aspect was explicitly addressed in the second phase, which anticipated that the problems of communication and deliberation within and across levels could be addressed by working at multiple levels and at the interfaces between them.

In the third phase (2008–2011), the aim was to link local level learning processes with national policy, and also to catalyze entrepreneurial innovations on forest products. For this, we expanded partnership base to facilitate ACA and sought to combine the strengths of various organizations including researchers, advocacy groups, business groups, and the media. Particularly, a coalition of five different organizations representing NGOs, CFUG networks, forest based enterprises, the media and academic institutions steered the project. The coalition worked with more than 60 CFUGs in three districts of Nepal and with a number of meso-level actors such as NGOs, networks, government agencies (District Forest Office, local governments), entrepreneurs and the media.

Adaptive collaborative approaches in practice

This section provides details of the key experiences of each of the three phases of applying ACA in the context of community forestry of Nepal.

Phase 1 (2000–2002): action research on facilitating community level planning and self-monitoring processes to promote equity and social learning

As part of the Center for International Forestry Research's (CIFOR) Adaptive Collaborative Management (ACM) project, we selected four CFUGs (two CFUGs each from two districts) from a sample of nine potential CFUGs (comprising five CFUGs from one district and four from another). These were selected in consultation with CFUG and district actors primarily on the basis of local socio-political heterogeneity and resource condition, as well as their willingness to participate. We visited each of the nine shortlisted CFUGs and collected data before making a final selection. Upon selection, we visited the selected four CFUGs several times, organized joint meeting with their executive committees and other key individuals,[4] and explained our research objectives and strategies. We explained then, as well as on several subsequent occasions, that the project had research focus, that local community members, if willing to participate, needed to commit some of their time to the project on a voluntary basis, and that the project does not have any provision for material incentives. In all cases, after hearing from us about the project, the CFUG representatives agreed to be a part of the project.

As a part of the 'traditional' research elements of the project, the project team conducted baseline studies covering socio-economic, biophysical and policy aspects and historical trends. To make these more 'objectively verifiable', a group of CF experts at the national level interpreted the CIFOR-developed criteria and indicators (C&I) for sustainable forest management (CIFOR 1999), to contextualize them for Nepal and provide the framework for C&I-related data collection. While researchers from within the team conducted a socio-economic analysis, a team of consultants undertook a biophysical assessment of forest resources. While the baseline studies did involve participatory methods, this stage was researcher-driven and was, as such, distinct from the 'participatory action research' (PAR) component, in which local people were to be actively engaged.

To conduct PAR for governance innovations, the project team developed a plan with the CFUG executive committees as to how they could promote more democratic and inclusive planning processes within the CFUGs. This led to a series of hamlet-level meetings, as well as separate meetings with women, *dalits*[5] and ethnic groups to identify and understand key issues from their perspectives, and to explore their opinions on possible solutions.

Information required for the research was derived from researchers' ongoing reflective documentation of the processes and outcomes, including cycles of planning, implementation, reflection and observation. After being oriented to take the reflective notes, research team members participated in all CFUG processes. These notes later became a valuable source of information. Researchers carried out a final assessment in each site, using a common framework across sites.

The most important experiences and insights of applying ACA in this phase are as follows:

1. Being political about forestry

When I joined the project in mid-2000, my ontology (perspectives, beliefs, disciplinary orientation) was shaped by two, at times conflicting, cultural traditions. First, my interest and involvement in student politics, itself linked to larger popular politics, motivated me to respect and protect the rights of people in relation to both larger political matters and control over local matters, including forest governance. Second was my government position, and my schooling of forestry, where foresters (Forest Rangers and Forest Officers) received knowledge and administrative power in the context of a colonial legacy transmitted via the Indian Forest Service.[6] This conventional forestry constructed and transmitted a deceitful image of local people where they were postulated as 'destroyers' and 'encroachers' of forest resources and thus primarily labelled as enemies of forest conservation. Therefore, solutions were sought to get rid of them through stringent 'fine and fence' strategies.

My political motivation to give more rights and responsibilities to local people in managing forests contradicted sharply with the attitudes of many other foresters at that time, who were reluctant to open up space to local people. My earlier work in helping to form CFUGs, and in providing backstopping support to them in group governance and forest management, had been key in my decision to accept a role in this project.

In fact, participating in learning and collaborative management initiatives helped me to undo the traditional forestry mindset, and to think about how a forestry professional can help bring different kinds of people together to effectively manage forest resources.

2. Understanding and internalizing ACA concepts

In 2000, the project which applied ACA principles had an ambitious goal of improving both the forest condition and the livelihoods of the people through community forestry. As described above, it combined traditional research strategies (e.g. in-depth socio-economic study and rigorous inventory of forest resource) with PAR where local people and researchers worked together to experiment with governance-related institutions and practices. In a project of less than three years, improving wood production was rather an unrealistic goal. It was perhaps the result of tensions typical of forest development projects, such as those within CIFOR in which social scientists had to accommodate the dominant reductionist and experimental biophysical research paradigm within the organization.[7]

Except for the motivation to respect local people, I had limited understanding of how governance and management of community forestry could be better instituted to address the issues of equity, livelihoods and resource sustainability. I entered the project with the aim of advancing my career as a researcher and an interest in learning new concepts and skills. Although I was broadly appreciative of the learning and collaboration aspects of forest management, I was not entirely sure what exactly we were doing and how that would lead to improvement in forest condition and livelihoods. The project provided limited opportunity for me to understand the ACA concepts[8] and how it linked to the local context, before I actually started undertaking the project works. Whilst academic and practitioner literature in English from a variety of fields including common property resource management, governance, social learning and collaborative governance were provided to me, it was still difficult for me to understand ACA concepts and to explain them convincingly to others. In the initial phase, our limited understanding of the concepts and ability to effectively communicate them was more challenging and often frustrating. Nevertheless, continuous sharing and reflections with local people, senior scientists and colleagues as wells as reading of related literature slowly helped me to understand the relevance of learning and collaboration aspects of ACA.

3. Dealing with community expectations

For the people in the CFUGs, a 'project' was always a bag of money for infrastructures such as drinking water schemes and roads. They did not think of a 'project' without such schemes. Despite our clear communication at the beginning, local communities continued to expect and even explicitly demanded material support.

For example, although as a field researcher, I participated in planning and reflective meetings within the CFUGs and clarified the project activities, local people suspected as if I were not disclosing the amount of funds that the project had as a material support to them. Local leaders personally approached me several times and inquired on the available fund and types of project that we could support. I repeatedly clarified that there was no money to support infrastructure development, and that I was there to facilitate their CFUG processes so that we could reflect upon the existing practices, envision the better future and devclop concrete plans of action to improve forest management and group governance for the benefit of them, with specific focus on the poor, women and disadvantaged groups. However, it seemed that they were not satisfied because they kept putting the same queries in front of other senior researchers visiting the sites.

In the first year, there was ample enthusiasm among local leaders and they participated in most of the meetings. But when they were fully convinced that there would be no material support from the project, they became less interested in the project. Thereafter, I and other researchers working in the field always faced difficulty in securing participation of local people in the project, particularly the leaders of the CFUGs.

In the initial stages, the research team encouraged local people to identify issues and plan through an inclusive and participatory process so that the team could help them identify possible funding sources when the planned activities required resources greater than the CFUGs could afford.

Field researchers met in Kathmandu several times, and reported the same issue for all the participating CFUGs.[9] Confronted with the demands for material support, field researchers including me continuously pressured the project's national level researchers to find some resources to meet at least some expectations of the CFUGs. Finally, the national teams, particularly Kaski Team and New ERA teams, agreed to raise the issue in the cross-country meeting involving researchers from Indonesia, the Philippines, Zimbabwe and Cameroon, as well as members of the project steering committee[10] and researchers from CIFOR who oversaw the research. The meeting agreed that the project raised the expectations of local people without adequate resources to meet them, and that the national teams needed to forge linkages with other funding sources at district and sub-district levels.

In another case, one of the key concerns raised from the CFUG in Kaski district was about establishing a sawmill to utilize the valuable Sal (*Shorea robusta*) and Chilaune (*Schima wallichii*) timber from the CF. Though the community forest was very productive, people were collecting timber only from locations close to the settlement because of high transportation costs caused by difficult terrain and lack of a road network. Consequently, the overall level of wood harvest from the CF was significantly lower than prescribed by the operational plan. The leaders of the CFUG had the opinion that, by establishing a sawmill, the CFUG income would increase sharply. According to them, the sawmill could transform timber for making diverse products including furniture, artifacts and construction materials. This would increase the demand of timber locally. In addition, with added value and transportability of certain products, they could also sell the products outside the CFUG. Once the establishment of the sawmill became a priority option for the CFUG, the whole discourse at the CFUG revolved around the sawmill, and local people kept pressuring the field researchers to provide support for establishing it. We also received threats to boycott the project. In this situation of intense and prolonged pressure, the project ultimately had to provide the matching fund for establishing the sawmill notwithstanding its longstanding position against it. Later, however, the sawmill could not operate as expected, despite the ambitious expectations of the leaders. Subsequently, in a reflection, one of the CFUG leaders revealed that the CFUG leaders strategically raised the issue of sawmill establishment repeatedly so that they could challenge the position of the 'project' and get some material support from it.

Despite these challenges, as we moved through cycles of visioning, experimentation and reflections, governance processes within the participating CFUGs improved significantly, also enhancing leadership capacity at the local level. Members of disadvantaged groups became aware of institutional procedures and their own rights to participate in and claim benefits from forest management. Similarly, once ACA was applied in the CFUGs, the awareness of the members of the CFUGs in general was increased on how they could be part of the CFUG governance and with what rights, and how deliberation in the planning and decision-making processes could improve decision outcomes. Institutional arrangements and benefit sharing mechanisms were made more equitable (details of such achievements are discussed in McDougall *et al.* 2002). Towards the middle of the project, we felt that a focus only in the CFUG level was limiting the advancement of community aspirations and the promotion of community autonomy and rights, then we slowly started helping to develop networks with neighbouring CFUGs to advance these objectives.

4. Conducting the participatory action research process

A crucial part of facilitating adaptive collaborative approach was the participatory action research (PAR) process, which started within the CFUGs with vision-based planning and self-monitoring workshops. The aim of the workshops was to develop long-term and short-term plans based on collective vision of representatives from each hamlet (locally called *tole*), and from other strata (in terms of gender and ethnicity) of the community. In addition, the workshops could arrive at a plan for reflection and learning. Biophysical and socio-economic assessment, hamlet level discussions and collective visioning by representatives of each hamlet and executive committee members were the key processes during this phase.

Although the plans were developed through an inclusive deliberative process, CFUGs lacked needed local resources and capacity to materialize them. Therefore, they needed to explore external funding opportunities, which was a challenging task. The community members and leaders had the expectation that the project would provide necessary funds or help link with external funding sources. Even when the research team tried to link the CFUGs with external funding agencies, the funds sought did not eventuate for a variety of reasons – such as the funding processes requiring additional work to meet funding conditions, timing, the stage of proposal development, or the need to secure matching funds. Consequently, the CFUG leaders showed their frustrations several times, and in some cases, members of the ACA project gave promises so as to cultivate some hope[11] within the PAR process – for instance giving information on potential funding and services.

5. Tensions within the team: international partners and local project hierarchy

We also experienced some tensions within the project team as we had different views on balancing roles across two high-demand areas, i.e. collecting information and maintaining relationships among different stakeholder groups. This was particularly between CIFOR based researchers who wanted more data, and the field researchers who had to develop close relationships with local stakeholders.

Project field researchers were asked to keep detailed records of local processes and observations of relationships among different stakeholder groups. The project's earlier assumption was that the field researchers would allocate more than 60 per cent of their time in the field so that they could facilitate the transition towards ACA, and keep adequate records and documentation of processes and relationships of interest. Later, however, the teams were asked by CIFOR to provide more information and produce additional reports, requiring more time in Kathmandu than in the field. Capturing the ongoing processes was a key task for the project since regular reflections on the processes and improving future actions based on learning from this reflection were considered crucial in PAR (Whyte *et al.* 1989; Patton 2002).

The field researchers thus felt it necessary to hire a local person in each CFUG for ongoing process documentation. After the initial process of recruitment was completed – including sharing the idea with CFUG executive committees, developing selection criteria for the position, soliciting interest from within each CFUG and interviewing potential candidates – the field researchers received a directive from senior project leaders of Nepal to annul the hiring. The cancellation had dire consequences in the CFUGs. Specifically, the CFUG leaders questioned the project transparency, potential candidates considered the outcome as cheating, and the CFUG leaders threatened to withdraw from the project. The researcher had to apologize to the CFUGs for confusion caused by the reverted decision, and tried hard to convince them that there was no sufficient fund to cover the costs of additional people.

6. Understanding and muddling through community power dynamics

Once processes were initiated to strengthen the principles of ACA (inclusion, deliberation and reflection) within the CFUGs, all participating CFUGs experienced significant changes in institutional arrangements and CFUG governance processes. These include greater representation of marginalized users in the decision-making bodies, changes in leadership, recognition of *toles* (hamlets) and thematic committees as part of planning and decision making, and more equitable benefit sharing arrangements. In general, there was a significant improvement in communities' capacity for management and governance, including awareness on inclusion and representation issues, planning and decision-making processes, and the rights of various groups of people over forest benefits and governance processes. However, the project team also experienced significant resistance from local elites. In some CFUGs, after completion of the project, these vested interests started to reverse provisions instituted for poverty reduction and equity. In other CFUGs, where the leadership was keen to continue ACA, they faced severe challenges from elites – mostly male from higher caste categories and rich – to continue inclusive and equity-oriented policies and practices.

7. Concerns about continuity of ACA

Towards the end of this phase, the researchers became aware that the local capacity to continue ACA beyond the project was not sufficient. Since the whole process was facilitated and supported by external actors, the field level PAR facilitators shared the concerns with the project team that the ACA process might not continue in the participating CFUGs. Likewise, both researchers and the participating CFUGs realized that working at the CFUG level alone was not sufficient, since changes in local institutions and plans subsequently needed to be endorsed by key stakeholders at other levels, such as Range Posts[12] and District Forest Offices (DFOs). Many local expectations could only be met only by collaborating with agencies beyond a CFUG. Linkages between participating CFUGs and other levels of governance – both meso and macro – were limited, which ultimately limited the opportunity of creating enabling conditions for local level innovations in livelihoods. The knowledge interface of different stakeholder groups was also a limiting factor in this phase. Though ACA facilitators had begun to bridge the gap between the CFUGs and other stakeholders, the situation was more a case of information transfer, than the co-creation of knowledge.

Researchers' awareness of this shortcoming in linkages and local facilitative capacity led to issues of continuity becoming a central design feature of the subsequent phases.

In the concluding national workshop of Phase 1, a few key forestry stakeholders in Nepal including the officials of Ministry of Forest and Soil Conservation (MFSC) raised a concern that the limited number of participating CFUGs was not adequate to represent the overall national context. Although we believed that our conclusions had important insights for policy, and we were clear that these concerns grew from a positivist (hypothesis-testing) research paradigm (see Colfer, Chapter 2 in this volume), we nonetheless wanted to respond to the concerns in our subsequent ACA initiatives.

Phase 2 (2004–2007): linking across scales of governance to foster inclusion and equity, deliberation and learning

The second phase research aimed to specifically address key design concerns raised in Phase 1, those raised by the key stakeholders as well as our own reflections. Specifically, the new project included new districts and CFUGs also from Terai region which was missing in the previous phase. Similarly, the second phase of research also responded to the concern regarding the need to link CFUGs with other levels of governance, and to build local capacity for facilitating ACA. In addition, selection of local facilitators[13] was an important step for the second phase, as they were involved in active facilitation; this role was carried out in the previous phase by the external researchers. Locally based facilitators were identified from CFUG and meso levels, and they received training and backstopping support from the external researchers.

Key points that emerged from the reflection on the second phase are the following:

1. Internalizing the ACA concept

We always felt that the ACA concept provided an accommodative framework to understand and act on the issues that characterized the complex social and ecological systems around CF, but we continued to experience difficulty in making local communities and stakeholders realize the importance of taking a learning-based approach and also taking a more system-wide view of the socio-ecological processes. One researcher[14] reflected it as:

> Though the concept of participation has been a buzzword in the CF, ACA as a concept and approach has been hard to understand and internalize. We as researchers had a hard time in explaining the concept to other stakeholders and to convince them.

There was also an issue raised by Pandit (personal comm.),[15] that though ACA borrows positive elements from various relevant fields, it is difficult to integrate them to realize an effective whole. We had to spend quite some time clarifying the scope of our work and the potential benefits that the communities can get eventually. Through successive phases of our work, we were keen to deepen the understanding of ACA especially on linking research with management decisions, and linking local level processes with meso and national level learning and decision-making. Ultimately, stakeholders including the CFUGs did realize that the approach was extremely relevant and different from many other development strategies they have experienced.

2. Legacies of the earlier phase

Some legacies of the earlier phase also constrained the new project. For instance, as described above, we were aware that the earlier project had raised several expectations about material benefits within the participating communities, which were left unmet. However, we did not expect opposition from the potential facilitators of ACA from the site's previous phase, where we assumed that ACA initiatives were largely successful.

In this phase, the first step was to provide training to the selected local facilitators (they were called change agents [CAs]) on how to facilitate ACA at both CFUG and meso (district and sub-district) levels. The training was organized in two regional clusters, East and West. In both the clusters, training events were jointly organized by ForestAction (responsible for new sites) and New ERA (responsible for sites from the previous phase) for CAs from the CFUGs and meso level. Our assumption was that when some leaders of the CFUGs from the previous phase[16] participated in the training, the new CAs would gain insights from their previous experience, and would thereby add enthusiasm for facilitating ACA in their respective communities and meso levels. While this did happen to some extent, some meso level CAs, who were the CFUG leaders during Phase 1, raised questions about the limited material outcomes of applying ACA within the CFUGs, and transparency and attitude of researchers involved. The transparency issue was related to the unmatched expectation of the community leaders and the kind of support that the project offered to them (as discussed in previous section). In this way, they brought into question the legitimacy of the new project.

In a meeting with the Project Leader from CIFOR, the district chapter leaders of FECOFUN, who eventually became the CAs in this phase and had previous experience of ACA, argued that they wanted to work directly with CIFOR. They requested her not to engage other organizations as lead partners in the project. Because the successor project was already designed in collaboration with ForestAction and New ERA, she politely declined their proposal. When these CAs understood that their request would not be met, they began to perceive these two organizations as rivals. In addition, they also learned that the project would not provide remuneration to them for facilitating ACA. Both of these concerns made them oppose the project during the training. However, after subsequent negotiations following the training,[17] which then secured resources to fund their participation in the project activities, they became active in later stages of the project. Although they subsequently successfully facilitated the ACA in their district, the concerns they aired during the training activities had a negative effect in the other sites. For example, in one of the new districts during the early stage

Table 7.2 Key strategies and processes applied at different governance levels in the period 2004–2007

Governance levels	*Key strategies*
CFUG and meso levels	Selection of facilitators from CFUGs and meso institutions Training and backstopping to facilitators to facilitate applying ACA within CFUGs and at meso level ACA-based planning and self-monitoring at CFUG and meso levels Helping meso level staff to be responsive to local level plans and concerns Implementation and reflection of planned activities at each of these levels
National level	Setting up national level policy learning group (NPLG) Further research through sub-groups on the prioritized issues Reflection on findings to assess application at policy level Reflect on the research at CFUG and meso levels
Linkages across levels	Linking the meso and CFUG levels by involving representatives from another level Bringing members of NPLG to meso and CFUG levels for interaction and learning Reflection meetings involving key representatives from each level Sharing key concerns, challenges and lessons through publication outlets

Source: Adapted from Banjade 2006.

of this phase, six of the seven trained CAs refused to facilitate the ACA, and, as mentioned above, the major reason they gave was the question against effectiveness of ACA in the previous phase.

3. Key interventions and diversity of outcomes at community, meso and national levels

Key ACA strategies and processes applied at community, meso and national levels, and at the interface of these levels, in this project period are summarized in McDougall *et al.* 2007 and 2008. Although processes at each of these levels were informed by the same principles of inclusion, deliberation and learning, the results of the interventions were different at each level.

At the CFUG level, most CFUGs engaged positively in a transition towards ACA, and contributed significantly in awareness raising, capacity building, social inclusion and equity. These CFUGs became very popular among CF actors within their district and beyond, for their empowerment of marginal groups, for development interventions oriented to improving equity and livelihoods, and for democratic governance, particularly greater inclusion in institutional arrangements and deliberation in decision-making processes. McDougall *et al.* (2007) have provided the details of outcomes and insights on applying ACA in the context of CF in Nepal during this phase. However, at least in three out of the five CFUGs we worked with, there was a significant resistance to this transition by the existing leadership. In some of these cases, external researchers decided to work as change agents along with local CAs in order to address the resistance through empowering other people to challenge the existing leadership. While taking an active role in one of the CFUGs in Eastern Hill District, one researcher even received threats of physical harm from existing leaders for 'interfering in their business'. By working with local CAs and other members of the CFUGs, the resistance could be successfully challenged. However, we found that at least in one CFUG some leaders resisted again after the project period.

We began to link community level issues and learning with meso and national levels, and the complexity grew as we moved up in the scale. At the meso level the processes and outcomes were more differentiated and erratic than at the CFUG level for a number of reasons. In the initial stages of working with the meso level, the intended processes of visioning, plan for collaborative action, and joint reflection and adjustment were not working properly. Despite limited resonance between researchers' assumptions and planned strategies for this level at the beginning of the PAR, we (the external researchers) were initially slow to review our assumptions and way of learning to effectively respond to the contextual specificities. Our limited reflexivity and re-strategizing on this often led to frustration among us. After several months of PAR, the teams began to thoughtfully reflect on differences on issues, processes, responses and challenges across the participating meso districts, and how we could address these. At this point, researchers (and CAs) then started facilitating a more organic, context-specific development of the meso-level multi-stakeholder forums in each district, i.e. encouraging actors to identify an issue of collective significance, deliberate on the issue, make joint action plans to understand (i.e. research) and deal (i.e. action) with the issue, and finally, to reflect on the process and outcomes. Some meso forums performed well with this strategy. For example, the meso group in Lalitpur District organized reflection meetings continuously to energize some of the less active groups, while also continuing with the primary objective of sharing of issues and lessons with the entire member CFUGs. In Dhankuta District, participants established their own 'learning group' and planned a few joint visit programmes to learn from innovative communities, so that they could use that knowledge in other CFUGs in the district. In Nawalparasi District, the government was resistant to handing over community forestry without addressing the issue of regional equity in forest product distribution (i.e. sharing of benefits between forests in the district's north with residents in the district's far south). The meso group focus in this district thus revolved around addressing spatial inequity and handover of more forests to local communities.

While we initially found it difficult to promote deliberative forums at the meso level, towards the middle of the second phase we were able to fairly effectively catalyze and support these by working with meso actors to devise unique strategies for each context. From a research perspective this meant that we could not identify a single appropriate mix of actors and institutions for effective ACA at the meso level or a specific type of forum (McDougall *et al.* 2008). Nevertheless, the research identified various challenges and potential strategies for working in different contexts. For example, districts with an earlier culture of organizing multi-stakeholder sharing events, with or without a support of external agency, easily accepted the concept of meso-level deliberative processes. They were able to progress with some minimal facilitation to identify common issues of concern. In contrast, in the districts where earlier experience of holding such forums was limited, issue-based networks might only act as a starting point. Further, while some actors saw benefits in these processes, others saw them as a burden because of the extra time required in addition to their other tasks (in case of some NGOs in Nawalparasi). Some others also saw the forum as a threat to their personal interest (more so when that interest extra-legal), with the potential to humiliate them. In the meso level of Nawalparasi District, issue based networks such as the Non-timber Forest Product (NTFP) network, were an effective strategy for some time. The diverse goals of different actors and their mandates would often limit the effectiveness of ACA processes at this level.

3. Frequent turnover of change agents (local facilitators)

Many of the trained CAs left the project after receiving training, the main reasons being the marriage of woman CAs to men from other areas, the departure of male CAs to other countries for work, and the withdrawal of CAs from the project because of the inability to manage time due to their other responsibilities in their village or district. This departure of CAs created a significant challenge for the ACA process.

In Nawalparasi District, for example, training was provided to seven CAs (two from the participating CFUG and five from meso organizations), all of whom had been selected and endorsed by the CFUG. It had been clarified from the very beginning of the project that CAs had to work on a voluntary basis, unless the CFUG and meso forum were able to compensate them. One of the criteria for the selection was that the CAs 'can and are willing to contribute time voluntarily'. In this site, however, all but one CA left before ACA started in the CFUG and meso levels. These difficulties created a rather disappointing situation for the project team. In the case of Nawalparasi District, we then discussed with the CFUG executive committee (EC) about CAs. Although the enthusiasm of the CFUG leaders to find volunteer youth CAs was limited, we agreed to train a number of youths, most of them undergraduate students motivated by the opportunities for learning and career development. We provided them with basic training in facilitation and ACA. Later, the EC covered the costs of lunch and snacks for the CAs while they were visiting users. In this site, and others in similar situations, it thus took some additional time before ACA could begin. Similarly, at the meso level, the departure of some of the originally selected and trained CAs appeared to temporarily reduce the interest of some members and thus slow down the institutionalization of the forums.

4. Transition of CFUG leadership

Ownership of the ACA was often reduced significantly with the change of leadership particularly in the CFUGs where existing leadership was positively leading the ACA initiatives. Tensions and conflicts between CAs and new leadership, as well as between the new leadership and the researchers, were observed frequently due mostly to misunderstanding but also with the personal and ideological differences. The project team responded with different strategies, depending on the case. For example, in some CFUGs, we engaged the disgruntled leaders in meso-level ACA processes (to allow them deeper understanding and ownership), while in others we talked with the identified local champions who had formal or informal power to motivate the CFUG leaders.

Difficulties associated with leadership change were often related to local power dynamics, i.e. when new leaders had a different party line or lobby group from their immediate predecessor (as is the case in the above mentioned condition), the problem was more severe than when new leaders came from the lobby group of previous leadership.

Change in the leadership of meso institutions also posed some challenge of keeping new leadership on board, requiring additional time and effort to update them about the ACA initiatives and the status of implementation. In these cases, we had to provide orientation to new members about ACA. In later stages, it became almost a regular practice for us to include a session on ACA in the beginning of meso level deliberations, at least once in every visit.

5. Attempts for social change not always welcomed

We put our every effort into sincerely facilitating ACA processes at local and meso levels, with the aim of contributing to positive socio-economic and political change within the CFUGs. We were aware of, and committed to, the fact that ACA recognized the role of the researchers as partners and agents of change in a social process. As such we sought to be proactive in understanding local knowledge systems and power asymmetries, and aimed to play a vital role in minimizing the domination of some actors over others, so that local processes of decision-making would fulfil the minimum condition for non-coercive exchanges during deliberative events. We also tried to create space for synergy between local and external knowledge systems and pulling other resources from various sources for effective management and governance of community forestry.

However, local people perceived the researchers as NGO employees, among thousands of others, whom they considered as having lucrative jobs and being interested only in 'dollar *pachaune*' (literally, 'digesting dollars coming from foreign sources'), rather than being committed to enhancing community well-being. As such, the researchers were victims of the dominant image of NGOs in Nepal, which are commonly blamed for (mis)using foreign support in the name of people. When people participated in the forums with a prior negative perception of the researchers, the outcomes of the deliberative process were not those that were anticipated, and instead were often highly frustrating for the researchers. For me personally, I had left the government position and joined ForestAction in part to make a contribution to larger society through research and advocacy – then found this potential to be undermined by the negative public perception against NGOs.

6. Organizational commitment in applying ACA

The nature of the organization leading the project in Nepal also had significant implications for the application of ACA in a context, and refining the implementation on the basis of learning. For example, ACA has been a core approach of ForestAction since 2000. ForestAction's main thrust was to establish itself as a learning organization, apply inclusion, deliberation and social learning as its key working principles, and promote linkages and partnerships with several organizations from community to international levels. It was therefore centrally concerned with promoting the ACA concept beyond the project boundaries and period. For example, ForestAction applied the concept indirectly in other non-related projects as well, such as 'Participatory Research on Harvesting Non-timber Forest Products' and 'Improving Livelihoods through Community Forestry'. Similarly, ForestAction led the conceptualization of the project in Phase 3, which is discussed in the following section.

Phase 3 (2008–2011): combining deliberative governance and social learning across scales and disciplinary boundaries

By our third phase of ACA work, we had gained substantive experience from the field, and we were engaged more deeply (and theoretically) on the issues related to deliberative governance, social learning, and cross-scale governance. We were also enthusiastic to have a strong component of entrepreneurship development benefiting the poor. A coalition at national level with a range of actors such as NGOs, networks, media, business groups and academia was formed during this phase to promote ACA. A project management team was recruited by the coalition to foresee the project activities. The team provided training to local facilitators on governance and forest management in all the participating districts. The Federation of Community Forestry Users, Nepal (FECOFUN) was one of the coalition partners and led activities related to awareness and capacity building at the CFUG and meso levels. We also conducted issue-based networking of CFUGs in the meso sphere. The explicit target of all these was to promote rural livelihoods through entrepreneurial innovation with vertical and horizontal linkages, fostering interactions and interconnections. During the course of applying ACA, we were also involved in resolving intra- and inter-institutional conflicts and tensions. Self- and collaborative monitoring and reflection by the coalition partners as well as jointly with CFUG and meso level actors formed a major part of the process.

In this phase, we tested whether ACA could be scaled out widely. We did so through working with 60 CFUGs from three districts representing three agro-ecological zones of Nepal (Western mid-hills, Central Terai[18] and Central mid-hills). Out-scaling strategies included organizing interaction at and across various levels with a wider range of stakeholders, developing and airing radio and television programmes on key initiatives and innovations, and establishing of community-based resource centers in each of the participating districts.

There were five key areas of research and intervention in this phase: i) applying ACA within CFUGs; ii) effective forest management by combining local knowledge and skills with external expert knowledge; iii) enhancing entrepreneurship skills within CFUGs and supporting development of enterprises and marketing of forest products and services; iv) applying ACA at meso and national levels; and v) developing effective links of CFUGs with multiple actors from multiple-governance levels (from local to national).

A member from each of the 60 CFUGs participating in this project period received ACA training with a focus on forest management and CFUG governance. The training content and resources were developed based on lessons and materials from the previous phases. After the training, the participants facilitated ACA within their CFUGs and helped develop links of their CFUGs with other CFUGs, and with other actors at meso and national levels. In addition, each facilitator reflected on the processes facilitated by them, and recorded reflective notes including their own facilitation approaches.

We also conducted a baseline study of socio-economic and ecological aspects of forest management in the 60 sites (Kattel *et al.* 2009). The analysis informed ACA planning and reflections at all levels from CFUGs to national level. The baseline study revealed that certain groups of people – based on gender, ethnicity and economic class – were excluded from the decision-making processes and benefits sharing in the participating CFUGs. Evidence also suggested that the CFUGs had limited access to important resources from outside the CFUG. For example, they could have benefited from forging linkages with other important actors and institutions such as entrepreneurs and business people, service providers and external networks. However, the potential for learning from other CFUGs and these external actors was not recognized by the CFUGs or other actors and institutions prior to the research (Kattel *et al.* 2009). Despite overwhelming enthusiasm on the part of the project team for applying ACA through overcoming several challenges, we continued facing challenges in implementing the project. I describe below two of the most important insights relating to these challenges from this phase: i) managing collaborative relationships with multiple actors, and ii) satisfying the mechanistic expectations of donors:

1. Managing collaborative relationships

I would like to briefly explain several challenges we encountered during this phase in relation to managing relationships with various actors including tension between various organizational levels within a coalition partner, relational problem of another partner and a CFUG continued from the earlier phase, and problems with the government forest authorities.

Two important relationship problems within FECOFUN hindered project implementation in one district. The district chapter of FECOFUN, the implementing leader of the project at the CFUG and meso levels, had problems with the center. According to the leaders holding key positions at the district chapter, the central body was less concerned with the issues of the Terai and was more concerned with the issues and discourses of public consumption at higher level rather than responding to the interests and concerns of CFUG and meso levels. In contrast, the central office apparently perceived the crux of the issue to be about the sharing of resources received from the donors (Indrawati Moktan, personal communication), although this was denied by the district chapter. In the situation of 'cold war' between the center and the district chapter, the district chairperson was not advancing the project inception process. The other four coalition partners remained concerned about the expected role and commitment of the district chapter.

To address the situation the coalition partners, specifically the central project management team, had to spend significant time and energy in discussing the issue with the FECOFUN leaders, individually and in groups, and to help carry out the project activities. The part of the project outcome was to reform the internal governance of FECOFUN to make it inclusive, deliberative and learning oriented so that it could effectively respond to the CFUG interests and concerns. Essentially, it would demand more deliberation in decision-making and transparency of financial management. Some members of the district chapter revealed that they believed the resistance from the chairperson of the district chapter was largely because his leadership was more authoritarian rather than deliberative. They argued that the project would eventually expect him to change his leadership style. Ultimately, a series of meetings between researchers and district level leaders, researchers and central leaders, and joint meeting of researchers, central and district leaders of FECOFUN made it possible to start the project activities after about a year. Finally, they agreed to continue the project while keeping doors open for discussion to resolve remaining problems.

In other cases, for example in Lalitpur, the district chapter leadership had not been functional, and individual leaders resisted taking on leadership of the project, because the resources available for the facilitators to facilitate the ACA from CFUG up to the meso level initiatives was not adequate.

Similarly, some resistance was observed from a leader of a CFUG, who was initially shortlisted to be part of the project but later refused by blaming a coalition partner (one field researcher in particular) for not being transparent in an earlier phase. However, there was no such complaint from the local people in the CFUG or the local change agents involved in the earlier phase. When the research team discussed the issue with the executive body of the CFUG and promised to provide them the necessary information from the earlier phases, the executive body agreed to participate in the project. However, in the inception workshop of all participating CFUGs, the researchers learnt that the CFUG was not participating. The researchers again discussed with the leaders in the CFUG and learnt that the reason given this time for not participating in the project was different – the chairperson informed them that the project demanded a lot of time, which he could not afford. When we explored the matter further by talking to local leaders, particularly CAs from previous phases and current EC members, they mentioned that the decision of not being part of the research was possibly made because: i) the chairperson did not want to be transparent and deliberative as ACA would require; ii) there were hidden economies, such as corruption and mismanagement in the CFUG, which could be revealed when the project empowered people to raise the issue; and iii) local government forestry staff were against the project. Four other CFUGs from the same district in the later stage of the phase also shared that they received pressure from government staff to defy the project. Repeated discussions with individual CFUGs and joint meetings with all the CFUGs were conducted to reduce misunderstandings due to communication gaps between the central project team and the CFUGs.

By analyzing media and reports from investigating committees from a parliamentary committee and the CIAA (Commission for Investigation for the Abuse of Authority), Banjade *et al.* (2011) have found that in many parts of the Terai, some local leaders, corrupt government officials and timber smugglers have formed a strong nexus to misappropriate forest resources. For example, more than 75 per cent of the 215 media stories related to forestry in 2010 in Nepal's three daily newspapers (*The Kathmandu Post*, *Kantipur* and *Gorakhapatra*) claimed such a nexus (Banjade *et al.* 2011). Obviously, the actors with corrupt attitudes would like to be concealed from public exposure. For them, the presence of external actors and their intervention on the CFUG could become a threat. The revelations of leaders of four CFUGs of threats from a government Ranger not to participate in this research project support such a contention. In addition, in one of the meetings of a participating CFUG, a forest officer warned the researchers that he could put them in the jail if they work with a CFUG without getting permission from DFO (Dharam Uprety, personal communication).

Although the project was conceptualized to include government forest service on the basis of specific needs of the participating CFUGs and their networks, the resistance from the forest officials was immense. This was partly rooted in the history of ForestAction's presence in the district. Particularly, early engagement of ForestAction in the district was to provide advocacy support to the community movement for CF, and one staff member in particular had had severe conflict with local government forest agencies (Range Post and Area Forest Office). The legacy of that relationship still remained despite our new strategy of constructively working with them. To engage local government forestry officials in the project, the ForestAction team maintained communicative interface with the officials at multiple levels – from the Range Post (a lowest unit under Department of Forest), Area Forest Office (a unit comprised of 5–8 Range Posts), District Forest Office, Department of Forest to Ministry of Forest and Soil Conservation.

Engagement with national and meso level government forest officials was important in the process of working on forest governance in Nepal. We rejected the monolithic perception of the government, which sees government techno-bureaucrats as essentially authoritative and autocratic. Rather, we took the view that government bureaucracy staff hold a range of views and attitudes, and we sought to collaborate with progressive foresters. Our first reason for this strategy was a practical one: working with the senior officials within the Ministry of Forest and Soil Conservation would help in securing support from lower level authorities. When district and sub-district level authorities cooperate in a project, it would certainly increase its performance. Second, upon reflection on Nepal's forest sector policy processes, we observed that the discursive practices of 'technocrats' had a pivotal influence on wider CF discourse. Applying the ideas of technocracy from Habermas (1975),[19] forestry technocrats refers to the foresters within government forest bureaucracy (i.e. techno-bureaucrats, c.f. Ojha *et al.* 2010) and other foresters having a similar perspective of expert superiority in development. Following this line of thought, we perceived engagement with the technocratic sphere as necessary to the transformation of their technocratic mindset and attitude so that the sphere could spread positive messages to the policy processes for CFUG autonomy and democratic forest governance. Researchers and policy analysts have often overlooked the technocratic sphere. Therefore, we decided to work with positive individuals from the government forest bureaucracy so that we could contribute to the process of shaping discourse and thereby the policy processes. Our purpose of being engaged with the technocratic sphere was to promote democratic forest governance, increasing motivation among technocrats on empowering local communities and providing more autonomy to the CFUGs. Finally, working with government forest officials was necessary to break the polarization between the government and the community and non-government institutions. Despite CF being a main government programme, many of the government forest staff perceive it as a programme of the non-governmental sector. Thus the polarization of discourse around forestry often diminished the ownership of government officials over the programme. We expected that increasing opportunities of deliberation among government and other agencies, including NGOs and networks, could help in reducing the division and relationship problems between them.

2. *Engaging with mechanistic expectations of donors*[20]

Immediately after the meeting with the external mid-term review team of the project, sponsored by a donor, the project team met for self-reflection and sharing regarding the meeting. (This was in keeping with our application of ACA to research project management, as ACA envisages reflection as a regular process conducted after every significant event or activity for learning and further improvement.) There was a unanimous agreement that the meeting with the evaluators had gone very well, and that we had been able to showcase the project's on-ground realities and its achievements and multi-actor innovations at the project sites. Our understanding, however, contradicted that of the evaluators. The evaluation reported that the aims and objectives of the project team did not fit with those of the project and that the project should be brought to an early closure. Moreover, the evaluation was of the opinion that the project should not have received funding and that the modalities and working methods are entirely different than those intended by the donor programme's concepts of putting research into use. The evaluators' perspectives sharply contradicted our understanding about applying ACA in practice.

Contrary to the evaluation report, the data collected from the field, our own field observations, and the reflections of local people and several other stakeholders demonstrate that the project was a success. For example, a journalist from UK came to visit several CFUGs in the three participating districts around the end of the first year of this project phase; she captured local innovations and successes, organized a photo exhibition in UK, and published an article highlighting the success of the project (Green 2009). Thereafter, several additional initiatives were applied at the project sites on the themes of enterprise development, governance reform, forest management, poverty reduction, and enhancing partnerships between the CFUGs, service providers and market forces. There was clearly a difference in how the team and the reviewers were measuring a 'success'. The project team was considering factors such as: all activities were conducted, plus additional ones; the project had successfully catalyzed process innovations, learning and partnerships in the CFUGs and districts, and some material well-being of improved forest management practices, and increased the flow of funds in livelihoods' improvement activities. We were able to conduct all of the activities specified in the project logframe, and several others in addition. In contrast, the evaluators' assessment was largely framed in terms of material products and conventional economics, where process-related innovations, learning, and broadening partnerships and collaboration are not counted as achievements. In contrast, our ACA-oriented worldview perceived procedural aspects of a system should be given equal emphasis to the material outcomes.

We wrote a report responding to the comments, highlighting the differences of ontology between the researchers and the evaluators. While the evaluators used what we perceived as a 'transfer of technology' view of the 1970s, our view was that complex cultural and structural barriers to change could not be penetrated by such a linear and mechanistic approach. Eventually, the research section of the programme from the donor appreciated our idea and encouraged us to bring forward the issue through a discussion paper (Giri *et al.* 2011).

Analysis and discussion

This section reflects on the principal learning over the three phases of the project.

Understanding reform processes in community forestry

Natural resource governance has shifted from a single agency's centralized control to multi-stakeholder collaboration. While networking, struggles and civil movements have all contributed to making the shift significant, there are still challenges in terms of some lack of genuine interest from the government in devolving rights and resources to local people. Similarly, the role of actors other than communities and government forest authorities has been under-recognized until recently. Our work has led us to the conclusion that the existing techno-bureaucratic paradigm should be brought into much wider stakeholder discourse to bring about agency and structural changes (Banjade *et al.* 2006b; Ojha *et al.* 2009b; 2010).

The governance of community forestry is inevitably a political process that involves knowledge and power, often differentiated by imbalances in socio-economic and symbolic resources. Our experience confirmed that the process of reconfiguration of power entails significant conflict and resistance. Even though the deliberative process is very important, in the given condition of inequalities between actors, it is prone to the exercise of coercive power, and contestations revealing deeper fissures that risk undermining the successes already achieved. When facilitated carefully, deliberative processes may provide spaces for both marginal and dominant stakeholders (Häusler 1993), but the results depend on how existing knowledge and power are reconfigured.

Deliberative processes at various levels sometimes benefit, but at other times limit, the autonomy and livelihoods of CFUGs. In many settings, seemingly participatory institutions are captured by a few powerful groups, and the attendance of marginalized members such as women, the poor and ethnic minorities in the decision-making processes are used merely to legitimize the interests of the powerful groups (Agarwal 2001; Nightingale 2002). Gaining an understanding of context-specific power dynamics can lead to a broader understanding of which approaches may best address the complexities inherent in CF in Nepal, as well as in natural resource governance regimes elsewhere.

As context-sensitive ACA practitioners, we had to pay an increasing attention to working with marginalized groups, trying to bring changes in planning and learning processes. In all these processes, the role of external actors and facilitators, and their influence in CF governance, is a crucial element in deliberative and adaptive learning processes (Banjade and Ojha 2005), first to make the local people aware of the complexities of management and governance, and then to catalyze adaptive and collaborative planning and management.

Contribution of ACA to community forestry

A number of governance reforms were possible when we applied ACA in community forestry in Nepal. We demonstrated substantial progress in institutional arrangements for conserving forest resources, awareness about sustainable harvesting of forest products, and investments in forest development (McDougall *et al.* 2007; ForestAction 2011). Significant achievements were made in relation to equitable representation in the decision-making bodies by gender and ethnicity, establishing deliberation as a key governance mechanism, more equitable benefit sharing mechanism, improving communication within the CFUG and with external actors, and making meso level actors more responsive to local interests and concerns (see Table 7.3 for key achievements in different phases of ACA).

New institutional arrangements, such as hamlet committees with some decision-making power, promoted deliberative democratic governance within the CFUGs by giving legitimate space to all members so that they could raise their concerns and engage in domination-free communication during the decision-making process (McDougall *et al.* 2008; Uprety *et al.* 2011).

Networks played important roles at different levels, from local, to meso, and to national levels, by strengthening solidarity among different groups of stakeholders, and enhancing empowerment, working relationships and two-way communication. This was accomplished through the establishment of several interactive platforms and self-monitoring by the respective collaborators. In addition, these networks emerged to be promising local institutions for managing disputes and for information sharing, contributing to policy processes and building a collective plan of action to improve management and governance of community forestry (Banjade *et al.* 2007; Uprety *et al.* 2011). This is in confirmation to Dryzek (2010), who highlighted the efficacy of the networked governance, including those operating at

Table 7.3 Key achievements in different phases of ACA projects in community forestry in Nepal

Phase I	*Phase II*	*Phase III*
Inclusive and deliberative local processes Equitable benefits sharing Increased awareness and voice among marginalized groups	Reform in CFUG level governance processes, including leadership change More inclusive and deliberative processes at all levels Improved local awareness, management capacity and governance Significant improvement in representation of marginalized groups Equitable distribution of forest benefits and contribution to pro-poor activities	Culture of partnership developed among CF actors at all levels Improved linkages of CFUGs with other spheres (meso level and other agencies such as business groups, NGOs) Improved governance at CFUG and meso level Meso level more responsive and deliberative to the CFUGs More effective networking at different levels

Source: McDougall *et al.* 2002, 2007; ForestAction 2011.

multiple levels, in processes less coercive than government hierarchies or less competitive than among market actors.

Similarly, at the national level, ACA was able to 'break the ice' in the troubled relationship between key CF actors by providing a forum for deliberation and breaking a deadlock of communication between actors. For example, one of the most important outcomes of the ACA initiative at the national level during the second phase was that actors with conflicting positions on the contentious Terai forestry issues began to engage in regular dialogue and increased their mutual understanding. In addition, engaging with the technocratic sphere has been vital in contributing to the broader policy process.

Despite an operational plan (OP) that specifies the amount of each forest product that can be harvested from the community forest, in practice the DFO retains the authority of issuing a permit for annual allowable harvest by manipulating OP. In almost all CFUGs they required to put the provision of mandatory prior approval by the CFUG from the respective DFO. When the CFUGs engaged in ACA, the members gradually became aware of their rights and often challenged the hierarchical relations between CFUGs and meso level actors. This has been particularly visible in the local forest users' increased questioning of, and even at times challenging, the 'techno-bureaucratic' domination in forest management in the meso level forums (Ojha *et al.* 2010).

Limited appreciation of formalized social learning processes

Social learning lies at the heart of ACA. Social learning can be understood as a 'process in which multiple stakeholders bring together their different knowledge, experiences, perspectives, values and capacities for a process of communication and critical reflections as a means of jointly understanding and addressing shared issues, challenges and potential options' (McDougall *et al.* 2008: 30). For example, with ACA members of CFUGs often informally or formally engaged in reflective processes after major initiatives or processes, they sometimes identified key learning questions from the issues they wished to address, and they routinely deliberated on key issues in hamlets and in hamlet-committee meetings.

We observed that, in practice, the most structured social learning process, i.e. the annual self-monitoring, received less appreciation by CFUG members. The relatively low appreciation was partly because CFUG members were initially unclear on its benefits. This likely reflected the researchers' difficulties in explaining adequately why CFUG members should spend their time and effort on institutionalizing reflection and learning processes. Moreover, the self-monitoring required a relatively demanding structured process (assessing each criteria) that local people felt alien with the informal and less structured local cultural practices. Also, using a written vision and indicators sometimes unintentionally disempowered illiterate or less literate people (a critical evaluation of this concept and practices of the monitoring has been provided by Ojha *et al.* 2002 and Paudel and Ojha 2007). To address the challenge, ACA facilitators allowed flexibility in using locally appropriate tools and techniques such as the use of local materials to represent the criteria and measuring the achievements.

Linkages across scales and levels of governance

The research suggests that it is necessary to make institutional arrangements that help connect community-level governance with that of meso and national levels. Community forestry provides the platform for bringing together and integrating wide and diverse experience, knowledge and wisdom related to community-based forest management for strengthening, improving and adapting innovations (Banjade 2006). There are, however, still several fundamental questions that need to be addressed. These include: how individual CFUGs who have successfully applied ACA and have generated substantial changes could transmit their experiences to the policy process to those at other levels; and how external stakeholders supporting CF could, in turn, learn more effectively from the CFUG level processes.

Although the research described in this chapter started with a narrower geographical and conceptual focus, it subsequently embraced broader spatial and conceptual coverage – from a local planning and 'self-monitoring' focus to larger geographical coverage and inclusion of issues of deliberative governance, cross-scale linkages and systemic comprehensions. Even though particular concepts or elements of ACA were emphasized during different phases, the basic focus remained consistent: how different actors and institutions could deliberate in the process so that they could learn to deal with the complex issues of exclusion, inequity, sustainable resource management and livelihood enhancement.

Challenges were greater than anticipated

Facilitating ACA in the context of CF in Nepal raised several major challenges. First, releasing prior norms to facilitate deliberative governance was a challenging task in itself. For example, existing power holders often resisted other members from participation in intra- and inter-CFUG deliberations, which usually led to the same individuals who were historically powerful playing strong roles in the ACA processes. Similarly, elite domination in local decision-making processes could not be fully resolved. This implies that the constraints against 'ideal' deliberation could not be removed fully so that the poor and marginalized groups could raise their concerns and reasons without any risk of coercion. It also means that significant efforts are required to create conducive environments for deliberation so that marginalized members perceive an incentive to participate in public deliberation, feel fully respected by others as legitimate members, and offer flexibility of date, time and venue of meetings, etc. Moreover, challenging powerful actors, including government officials, is still an occasional rather than a consistent behaviour for most CFUG members.

Second, the role of facilitators in breaking the structural and cultural barriers of deliberation is important. However, a major problem in the latter phases was the turnover of change agents. Consequently, there have always been challenges in identifying the right facilitators (change agents) and ensuring their commitment to facilitating processes within the CFUGs. Further, many facilitators lacked basic skills and capacity to facilitate the deliberative processes, even after providing training and backstopping support.

Third, as discussed previously, partnership and collaboration were found to be more complex than initially assumed in the project, especially because different partner organizations differ in capacities, working modalities, levels and forms of staff motivation, organizational hierarchies and approaches to development and innovation. In the third phase, while some organizations within the coalition had well-developed professionalism in organizational management, such as book-keeping, staff recruitment and mobilization, or project management and reporting, others were good at advocacy and political action but lacked other capacities. In such a situation, working in a team to deliver a joint outcome was a challenge, and it demanded considerable patience and continual effort. In this phase, we addressed these issues through a series of formal and informal processes such as peer feedback mechanisms, informal procedures for empowerment, joint field visits, joint reflection and planning, continuous follow-up, mentoring, and layers of institutional communications. Given that an a priori fixed design may not fit actual circumstances, ACA methodology was adapted to address the real problems on the ground, which required extra time and efforts.

Fourth, the assessment of 'success' of innovations and innovation systems through the lens of classical economics (as some donors and public officials tend to see) undermine the overall change process. Our view is that success should not be judged merely by the adoption of a technology or some form of immediate economic improvement, but rather also on the basis of institutionalizing processes that are likely to support long-term equity and sustainability of livelihoods and resources, such as adaptive and collaborative approaches. ACA views that knowledge and technologies are co-created through the processes of innovation, and that appropriate frontiers of deliberation between local and various forms of external knowledge are essential to promote inclusive and deliberative governance.

Fifth, as our experiences indicate, overcoming the hurdles associated with improving the linkages with external markets and harnessing economies of scale remained a formidable task for the CFUGs. Some corrupt leaders would like to strengthen their nexus with corrupt government officials and business people. As such, ACA may risk being an additional instrument for these fraudulent interests to foster. Conversely, when it is facilitated in an appropriate way, it can help in enhancing transparency, stakeholder vigilance and action against these malpractices.

Sixth, it is clear that the design phase of an action research project needs to understand communities' expectations of the project so that the problems arising from undue expectations can be minimized. Finding the right mix of 'action' and 'research' always remains an issue within action research. In our context, we found that local people's emphasis was on material assets, while we wanted to focus on processes of socio-institutional change, from which improved material well-being could emerge. It took a long time for local people to appreciate the significance and value of reflection and inquiry in their development process, because they were exposed to the mechanistic views of a development industry in which only material well-being is counted as development.

Finally, on a number of occasions, the personal attributes of ACA facilitators had a large bearing on the outcomes of the processes with which they were engaged. For example, some facilitators seemed to be too research-centric, and others too advocacy-oriented, rather than ACA-oriented. In the former case, facilitators were relatively passive in facilitation and active in documentation, whereas in the latter case, they were concerned about outcomes and did not appreciate the value of failures as an underlying source of learning. Limited institutional support for reflection and writing, as well as frequent changes in staff from leading research organizations, also hindered the process of developing the capacity of such facilitators.

Conclusion and ways forward

This chapter has provided a retrospective analysis and reflection on over 10 years' experience of applying various ACA related initiatives in Nepal's CF. We have been promoting ACA in the context of the long history of CF and NRM in Nepal to address the issues of inclusion, democratization, livelihood improvement and resource conservation. Our focus has been primarily to uncover the challenges of applying ACA in the context of multi-scale resource governance involving multiple stakeholders, and describe our strategies and interventions in addressing these challenges.

From my experience of the three phases, the main conclusion is that, in the context of forestry and natural resource governance in Nepal, ACA has the potential to uncover deeply hidden interests and concerns, engage actors in deliberation and reflection, and provide new ways of learning and innovation. We also acknowledge that the transition to more adaptive and collaborative approaches is fraught with many challenges. These challenges are in sustaining the achievements already realized, as well as in advancing further change. We also observed how regressive forces often tried to revoke the inclusive and deliberative institutional changes that followed from applying ACA.

Integrating learning into everyday practices has been the most difficult part of applying ACA. This reinforces the classic challenges faced by those attempting to embed research within an agenda of change. Similarly, as I have recounted in this chapter, deliberation within a governance process is usually subject to existing power asymmetries caused by the differential possession of economic, cultural and symbolic resources.

Establishing links between CF spheres (levels) through networked governance or deliberative multi-level forums often proved very useful in improving CF governance. Skilled facilitation and coordination functions were central to the success of this process and demanded additional resources and institutional commitment to sustain ACA at the meso and national levels. Similarly, developing capacity to integrate research into the practice of change, and to effectively communicate research, are other areas of practice that need attention if ACA is to be scaled up within the forest sector, and also into other sectors.

Our experience of applying ACA shows that challenges increase with promoting and sustaining deliberative forums at the meso and national levels. Nonetheless, it also has shown the promise of forging effective linkages between communities and meso and national levels. Recognition of multi-stakeholder processes and policy feedback from local practices are gradually being appreciated by the policy-makers. However, they are yet to institutionalize the essence of ACA in light of limited institutional responses to these new insights. This implies a need for change in governance frameworks alongside ACA processes.

Acknowledgements

I would like to acknowledge Peter J. Kanowski and Cynthia McDougall for their comments and editorial assistance, and John S. Dryzek, Hemant R. Ojha, Kalpana Giri and Dharam Uprety for their valuable comments that helped improve the chapter. Similarly, I would also like to acknowledge: the Center for International Forestry Research (CIFOR)-led Community Forestry Research Project (Nepal, 2000–2002) that was a part of the ADB RETA 5812; the CIFOR, ForestAction and New ERA project Improving Livelihoods and Equity in Community Forestry of Nepal (2004–2007) which was supported by IDRC; the ForestAction-led Reducing Poverty through Innovation Systems in Forestry project (2008–2011), which was supported by DFID and IDRC; and, the ForestAction and Nepal Australia Community Resource Management and Livelihoods Project (NACRMLP) collaborative project on Reforming Governance of Forest User Groups: Facilitating Self-Monitoring Process (2003–2004).

Notes

1 I worked as field researcher with the Center for International Forestry Research (CIFOR) during the project phase of 2000–2002 and joined ForestAction in 2003. ForestAction is an NGO working in forestry, livelihoods and environment in Nepal. It was set up in 2000 by a group of professionals from forestry, agriculture and social sciences with their shared passion for triggering community empowerment

in forestry and natural resource management. I led the ForestAction component during 2004–2007 and led the first half of the project during 2008–2011.

2 These are the short versions of the projects. Full names are listed in the acknowledgement section.

3 These low caste people are called *dalits* in hierarchical Hindu caste system.

4 Key individuals here refer to the local politicians, social workers, teachers and other influential persons who are locally called *buddhijibis*.

5 *Dalits* are often known as *Achhut* (i.e. untouchables) in hierarchical Hindu caste system. They are the lowest caste group in the hierarchy. Most of them are at margins in terms of education, and social and economic statuses. Untouchables are subject to discrimination by other higher caste/ethnic groups. They are mostly marginalized in decision-making and in access to resources as well.

6 Until late 1980s, forest officials from Nepal government received training from Indian Forest College, Dehradun, India, which provides technical and managerial knowledge and skills about forest conservation to the officials from the Indian Forest Service.

7 Colfer (in this volume) describes this tension clearly.

8 When the author talked to other colleagues who joined the project several months earlier than me and got the opportunity to participate in the inception workshop, revealed that the project concept was evolving and many of the senior researchers had also the same problem of digesting the concept and grounding it in the local context.

9 There were four CFUGs participating in the research – two each from Kaski and Sankhuwasabha districts. In each CFUG there was one field researcher. There were separate teams working in Kaski and Sankhuwasabha districts and all the researchers from each team were meeting occasionally to discuss the field level issues, and for planning.

10 The ACM project at that time was introduced in 11 different countries including Nepal, the Philippines, Indonesia, Bolivia, Cameroon and Zimbabwe

11 In many cases, we offered them conditional promises. We kept assuring them that we were with them to secure additional funding. Narratives used to receive their faith to the project include, for example: i) there are several funding sources available at higher levels where we could join you to explore and push, ii) let us ask higher levels of the project whether there is some matching fund available by chopping other project expenses, and iii) you will also receive some opportunities of training, etc. Preconditions set for all these provisional promises were that the proposal should come through the inclusive and deliberative process and that the leadership showed full commitment for deliberation, inclusion and equity.

12 The lowest administrative unit of government forest authority often comprising more than one village development committee, which is the lowest political unit of the local government.

13 The facilitators were conceptualized as change agents (CAs) in this phase. Rather than sticking with the conventional role of facilitators where they play unbiased or neutral roles only, they were encouraged to engage in their regular roles that they used to perform in their respective positions in the CFUG or meso level. Later, most of the CAs pursued, the role of an agent of change for pro-poor and equity oriented deliberative governance.

14 Sushila Rana (personal communication, October 2010).

15 Bishnu Hari Pandit, Lead Researcher for New ERA sites for the phase 2004–2007 (talked to December 2010).

16 One of the CAs for the second phase was from a case study CFUG of Phase 1, who later became the leader of FECOFUN district chapter. Since researchers were regularly sharing information with FECOFUN district chapter, many of the executive members were familiar with the concept and progress of the project. Another CA remained the leader of FECOFUN from the early stage of Phase 1.

17 Based on the discussion with New ERA team during the inter-team meetings.
18 Low land plain area of Nepal which extends from east to west on the southern part and covers about 17 per cent of land area.
19 Habermas uses the term technocracy to refer to the influence of external experts in the process of governing people. We would like to use the term technocrats and technocratic sphere to refer to the technical experts within and outside of government systems and the communicative sphere around them, respectively.
20 Draws on Giri *et al.* 2011.

References

Agarwal, B. (2001) *Participatory Exclusions, Community Forestry, and Gender: An Analysis for South Asia and a Conceptual Framework.* World Development 29(10): 1623–1648.

Agrawal, A. (1995) *Dismantling the Divide between Indigenous and Scientific Knowledge.* Development and Change 26(3): 413–440.

Banjade, M. R. (2006) *Transforming Policies and Institutions in Community Forestry of Nepal: The Role of Participatory Action Research.* Exploring Regional CBNRM Policy and Policy Advocacy. International Institute of Rural Reconstruction (IIRR), the Philippines.

Banjade, M. and H. Ojha (2005) *Facilitating Deliberative Governance: Innovations from Nepal's Community Forestry – A Case of Karmapunaya.* The Forestry Chronicle 81(3): 403–408.

Banjade, M. R., H. Schanz and C. Leeuwis (2006a) *Discourses of Information in Community Forest User Groups in Nepal.* International Forestry Review 8(2): 229–240.

Banjade, M. R., N. P. Timsina, H. R. Neupane, K. Bhandari, T. Bhattarai and S. K. Rana (2006b) *Transforming Agency and Structure for Facilitating Pro-Poor Governance in Community Forestry.* Journal of Forest and Livelihood 5(1): 22–33.

Banjade, M. R., N. S. Paudel, H. Ojha and C. McDougall (2007) *Conceptualising Meso Level Governance in the Management of Commons: Lessons from Nepal's Community Forestry.* Journal of Forest and Livelihoods 6(1): 48–58.

Banjade, M. R., N. S. Paudel, R. Karki, R. Sunam and B. R. Paudyal (2011) *Putting Timber in the Hot Seat: Discourse, Policy and Contestations over Timber in Nepal.* Discussion Paper Series. Kathmandu, ForestAction. 11:2.

Bengston, D. N. (1994) *Changing Forest Values and Ecosystem Management.* Society and Natural Resource 7: 515–533.

Berkes, F. (1993) *Traditional Ecological Knowledge in Persepctive.* Traditional Ecological Knowledge: Concepts and Cases. J. T. Inglis. IDRC.

Blaikie, P., K. Brown, M. Stocking, L. Tang, P. Dixon and P. Sillitoe (1997) *Knowledge in Action: Local Knowledge as a Development Resource and Barriers to its Incorporation in Natural Resource Research and Development.* Agricultural Systems 55(2): 217–237.

Brokensha, D., D. Warren and E. Werner, Eds (1980) *Indigenous Knowledge Systems and Development.* Lanham, MD, University Press America.

Bryant, R. (1998) *Power, Knowledge and Political Ecology in the Third World: A Review.* Progress in Physical Geography 22(1): 79–94.

Chambers, R. (1997) *Whose Reality Counts? Putting the First Last.* London, Intermediate Technology Publications.

CIFOR (1999) *The CIFOR Criteria and Indicators Generic Template.* Bogor, Center for International Forestry Research.

Dryzek, J. S. (2000) *Deliberative Democracy and Beyond: Liberals, Critics, Contestations.* Oxford, Oxford University Press.

Dryzek, J. S. (2010) *Foundations and Frontiers of Deliberative Governance.* Oxford, Oxford University Press.

Escobar, A. (1998) *Whose Knowledge, Whose Nature? Biodiversity, Conservation and the Political Ecology of Social Movements.* Journal of Political Ecology 5: 53–82.

Fals-Borda, O. and M. A. Rahman (1991) *Action and Knowledge: Breaking the Monopoly with Participatory Action-Research.* London, Intermediate Technology Publications, The Apex Press.

FAO (1978) *Forestry for Local Community Development.* Rome, Food and Agriculture Organization of the United Nations.

Finger, M. F. and P. Schuler, Eds (2003) *Poor People's Knowledge: Promoting Intellectual Property in Developing Countries.* Washington DC, The World Bank.

Fisher, R., R. Prabhu and C. McDougall, Eds (2007) *Adaptive Collaborative Management of Community Forests in Asia: Experiences from Nepal, Indonesia and the Philippines.* Bogor, Center for International Forestry Research (CIFOR).

ForestAction (2011) *Reducing Poverty Through Innovation System in Forestry: End of Project Report.* Kathmandu, ForestAction Nepal.

Forester, J. (1999) *The Deliberative Practitioner: Encouraging Participatory Planning Processes.* Cambridge, MA and London, MIT Press.

Gilmour, D. A. and R. J. Fisher (1991) *Villagers, Forests and Foresters: The Philosophy, Process and Practice of Community Forestry in Nepal.* Kathmandu, Sahayogi Press.

Giri, K. and H. R. Ojha (2010) *Enhancing Livelihoods from Community Forestry in Nepal: Can Technobureaucratic Behaviour Allow Innovation Systems to Work?* 9th European IFSA Symposium Vienna, Austria.

Giri, K., D. Uprety, H. Ojha and M. R. Banjade (2011) *What Does Research into Use Actually Mean: A View from the Reality of Practice.* Draft Discussion Paper. Kathmandu, ForestAction.

Green, D. (2009) *Turning Spotlight on Nepal's Hidden Battles.* East Anglian Daily Times. East Anglia.

Habermas, J. (1975) *Legitimation Crisis.* Boston, Beacon.

Habermas, J. (1987) *The Theory of Communicative Action: Lifeworld and System – A Critique of Functionalist Reason.* Cambridge, UK, Polity Press.

Habermas, J. (1996) *Between Facts and Norms: Contributions to a Discourse Theory of Law and Democracy.* Massachusetts, MIT Press.

Häusler, S. (1993) *Community Forestry: A Critical Assessment: the Case of Nepal.* The Ecologist 23(3): 84–90.

Hendriks, C. M. (2009) *Deliberative Governance in the Context of Power.* Policy and Society 28(3): 173–184.

Holling, C. S. (1995) *What Barriers? What Bridges?* Barriers and Bridges to the Renewal of Ecosystems and Institutions. L. H. Gunderson, C. S. Holling and S. S. Light. New York, Columbia University Press: 3–37.

Kant, S. (2007) *Economic Perspectives and Analyses of Multiple Forest Values and Sustainable Forest Management.* Forest Policy and Economics 9(7): 733–740.

Kattel, S. P., M. R. Banjade and D. Uprety (2009) *Baseline Study: Reducing Poverty through Innovation Systems in Forestry.* Kathmandu, ForestAction Nepal.

Khadka, S. R. and D. Schmidt-Vogt (2008) *Integrating Biodiversity Conservation and Addressing Economic Needs: An Experience with Nepal's Community Forestry.* Local Environment 13(1): 1–13.

Kumar, N. (2002) *The Challenge of Community Participation in Forest Development in Nepal.* Washington DC, World Bank.

Lee, K. N. (1999) *Appraising Adaptive Management.* Conservation Ecology 3(2): 3.

Leslie, A. J. (1987) *Foreword.* The Purpose of Forests. J. C. Westoby. Oxford, Basil Blackwell: vii–xiii.

Mahesh, G. and G. Hemlata (2004) *Combating Desertification: Building on Traditional Knowledge Systems of the Thar Desert Communities.* Environmental Monitoring and Assessment 99(1–3): 89–103.

Malla, Y. B. (2000) *Impact of Community Forestry Policy on Rural Livelihoods and Food Security in Nepal.* Unasylva 51(202): 37–45.

Malla, Y. B. (2001) *Changing Policies and the Persistence of Patron–Client Relations in Nepal: Stakeholders' Responses to Changes in Forest Policies.* Environmental History 6(2): 287–307.

McDougall, C. and A. Braun (2003) *Navigating Complexity, Diversity and Dynamism: Reflections on Research for Natural Resource Management.* Managing Natural Resources for Sustainable Livelihoods: Uniting Science and Participation. B. Pound, S. Snapp, C. McDougall and A. Braun. London & Ottawa, Earthscan Publications Ltd and The International Development Research Centre: 20–47.

McDougall, C., K. A. Team, N. E. A. Team and ForestAction (2002) *Planning for the Sustainability of Forests through Adaptive Co-Management: Nepal Country Report.* Bogor, Indonesia, CIFOR.

McDougall, C., F. Team, N. Team and E. Team (2007) *Final Technical Report Enhancing Livelihoods and Equity in Community Forestry in Nepal: The Role of Adaptive Collaborative Management.* Bogor, CIFOR.

McDougall, C., H. Ojha, M. R. Banjade, B. H. Pandit, T. Bhattarai and M. Maharjan (2008) *Forests of Learning: Experiences from Research on Adaptive Collaborative Approach to Community Forestry in Nepal.* Bogor, CIFOR.

Messerschmidt, D. A. and A. L. Hammett (1998) *Local Knowledge of Alternative Forest Resources: Its Relevance for Resource Management and Economic Development.* Journal of Sustainable Forestry 7(1–2): 21.

Nightingale, A. (2002) *Participating or Just Sitting In? The Dynamics of Gender and Caste in Community Forestry.* Journal of Forest and Livelihood 2(1): 17–24.

Nightingale, A. J. (2005) *'The Experts Taught us All we Know': Professionalisation and Knowledge in Nepalese Community Forestry.* Antipode 37(3): 581.

Ojha, H. (2006) *Techno-Bureuacratic Doxa and the Challenges of Deliberative Governance: The Case of Community Forestry Policy and Practice in Nepal.* Policy and Society 25(2): 131–175.

Ojha, H. R. (2008) *Reframing Governance: Understanding Deliberative Politics in Nepal's Terai Forestry.* New Delhi, Adroit Publishers.

Ojha, H., B. K. Pokharel and K. Paudel (2002) *Adaptive Collaborative Management: Towards a New Approach to Community Forestry.* Seminar on Human-Institutional-Natural Resources Interaction, Pokhara, Nepal, Institute of Forestry, Pokhara in collaboration with Indiana University, USA.

Ojha, H., J. Cameron and C. Kumar (2009a) *Deliberation or Symbolic Violence? The Governance of Community Forestry in Nepal.* Forest Policy and Economics 11(5–6): 365–374.

Ojha, H., L. Persha and A. Chhatre (2009b) *Community Forestry in Nepal: A Policy Innovation for Local Livelihoods.* IFPRI Discussion Paper, IFPRI and Vision 2020.

Ojha, H. R., N. S. Paudel, M. R. Banjade, C. McDougall and J. Cameron (2010) *The Deliberative Scientist: Integrating Science and Politics in Forest Resource Governance in Nepal.* Beyond the Biophysical: Knowledge, Culture, and Politics in Agriculture and Natural Resource Management. L. German, J. J. Ramisch and R. Verma. London, Springer: 167–191.

Patton, M. Q. (2002) *Qualitative Evaluation and Research Methods.* London, New Delhi, Sage.

Paudel, K. P. and H. Ojha (2007) *Imposing Indicators or Co-Creating Meanings Through Joint Reflection? Lessons from Community Forestry in Nepal.* Negotiated Learning: Collaborative

Monitoring in Forest Resource Management. I. Guijt. Washington DC, Resources for the Future.

Paudel, N. S., M. R. Banjade and G. R. Dahal (2008) *Handover of Community Forestry: A Political Decision or a Technical Process?* Journal of Forest and Livelihoods 7(1): 27–35.

Pokharel, B., P. Branney, M. Nurse and Y. Malla (2007) *Community Forestry: Conserving Forests, Sustaining Livelihoods and Strengthening Democracy*. Forest and Livelihood 6(2): 8–19.

Pokharel, B. K. (1997) *Foresters and Villagers in Contention and Compact: The Case of Community Forestry in Nepal.* Ph.D., East Anglia.

Shivaramakrishnan, K. (2000) *State Science and Development Histories: Encoding Local Forestry Knowledge in Bengal.* Development and Change 31: 61–89.

Thapa, B., F. L. Sinclair and D. H. Walker (1995) *Incorporation of Indigenous Knowledge and Perspectives in Agroforestry Development. Part 2: Case-study on the Impact of Explicit Representation of Farmers' Knowledge.* Agroforestry Systems 30(1–2): 249.

Timsina, N. P. (2003) *Promoting Social Justice and Conserving Montane Forest Environments: A Case Study of Nepal's Community Forestry Programme.* Geographical Journal 169(3): 236.

Timsina, N., H. Luintel, K. Bhandari and A. Thapaliya (2004) *Action and Learning: An Approach for Facilitating Change in Knowledge-Power Relationship in Community Forestry.* Journal of Forest and Livelihood 4(1): 5–12.

UN (2011) *International Year of Forests 2011.* Online at http://www.un.org/en/events/iyof2011/.

Uprety, D., K. Giri, M. R. Banjade, K. Bhandari and R. Bista (2011) *Innovation System for Pro-poor Community Based Forest Governance: Lessons From the Studies of 60 Groups of Community Forests.* Kathmandu, ForestAction.

Walker, D. H., F. L. Sinclair and B. Thapa (1995) *Incorporation of Indigenous Knowledge and Perspectives in Agroforestry Development. Part 1: Review of Methods and Their Application.* Agroforestry Systems 30(1–2): 235.

Westoby, J. (1987) *The Purpose of Forests.* Oxford, Basil Blackwell.

Whyte, W. F., D. J. Greenwood and P. Lazes (1989) *Participatory Action Research: Through Practice to Science in Social Research.* American Behavioral Scientist 32(5): 513–551.

8 Learning through action

Reflections on action research in natural resource management

Robert Fisher

Introduction

The aims of this chapter are to present a personal reflection on a variety of experiences involving action research (AR) and adaptive collaborative management dating back to the 1980s and to highlight some of the challenges involved in practising action research. The experiences covered include an early period working in a community forestry project in Nepal, participation as supervisor of several action research oriented Ph.D. studies, and training and mentoring action research/ACM (adaptive collaborative management) teams in a number of field-based projects concerned broadly with natural resource management and development.

As a personal reflection, the chapter necessarily covers a wide range of experiences in various levels of detail. I was trained as an anthropologist and have strong interests in human ecology and development. My experience has been somewhat unusual in that I have been engaged in action research and adaptive collaborative management both as an academic and as an active development practitioner. This has enabled me to experience both the opportunities and challenges of these approaches in both academic and applied contexts.

Because the chapter is based on personal reflection, there is a risk that it may seem self-indulgent. I have tried to avoid that by highlighting the challenges and lessons arising from various experiences.

Concepts

All the work discussed in this chapter deals with two generic elements in various ways. The first of these is action learning, or adaptive learning. The second is participation and collaboration. Although the experiences were sometimes labelled action research and other times adaptive collaborative management (or variations on these), the common link was a concern with learning for action and learning from action. The collaborative or participatory element was much more variable, ranging from fairly elaborate arrangements to enable participation by a wide range of stakeholders to more limited participation by a small group in the actual research activity. One of the main points I wish to make in the chapter is that there is a variety of different ways that stakeholders can be engaged in participatory or collaborative activities and also a variety of ways of structuring participation.

Defining the various approaches can lead to something of a quagmire, as terms are used by different authors in a variety of different and sometimes contradictory ways. I prefer to take a more eclectic approach and to use definitions in a pragmatic fashion recognizing the underlying common features of different approaches and specifying different features. In this chapter, except where I am referring to the terminology used by others, I use the following terminology:

- Learning refers to any approach that involves learning from actions carried out, particularly when this involves learning from one set of actions to improve future actions. It has essentially the same meaning as adaptive learning.
- Action research is a form of action learning that aims to assist improved action in a particular context and to contribute in some way to knowledge beyond that context. In other words it aims to identify lessons or insights that may be relevant to people working in different but similar contexts. Insights learned from implementing community forestry in a particular village in Nepal may be useful to people working in participatory forest management, or possibly even fisheries management, elsewhere. Drawing lessons of wider relevance than the immediate context is the 'research' element in action research. The insights/lessons may be published in some form or circulated by other means. The research aspect of action research is a concern with producing 'public' knowledge.
- Participatory action research (PAR) is action research that involves actors beyond the original researchers in the action research approach. Many practitioners and theorists of action research argue that action research is by definition participatory, that it must be participatory to be called action research (see for example Kemmis and McTaggart 1988a). I feel it is useful to distinguish between action research and participatory action research as a basis for understanding the differing levels of decision-making power in action research teams. In practice the roles of different stakeholders in action research differ widely and there is little point in being pedantic about what action research should be. Rather, 'the question is not so much about being pedantic about definitions, but about the need to specify just what sort of participation occurred and just who was involved' (Fisher 2003).
- Applied research is research undertaken to solve a particular problem. Sometimes applied research is carried out as a separate activity for application in the future (research precedes implementation); sometimes it is undertaken jointly with action. Action research applies when the

research and action are part of the same process. It is therefore a subtype of applied research.

- Adaptive management is an approach to natural resource management that operates on the assumption that the social and biophysical contexts of the natural resource system are not fully understood in advance and that, as a result, management must be based on steps that are subject to modification based on observation of the results of earlier steps (this tentative and iterative approach is very similar to action learning and action research). Adaptive collaborative management involves multiple actors and stakeholders in the process.

This chapter is structured around discussions of experiences in action research (and, to a lesser extent, adaptive collaborative management) in a number of different contexts. The next section discusses action research in the context of a long-term development/resource management project involving community forestry in Nepal. The following section shifts to experiences in the context of a role as an academic working in a university. It begins by discussing the ideas about action research that guided what happened in the university and then explores, in turn, a formal action research project in Papua New Guinea and action research in several research theses written by Masters and Ph.D. students in which I had a supervisory and sometimes collaborative role. The following section deals with a role in conducting training workshops for research and development practitioners. This is followed by some observations arising from collaboration in the Center for International Forestry Research (CIFOR)'s Adaptive Collaborative Management Project and The World Conservation Union (IUCN)'s Livelihoods and Landscapes Strategy. The discussion section looks at some of the broader challenges involved in applying ACM, based on these varied experiences, and attempts to suggest some strategies to deal with some of these challenges.

Action research experiences

The experiences covered include an early period working in a community forestry project in Nepal, participation as supervisor of several action research-oriented Ph.D. studies, and training and mentoring action research teams in a number of field-based projects concerned broadly with natural resource management and development.

Action research in the context of a 'development' project: community forestry in Nepal

I first became aware of 'action research' when I worked as project anthropologist with the Nepal–Australia Forestry Project (NAFP) in the period 1987–1989.[1] (My position was officially labelled 'Community

Worker', reflecting the reluctance of the donor and/or the Government of Nepal to see social science as a legitimate need in the project.)

A number of elements that it is hard to see as anything other than fortuitous accidents allowed the project to operate with a high degree of flexibility which, I believe, made an 'action research' approach possible. The first of these was that the project had evolved out of a long and relatively informal involvement of Australian foresters as advisors to forestry authorities in Nepal. For many years Australian forest scientists had been involved in providing technical advice on forestry matters, first during short visits and later through advisors posted in Nepal. Forest scientists from the Department of Forestry at the Australian National University were among the short-term advisors. Eventually this evolved into a bilateral project. From 1978 the project ('Phase 2') was managed and implemented by the Department of Forestry at Australian National University (ANU). The second element was that, at least in the earlier years, the project was not over-designed, giving the project team room for flexibility. A third element was that, although the project was not a research project, there was always a fairly strong learning culture. The project Director, Professor David Griffin, Head of the Department of Forestry at ANU, was himself a creative thinker, as was Dr Don Gilmour the Team Leader for much of the 1980s. Both of these had a commitment to achieving things on the ground and, at the same time, documenting the research process. The tradition of linking research and practice was exemplified by Ph.D. studies carried out at ANU by Tej Bahadur Mahat, a Divisional Forest Officer who inspired much of the innovative thinking on community forestry in Nepal in the 1970s and early 1980s. The tradition of learning from the project context and using what was learned to inform future action emerged early during the ANU involvement. The fourth facilitating element was that senior officers in the Department of Forestry gave the project considerable room to move in experimental activities.

By the time I joined the project in early 1987, the term 'action research' was used in the project. Griffin (1988), in a book that reflected on the experiences of the project, wrote explicitly about action research as an inspiration. He describes the way that the work of the project seemed to be 'undertaken in the existence of scant facts' (p. 70), yet it was not carrying out traditional scientific or applied research (defined as research that precedes implementation). The discovery of literature on action research helped him, as he describes it, 'to view the Project in a new light and to place it within a far more satisfying conceptual framework' (p. 71). For him the essence of 'participatory action research' involved making 'an irreducible minimum of initial assumptions' and working with communities to define and clarify issues. He saw that this approach made a 'grand design approach' impossible.

Interestingly it seems that Griffin's rethinking was not about discovering action research from the literature and then applying it. It was more about finding a framework to describe what was already going on in some sort of instinctive way. As I see it, the important thing is that, once the framework is identified, the 'action research' or 'action learning' approach (whichever you want to call it) becomes conscious, disciplined and a routine process.

At the country level, the project operated on a flexible learning based approach, with a strong culture of review and reflection as a basis for planning activities. Project staff, including the counterpart District Forest Officers (DFOs), routinely met to discuss ongoing activities and to plan new activities. As part of this process activities were reviewed and revised according to outcomes. New activities were identified. A key example of this relates to early efforts to develop a community forest management plan at *Chaap al Danada* forest (see Gilmour and Fisher 1991 for a discussion of this case). Community level meetings had been held to identify users and prepare a plan, and arrangements were made for people from particular villages to collect wood for fuel according to a schedule included in the plan. However some villagers did not turn up on the scheduled day. It was at this point that flexible planning and adaptation came in. After discussions by the project staff (including the DFO), the project's Women's Coordinator carried out an informal survey of the relevant villages to find out why people did not collect the fuel-wood allocated. She found that the villagers did not see the forest patches as being part of their traditional forest patches. It became clear that the community forest management planning process (one of the first exercises carried out in Nepal) had been based on a lack of knowledge of traditional use patterns and a totally different approach was developed for future cases.

In some ways the original forest management planning exercise seems to have been rather naive. In retrospect it was. However, the complexities of community use rights and community planning were new to forestry in Nepal. The important point is that critical reflection led to rethinking of what needed to be done.

Another outcome of this process was the identification of the need to learn much more about 'traditional' or 'indigenous' forest management practices. Such practices had often been discussed anecdotally, but relatively little was known about them. As a result I became involved in a separate study of this subject, in cooperation with the International Centre for Integrated Mountain Development (ICIMOD).

This example illustrates how a flexible learning based approach can work. A culture that values review and reflection as a basis for planning future activities is the essence of action research.

It is important to stress that, while there was a culture of review and reflection, and while there was a broad notion of action research as a framework for action, the application of action research was not based on any very detailed grounding in action research literature and methodology. The key elements were a strong culture of reflection and a recognition that the project was operating within a context of considerable uncertainty. There was also a recognition that new challenges and problems would emerge progressively. A paper on action research presented at the 1987 International Union of Forestry Research Organizations (IUFRO) Symposium (Gilmour *et al.* 1988) specifically discussed action research as a participatory approach involving problem identification though close collaboration with villagers. It focused on a role for foresters as 'new professionals' (Chambers 1986) working through a village centred approach. However it was not at all explicit about what the term 'action research' meant beyond the idea that it would link action and research.

For me, the most striking lesson about action research from the NAFP experience is the importance of a culture of 'review and reflection'. Theoretical literature on action research stresses the importance of a *conscious* cyclical process of review and reflection. (I will discuss this further in a later section.) In NAFP there was no really explicit model mandating such a process. It simply happened routinely. The project office was in Kathmandu. Every day those who were in the office about morning or afternoon tea time (including the DFOs when they were in town) met and discussed activities and plans over tea or coffee. Quite major decisions about activities and strategies were made at these informal meetings. There were, of course, many formal project meetings, but it seems to me that it was the culture that was crucial rather than any formal process. From subsequent experience I suspect that, once a formal process of meetings (including 'minute-taking') becomes a routine, a more informal culture of reflection will develop, given supportive leadership.

Another important observation that arises out of the NAFP experience is that action research was an overall project modus operandi that allowed progressive identification of issues and areas of focus. Plans to address these areas of focus often led to more 'conventional' research to address specific issues, either on biophysical issues or social issues. For example, concerns about sustainable harvesting of shrublands led to trials in silvicultural treatment of shrublands through forms of 'appropriate silviculture'. In other cases social research was undertaken to address social questions that had been identified.

It is worth commenting on the nature of participation within the action research. Although the project involved a high level of community participation at the local level, the action research process itself essentially involved the project team, including government staff, but did not extend to the communities. I would describe it as action research, but not broad based participatory action research.

The NAFP experience raised a very important challenge for action research. This is the difficulty of fitting a flexible learning approach into a large-scale aid framework, with governments and donors emphasizing standardized programmes and donor requirements for detailed log frames and specified inputs and outputs. While project designs sometimes talk about learning approaches, the same documents generally include targeted inputs and outputs that create a momentum that neutralizes any real adaptive or action learning. Griffin's concerns about avoiding 'a grand design approach' bring this tension into focus.

One issue that frequently arises in discussions about action research is the challenge it may present to researchers with traditional scientific backgrounds. This applies to many biophysical scientists, but may often apply to social scientists who have their own concerns with scientific rigour and credibility, perhaps particularly to quantitative social scientists. Although I had not heard of action research before I worked in Nepal, this never presented itself as a challenge to me. Perhaps the main reason for this is that action research is rather similar to the participant observation approach of social anthropology. It is characterized by extended periods of rapport building, informal discussion and systematic observation (Gilmour *et al.* 1988).

I see participant observation as involving a process of observing, questioning, analyzing and interpreting, followed by further cycles of follow-up questions and observations. This is very much like action research.

Practising and mentoring action research in an academic setting: University of Western Sydney – Hawkesbury

A year or so after I left NAFP, I was appointed as a lecturer in the Faculty of Agriculture and Rural Development (subsequently renamed several times under a series of restructures) at the University of Western Sydney – Hawkesbury (UWS-H) where I taught from 1991 to 1996. This was an immensely exciting period as the faculty (but not the university as a whole) was experimenting with a radically reorganized curriculum based on what was called 'systems agriculture'. The educational approach was heavily based on 'experiential learning' and action research was a crucial element,[2] especially for Masters and Ph.D. students whose thesis work was, in most cases, action research based. The approach was strongly interdisciplinary. Although I was appointed as a social scientist, most of my colleagues came from backgrounds in agriculture, biological science or agricultural extension. Although few had any specific disciplinary background in social sciences, all had become well-versed in learning theory and systems theory and there was a strong vision of agriculture as a 'human activity system', rather than a focus on agricultural science and technology. Action research was seen as both an approach to research and a means of achieving change and development, and thesis projects were heavily oriented towards change in agricultural and development contexts.

At UWS-H I had my first exposure to a more theorized and structured approach to action research. This included a strong emphasis on action research in education. Much of the inspiration came from work done in an education context by Kemmis and McTaggart at Deakin University (see Kemmis and McTaggart 1988a and 1988b).

There were a number of key elements in the more structured and theorized approach to action research that I was exposed to at UWS-H. Some of these were:

1 The grounding of action research in experiential learning (Kolb 1984).
2 The presentation of the action research as a cyclical process derived from Kolb's 'learning cycle'. This action research cycle was memorably presented by Kemmis and McTaggart (1988a) as a repeated series of cycles of planning, acting, observing, reflecting and re-planning (Figure 8.1). The action research cycle became a useful way of representing the structure for routinizing action research. In practice, much discussion went into developing diagrams that represented the complexity of the cycle in practice, with diagrams showing parallel cycles dealing with different issues concurrently, and cycles with sub-cycles as sub-issues were pursued.
3 The importance of action research as a participatory or collaborative activity involving a group of people with a shared 'thematic concern' rather than a research question. An example of a thematic concern might be that there is tension in the relationships between nomads and officials.

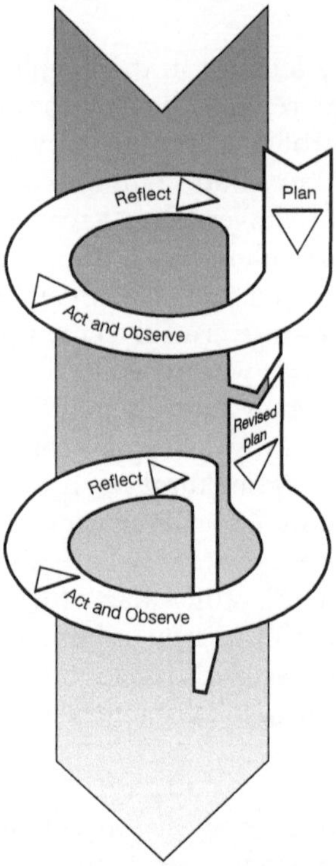

Figure 8.1 The action research spiral (source: from Kemmis and McTaggart, 1988a).

This is a general area of concern, but different actors may see the underlying problem differently. They have different ways of framing the problem and refining the problem or problems and developing solutions can only occur progressively (this example is discussed below). The important point here is that there is no necessary shared definition of the problem or clear objective of the outcome of the research at the beginning. Clarity about what needs to be done may be developed progressively as the action research progresses.

4 The existence of different traditions of participatory action research. According to Brown (1993, p. 249), one involves participation for problem solving and the other involves empowerment:

> 'Participatory action research' has been acclaimed for producing new knowledge as well as solutions to organizational and social problems. But the term has quite a different meaning to different audiences; The 'Southern' tradition is committed to community transformation through empowering disenfranchised groups; the 'Northern' tradition is concerned with reforming organizations through problem solving.

At the time I had, and still have, a problem with this type of view which sees action research as being always about social change, either at the organizational level or at a broader social level. While it is clear that action research is a *social process*, and while it is often about organizational and social change, my experiences in Nepal suggested to me that action research could also incorporate cycles of 'technical' research (such as research into appropriate silviculture arising from action research into project implementation). There were other examples of research involving agriculture and natural resource management. For example, a collection of papers by Whyte (1991) includes papers on agriculture and the work of Rhoades at the International Potato Center in Peru (Rhoades 1984) is essentially about PAR to develop appropriate agricultural innovations.

There is, nevertheless, a strong sense in the history of action research that it evolved very much as a method of social action. This is evident in the pioneering work of Lewin (1946), who seems to have invented the term. The more 'emancipatory' social change tradition is linked with Friere (1982). The excellent book *Introduction to Action Research: Social Research for Social Change* by Greenwood and Levin (1998) strongly emphasizes the role of action research in social change. They also stress that action research is 'a form of *social* science' (Fisher 2002). In a review of their book I pointed out that this

> may mask its potential relevance in the application to the physical sciences. There are many examples of action research applied to agriculture or natural resource management (such as in the case of community forestry), where social change and changing technical

> practices go hand in hand. In addition to eclectic use of social science tools and methods, Action Research can apply tools and methods associated with physical sciences in such cases. Transdisciplinary applications of AR are essentially ignored in the book.
>
> (Fisher 2002, p. 41)

Although my background is in social science, my experience in Nepal had strongly impressed on me the potential for action research to address both social and 'technical' change in a rather eclectic manner. Some of the action research experiences during my Hawkesbury experience demonstrated this. In the remainder of this discussion of the UWS-H experience, I will first discuss an action research project carried out in Papua New Guinea. I will then discuss some of the work done by research students who I supervised or for whom I was associate supervisor.

Action research in agricultural extension in Papua New Guinea

In 1991 and 1992 a colleague from Hawkesbury and I became involved in an action research project concerned with improving rural livelihoods in Goilala District of Papua New Guinea (Sriskandarajah and Fisher 1992).[3] The participatory action research was carried out in collaboration with extension staff in Goilala District in the context of the Smallholders Market Access and Food Supply Project (SMAFSP).

> The rationale for this [action research] project was a perception that lack of initiative, appropriate expertise and appropriate approaches to planning, had delayed implementation of SMAFSP in Goilala. The role of the Hawkesbury researchers was seen as a way of assisting staff to overcome the inertia and activating the SMAFSP project towards meeting its objectives. On the assumption that inactivity prior to the action research project was largely the result of a feeling of powerlessness among project staff as to their ability to implement project activities, we saw our role as being to facilitate a reflective learning process which would increase staff confidence and commitment.
>
> (Sriskandarajah and Fisher 1992, p. 1)

At the action level the project aimed to assist 'staff to learn how to use Participatory Action Research or PAR [as a learning approach] as a "routine" approach to their normal activities'. In terms of research the project was 'concerned with evaluating the utility of participatory action research as a methodology applicable to rural development projects in general' (Sriskandarajah and Fisher 1992, p. 1). In other words, the project was concerned with social action/change and was also explicitly a social research project.

It is important to understand that Goilala District, although only about half an hour away as the Cessna flies from Port Moresby, is very mountainous and is difficult to access by road. Travel within the district was possible mainly by walking or flying between small airstrips (of which there were quite a few). The district was depressed economically. At the time the project was undertaken it was suffering from severe law and order problems and the population faced high levels of malnutrition. District extension staff were not well supported by the national and provincial levels and were expected to implement a complex project in an inaccessible area with inadequate guidance. There had been cases of staff commissioning local people to carry out work included in the workplan, only to have commitments for payment dishonoured by administration. In the context of an area with a considerable reputation for violence this was both disheartening and dangerous. The result was that little activity was occurring.

The rationale for using PAR in Goilala was that:

- Involvement by actors in institutional change gives them a sense of ownership. This occurs when they identify goals, constraints and opportunities and plan and negotiate for the future.
- By enabling individuals to take responsibility for small achievable tasks, PAR helps people to develop confidence.
- By encouraging flexibility and continuing evaluation, PAR enables actors to prioritize activities, to focus on realistic action and to abandon or alter inappropriate or counterproductive activities.

The overall approach involved three visits by the Core Action Research Team (the two Hawkesbury researchers and the Rural Development Officer from Goilala District). Each visit involved reviewing the situation and identifying tasks in the form of action plans specifying individual responsibilities. The first visit began with a planning meeting in Port Moresby involving national and provincial level project staff as well as district and sub-district staff. This meeting introduced the concept of PAR and went through a process of review and planning for activities. This meeting was followed by a major workshop of district staff at the District Headquarters. It included local representatives of other government departments, some local leaders and the resident Catholic priest. Shorter visits were made to each of the three sub-districts, where meetings were held involving officials and community leaders. At all of the meetings, PAR was introduced and action plans were produced.

The action plans were, in my view, the key to the process. An example of how this worked is the plan to sell vegetables to the market in Port Moresby. This was one of the ideas in the SMAFSP plan, intended to encourage the production of vegetables for local consumption and income generation. The problem is that the project document didn't contain any real plan to achieve this and what might seem quite simple at first sight involved a complex variety of different tasks. First, there was a need to get the vegetables to the market. As there was virtually no road access, this meant that the vegetables had to be flown by light aircraft to Port Moresby. It meant that cold storage facilities would be needed at local airstrips, that there should be a process for collecting the vegetables from Port Moresby airport and that agreements be made with potential buyers. Each of these steps required smaller steps. For example constructing cold storage sheds required purchase of cement in Moresby and transportation to the district. Some of these steps could be carried out locally, but others had to be managed in Port Moresby. Communications were not easy before the development of mobile phone networks.

The idea of action plans was to break down the larger activity (vegetable marketing) into smaller achievable steps. The action research groups discussed what was needed and then negotiated who would do what. For example, person X agreed to arrange for cement to be delivered to Port Moresby airport for transport to one of the Goilala airstrips. Person Y agreed to arrange construction of the cold store. Allocation of tasks depended, of course, on things such as contacts and where people were located. Action plans listed planned activities against responsible individuals. It was understood that the obligation was to attempt to complete the task. If it could not be achieved for some reason, then the responsible person had to report this at the next meeting (that is, during the next action research cycle). The subsequent action plan would try to address whatever the barrier had been.

During the second visit to PNG the core team visited the District Headquarters and the sub-district centres. At each location meetings were held with members of the local action research groups to review progress on action plans, explore new developments and decide on next steps or revised steps. The action research cycle was followed and new action plans were developed.

During the third visit the core team again visited the district and sub-districts to review progress and facilitate the development of further action plans. This was followed by a wind-up meeting in Port Moresby.

So, what did the PAR project achieve?

At the action level there was a very clear improvement in morale and focus at the field level. There were modest achievements in project tasks, which began to be achieved for the first time. Unfortunately, long-term change was subverted because funding for SMAFSP in Goilala was reduced due to a lack of support from the central level. This seems to have partly been the legacy of a lack of confidence in local staff. If so, it was a tragic outcome that failed to take account of the changes that had occurred.

From a research point of view, the main outcome was that the value of PAR as a way to implement complex activities and to build confidence was supported. The importance of following a routinized learning cycle and the use of action plans to identify and implement small achievable steps were also demonstrated, thus providing a model for other action research projects. The research also demonstrated the need for genuine devolution of decision-making authority and the importance of confidence in staff in order to achieve efficient field implementation of development projects.

I have described this process in considerable detail, because it is crucial to understanding how a complex action research project can be structured and, importantly I think, it shows just how useful it can be to consciously follow a cyclical process in order for it to become a routine practice. The process was structured with a number of teams or groups operating at different locations. There was a 'core team', responsible for the overall research project, and groups with differing membership at various locations (national, district and sub-district). Membership of these groups overlapped as some people were members of both district and sub-district teams.

The project involved a large number of people as active participants, all aware of the methodology. These people did not just carry out planned activities, but participated in the action research cycle as a whole, consciously following the process of reflecting, observing, planning and acting. The project is an example of highly participatory action research, based on a complex organizational structure.

Following the somewhat loose notion of action research I had experienced in Nepal, this project provided insights into ways to integrate participation into a fairly large development project through a structure based on 'embedded' and overlapping teams and through the use of clear and documented action plans.

Hawkesbury Action Research Theses

This section discusses aspects of some of the action research-based Ph.D. and Masters theses with which I was associated as supervisor or associate supervisor at UWS-Hawkesbury. As mentioned earlier, most theses in the Faculty of Agriculture and Rural Development had substantial action research components. The theses covered a rich variety of subjects and the action research components were structured in a variety of ways.

Among the theses concerned with action research related to natural resource management or agriculture was a Master's thesis by Tevita Toafa (1994) who aimed to improve the marketing of pumpkins from Tonga for the seasonal Japanese niche market. In order meet the requirements of this market there was a need for a guaranteed supply of high quality pumpkins in a timely manner. This required cooperation between various actors, including merchants who generally competed against each other. Tevita established an action research team with these stakeholders to explore the needs of the market and to develop processes to meet these needs. In this sense his project was action oriented and concerned with organizational change. These combined action research with approaches based on marketing theories related to niche marketing.

Paul Weekley (1999), who had worked for many years as an agricultural advisor working with Fulani nomads in Burkina Faso, wrote his Ph.D. thesis on participatory action research concerned with food security. Helen Ritchie (1998) worked with farmers in New Zealand as an action research group on collective action for stream management. Bill Jackson (Jackson 1999), who I had worked with in Nepal, wrote his Ph.D. on high altitude forest use in Nepal, combining action research with an examination of 'the use and management of *lekh* [high altitude] forests' (p. 312). The descriptive examination of forest management can, I suggest, be thought of as more or less ethnographic research, although it has substantial elements of forestry research. The 'action research approach was used to guide the introduction of community forestry into Dhungkharka and Jethal' (p. 312). This process, combining an 'ethnographic' element or phase with an action research phase, was a common feature of several of the theses and can be a useful model for action research around natural resource management or agriculture. An important lesson here is that overly purist notions about what action research should be can be very limiting.

Likewise, another thesis was a research to improve rainfed lowland rice production in Cambodia (Solieng Mak). Solieng Mak was an agriculture extension specialist working with the Cambodia-IRRI-Australia Project (CIAP). Her Ph.D. research (Mak 1997) explored ways to improve rainfed lowland rice (RLR) production in a context where there was limited potential for change in agricultural practices. The research combined ethnographic research and action research (see Figure 8.2). It began with an ethnographic phase, which involved a detailed study of an RLR system in one village focusing on decision-making about agricultural practices.

The ethnographic research described the rainfed lowland rice farming system, with particular focus on the way the farmers coped with uncertain farming conditions to improve production. It also showed how political upheaval and uncertainty had negatively affected agricultural innovation. The research showed that shortages of labour affected the ability of farmers to use the green manuring technology being promoted by the International Rice Research Institute (IRRI) project. Although the technology, which involves the planting of sesbania, a nitrogen rich plant, in fields prior to the rice season, had been proven to be a successful way to improve rice production, farmers were reluctant to apply it.

The ethnographic phase was followed by two action research projects. The first of these involved villagers experimenting with green manuring as a technical innovation to improve soil fertility. Green manuring requires additional labour and changed farming arrangements, thus increasing complexity. Participatory action research showed how the farmers performed the trial in actual farm conditions. An important finding was an understanding of why some farmers followed and other did not follow the technical guidelines provided by the project (for issues such as spacing and timing of incorporation of chopped sesbania into the soil before transplanting rice).

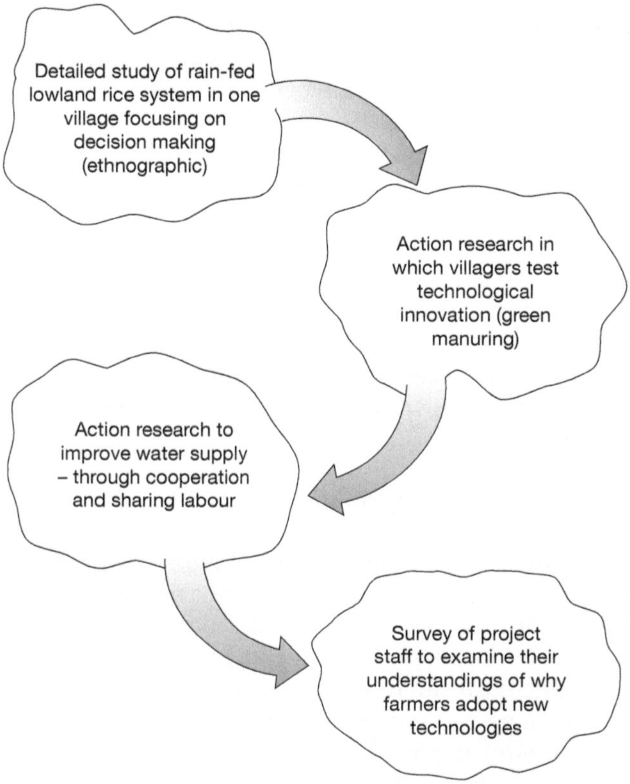

Figure 8.2 Research to improve rainfed lowland rice production in Cambodia (source: Solieng Mak).

Instead farmers re-experimented to find alternative processes that fitted practical requirements. Reflection by the farmers on green manuring indicated difficulties in managing agriculture due to lack of seasonal labour, draft animals and seasonal water availability.

The second action research project involved construction of a rainwater storage pond through cooperation and shared labour. This project emerged from the first action research project, which identified shortage of water as the priority problem to be addressed for the village as a whole. In the various stages of this project it became clear that willingness to cooperate varied according to the extent that benefits were perceived as immediate and meaningful to individuals. Longer term approaches with unclear benefits attracted limited cooperation. In conversations with Solieng about the field site (which I had visited) I was struck by her understanding of the reluctance of rural Cambodians at that time (1994) to work in groups. Beside the issues of perceived benefits, she attributed this to the legacy of the comprehensive system of forced labour established during the Khmer Rouge period (something she had personally experienced as a teenager).

A final phase of the research was a survey of CIAP staff to assess their understanding of why farmers adopt new technologies and dialogue with the farmer participants to reflect on the process and what they liked about it. In terms of action, an outcome of the project was that there was some adoption of new technology, with a resulting increased potential for increased rice production. The research outcomes were:

- a clear understanding of the RLR system and reasons for the 'conservatism' of farmers;
- an understanding of the process of adoption of innovations, as a process of re-experimentation and adaptation to local conditions;
- an understanding of the limited potential for group extension approaches in Cambodia at the time.

The project outcomes were a useful contribution to agricultural extension practice.

The project is a fine example of the way that social and organizational processes and technical research can be combined. Strikingly the PAR around green manuring was itself a form of technical action research.

Still another example of Ph.D. thesis I am highlighting here is about pastoralists, government and natural resources in Iran (M.H. Emadi). Mohammad Emadi's work on pastoral nomads in Iran (Emadi 1995, 2005; Emadi *et al.* 1992) was another example of a multi-phased research project (Figure 8.3). It combined an 'ethnographic' first phase with an action-oriented phase and a final phase of organizational change. The thematic concern behind Emadi's work was that development efforts among Iranian nomads had been largely unsuccessful and relationships between nomads and government agencies had been poor. While the Iranian economy depended (and still depends) heavily on nomads for production of meat, government officials blamed nomads for causing desertification.[4]

In the first phase ('ethnographic exploration') Emadi carried out a study that looked at the economic and social system of a nomadic group (the Ghareghani sub-clan of the Qashqai tribal confederation). He was particularly concerned with understanding the nomads' perceptions of their situation in Iran. These perceptions were:

> - the complexity of the current situation as perceived by the nomads themselves;
> - the unease of the nomads at their present situation;
> - the lack of any signs of improvement in future trends as they saw them;
> - an essential need for a mutual understanding between government officials and the nomads.
>
> (Emadi 2005)

The second phase was an action research phase in which Emadi formed an action research team made up of staff of various government departments concerned with nomadic issues. These departments were the Forest and Range Organization (FRO), the Organization for Nomadic People (ONPI) and the Planning and Budget Organization (PBO), the last represented by an ex-PBO official now working with the FRO. At the beginning, members of the team had different perspectives on these issues, just as their organizations had different perspectives, and each member felt that his perspective was the correct one (Emadi 1995). It was agreed that the team needed to understand the issues from the point of view of nomads, so the team moved to nomadic areas and met with 'some nomads who were invited to share their views and perspectives on the various projects with the government officers' (Emadi 2005). The result was a greater degree of understanding by the officials of the various perspectives of nomads and a greater understanding by nomads of the government's projects.

The members of the action research team had regular discussions to reflect collectively on their observations. (The nomads themselves were not members of the action research team.) The action research team identified a number of actions to address some of the issues facing the nomads. It also led to a

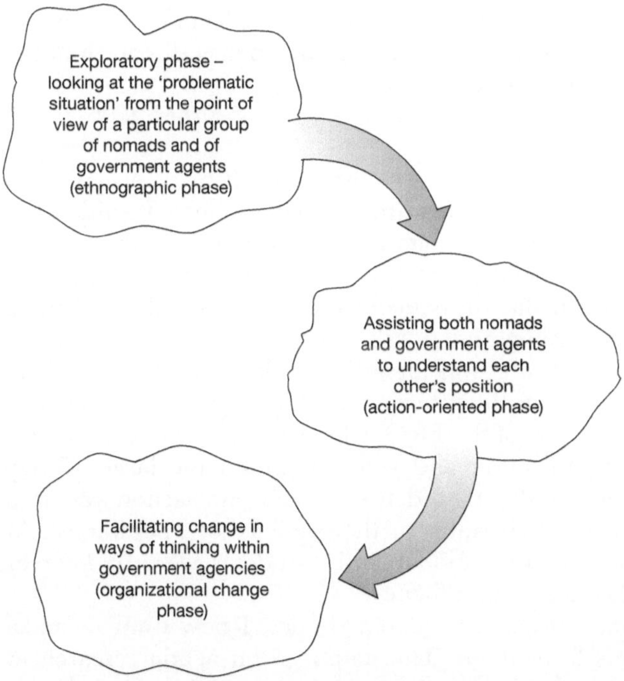

Figure 8.3 Pastoralists, government and natural resources in Iran (source: M.H. Emadi).

third phase, concerned with organizational change within the various government departments. One of the most important components of this was an intensive 13-day workshop with government staff which focused on experiential learning, systems thinking and people's participation in natural resource management.[5] Although discussions from diverse viewpoints led to clashes about ways of looking at the problem, including perspectives deeply enmeshed in the disciplinary training of participants, importantly:

> Feedback from the participants confirmed that there had been significant transformations in ways of thinking about the complex relationships between nomads in Iran, the environments in which they live and work, the technologies that they use as pastoralists, the agents of government departments concerned with these aspects of sustainable development, and Iranian society at large
>
> (Emadi 2005, p. 259)

The project had practical outcomes such as improved relationships between nomads and some government staff and improved understanding of nomadic perspectives by government staff. While the organizational changes were limited and many of the issues remain current, a visit to Iran more than 10 years after the research made it clear to me

> that the policy dialogue about nomadism has been significantly altered and remains so. At least there is an alternative policy discourse about nomadism within government agencies.
>
> (Fisher 2006, p. 20)

The project also had important research outcomes. It made a significant contribution to knowledge about nomadism in contemporary Iran that has been widely recognized. This knowledge was about both the ethnography of pastoral life and the complexities of nomadic life within the nation state.

It is worth mentioning one factor that contributed to the ability of Emadi to have such a large organizational influence. At the time when he commenced his Ph.D. studies he was already a senior government official with important contacts within the government. One of these contacts was the Deputy Minister in charge of the FRO, who was very keen to see a new approach within his organization and who supported the large 13-day workshop. The capacity of Ph.D. students to 'convene' action research activities often depends on their status or the presence of 'champions'. At Hawkesbury, many of our international Ph.D. students were mid-career professionals with high status and influence.

Having discussed various action research studies, I now want to make some overall reflections from these. The emphasis on action research at Hawkesbury was, I think, very unusual in a university setting. In the fields of natural resource management and agriculture some excellent work was done with good practical (action) outcomes leading to change as well as valuable research outcomes.[6] The examples discussed above illustrate some of the great diversity of postgraduate research projects carried out. Apart from research ('knowledge') outcomes related to specific subjects (such as nomadism, RLR farming systems, and high altitude forest management), there were also research outcomes in terms of the development and validation of methodologies for organizational development and change. While there may seem to be an element of 'navel gazing' in this (in the idea of action research researching about action research), I believe that it is quite understandable at early stages in the utilization of a new form of research, that methodological concerns are given considerable attention.

Methodologically, the experiences with action research were very valuable for me in establishing the importance of structured action research, rather than just a broad concept of 'learning by doing'. The idea of a structured approach does not mean that all action research should follow a standard pattern. On the contrary, I see great value in an eclectic notion of action research, with many different organizational features and approaches, often in combination. What it means is, instead, that clear ideas about process and the modes and structure of participation can assist in making action research rigorous. These processes and structures can be revised, but a conscious and deliberate, if provisional, plan is essential.

One of the features of discussion about research at Hawkesbury, especially early in the period of advocacy of action research, was what I considered to be an overly simplistic and idealistic notion of 'participation'. It was often said that participation in action research should include all participants in all aspects of the research. What this meant in practice was, of course, quite unclear and problematic. In the case of community based natural resource management or community development, did this mean that all members of the community should actively participate in preparing research reports, or planning activities? Would they necessarily want to? Do all 'participants' want to be involved in esoteric discussions of methodology? How could you organize meaningful participation in a project operating at multiple locations? It is, I think, clear from the discussion of projects above that the modes of participation can vary while still remaining valid.

I think there was sometimes a failure to recognize the power dimensions of participatory projects. The idea, perhaps not expressed very explicitly, was that participatory processes would lead to agreed outcomes. In a workshop on action research early in my period at Hawkesbury, I suggested, somewhat tongue in cheek, that there might be a case for 'partisan action research', that is action research that actively takes sides (Fisher 1991). I suppose this could also be called emancipatory action research.

This section has discussed experiences with action research for Ph.D. and Masters degree research projects.[7] There is something of a challenge in using action research in an academic context.[8] The issue is one of academic acceptability and standards. Would examiners accept a thesis based on action research as a valid form of research when it does not follow typical scientific research standards? Colleagues at Hawkesbury often discussed this as a concern. My experiences at Hawkesbury were that rejection of action research by examiners was not a major issue in practice. I don't think I ever saw an examiner's report that rejected the methodology in principle, although there were cases where the actual application may have been negatively commented on. I think it is important in action research to make a clear distinction between action outcomes and research outcomes. An action research project that contributes to change but does not contribute to knowledge is not acceptable as a Ph.D. project.

Conducting training workshops in action research for field practitioners

Between 1997 and 2001 I worked as Head of Programme Development/ Deputy Director at the Regional Community Forestry Center for Asia and the Pacific (RECOFTC) in Bangkok. While at RECOFTC, I facilitated or jointly facilitated a number of training workshops on action research for a number of organizations in Asia. (Since leaving RECOFTC I have been involved in several similar workshops.) This provided an opportunity to organize my ideas on action research more systematically.[9] The workshops were held with some development organizations as well as with groups of researchers from ICIMOD and CIFOR. The structure of these workshops was to introduce the concept of action research, provide examples of action research projects and assist the participants to plan future work involving action research. In each case the action research cycle was introduced and, as far as possible, the workshop was organized as an action research cycle, including reflection on the context in which the organization worked and planning for future action. Again as far as possible, an action cycle was carried out in the field, although this often tended to be artificial as participants were frequently operating outside their normal field location and in groups that functioned only for the duration of the workshop.

I do not intend to discuss any of these workshops in detail here, but here are a number of key points I would like to make:

- The best way to introduce action research is to structure the introduction around an action research or action learning cycle, following a process of reflection/analysis, planning, action and reflection. Ideally this should be linked to the work context of the participants. They should leave the workshop with some sort of action plan for future implementation, either as groups (if participants normally work together) or as individuals (if they don't).
- Despite an understandable tendency of some workshop participants to be wary of such an unscientific approach (by the standards of their discipline), many professionals working in a field context, whether as

development workers or researchers, are open to the new approach, especially if they can see how it applies in practice. The much more difficult problem is being able to apply action research in practice after the workshop, as they may need organizational support and often need to work to relatively inflexible workplans.

- Repeat workshops are desirable if possible. This means that people leave the first workshop with action plans and report back on these at a follow-up workshop, thus reinforcing the idea of repeated cycles and routine reflection.

A mentoring role with CIFOR's Adaptive Collaborative Management Project

One of the workshops facilitated while I was at RECOFTC was for the CIFOR team who subsequently developed and implemented the ACM project. My involvement began with the facilitation of a workshop on action research, continued with a mentoring role with the project and facilitation of a writing workshop and included a role as joint editor of the final book published out of the ACM Asia component (Fisher *et al.* 2007). I do not intend to discuss the ACM project in detail as it is covered in a separate chapter in this book (Colfer, Chapter 3). However there are a few observations that I think are important.

- Early in the history of the ACM project team members expressed considerable concern that they were having difficulty in arguing the scientific validity of the process with their colleagues at CIFOR. At a meeting of the project steering committee held in Kathmandu this was a topic of discussion and concern.[10] Ultimately what emerged from discussion was the idea that validity was about establishing 'plausible causal connections' between interventions and effects (this is discussed in Chapter 1 of Fisher *et al.* 2007, pp. 8–9). The issue of the credibility of research findings vis-à-vis scientist colleagues tends to be a problem for people working in action research and ACM. A similar challenge applies to any qualitative research.
- The project was specifically concerned with applying and testing ACM. Participatory action research was part of the nature of the project as a research project. The idea was that some activities would be undertaken using PAR. I always felt that, given the similarity of action research and ACM as learning approaches, the relationship between the two was very confusing. Was action research the overall research approach to testing ACM as a management tool, or was it used within ACM to address certain specific problems? I still struggle to get this clear and I think that the project, to some extent, was plagued by the terminological confusion so often associated with the field.

IUCN's Livelihoods and Landscape Strategy

The last set of experiences that I wish to discuss briefly is the experience of action learning/action research within the Livelihoods and Landscapes Strategy (LLS) of IUCN's Global Forest Conservation Programme. LLS is a major programme that operates in 27 landscapes in 23 countries in Africa, Asia and Latin America. In March–April 2008, I facilitated an action learning workshop held in Tanzania to introduce action learning/research to members of the LLS team working in various countries in Africa. The workshop followed the workshop approach discussed above. Workshop participants left the project with commitments to implement action learning in their landscapes. As an advisor to LLS I have mentored action leaning in several of the African landscapes.

One of the issues around the application of action learning in LLS has been a certain squeamishness about using the term 'action research' on the grounds that field practitioners are alienated by the term research, which is thought to be 'academic' and something carried out by outsiders. While I recognize this concern, I have argued that part of what we are trying to do in LLS is precisely to generate knowledge about landscape management and landscape approaches for a wider audience – in other words, research. Nevertheless, at a practical level, this learning can occur whatever terminology is used and I would have to say that the use of terminology has been inconsistent within LLS.

The adoption of action research within LLS Africa has been partial. Staff from a number of landscapes report that they have adopted action learning, at least informally. In at least two landscapes it has been applied in a more structured way.

In the landscape in south-western Ghana, LLS implementation has been managed through a conscious action research process by a small team representing IUCN and some local government departments and others. The team has met on a number of occasions to go through an action research cycle in a structured way, reviewing activities and planning new steps. In addition to occasional formal meetings the team members report that they routinely follow the action research process in the course of their normal activities. Thus, the process, while not very structured, has been very useful for implementation. At community level the team has encouraged smaller task based groups to follow similar reflection/action processes, although they have not introduced the action learning or action research cycle as a conscious tool at this level. In this case a more or less structured team follows an action research process consciously and there is a wider application of a learning approach. This all seems to work quite effectively and is a reminder that the key to effective action research is the learning culture.

In Uganda, an action research process has been used in the LLS landscape at Mount Elgon. In this case action research was introduced to a group of local officials, NGO workers and other stakeholders at a two-day workshop.

The group formed several issue-specific sub-teams and developed action plans, which have been substantially followed up. This was a much larger group than the Ghana action research team. I have been impressed by the extent to which people with a diverse range of jobs and professions were open to the approach. They expressed little scepticism about the scientific validity and seemed very excited by the approach to problem solving by breaking tasks down into achievable bits, just as had happened in Goilala.

Discussion

This section presents a summary of a few broad lessons about what action research can do and where it is appropriate. I will then discuss some of the challenges involved in applying action research in different contexts, such as development projects and academic research activity. Finally I will suggest some strategies for addressing these challenges, or at least some of them. The strategies are necessarily tentative. While they are somewhat based on experience, they are certainly not intended to be very authoritative.

This chapter has been a somewhat meandering journey through personal experiences with action research and adaptive collaborative management (mainly action research). The various experiences do lead to some conclusions.

- It is clear that action research can contribute to better implementation in natural resource management and agriculture projects. Natural resource management generally occurs in complex situations, where social and biophysical issues interact in complex ways. In such cases even clear identification of problems can be difficult and tends to be difficult to know where to start. Action research is a useful way to work when full knowledge doesn't exist and when the context is changing. By working in small incremental steps it enables action instead of paralysis. It is also particularly useful when stakeholders do not even agree with the framing of an issue. They can at least start with what they agree on.
- ACM and action research projects may sometimes have a primary concern with implementation (using learning to implement an activity better), but they can also have an important research role. In this sense research is defined as learning which is intended to inform a wider audience beyond the immediate project context, including providing insights that will assist more effective action in other activities and projects. It is sometimes important to differentiate between action learning and action research because field practitioners and administrators are sometimes alienated by the term research (as being 'just academic'). Nevertheless, it is important to stress the importance of learning being important beyond the immediate context.

- A degree of structure in terms of the way participants are involved can be very useful. Nevertheless there is no single model and the organization structure of participation may vary to fit the action research context.
- The discipline of regular reflection and revised planning is the essential basis for adaptive management and action learning/research. Following a structured process (such as one based on the action research cycle) is often useful, but a culture of learning is essential.
- ACM and action research can contribute effectively to policy change (as illustrated both by the Nepal–Australia Forestry Project and by the project in Iran).
- Meaningful participatory adaptive learning can be a major motivating force in projects. For example, the PNG project contributed to disempowered and demotivated extension staff becoming actively engaged with rural communities and project activities. On the other hand, bureaucratic decisions ultimately undermined the project, demonstrating the importance of long-term commitments by donors and governments.

In this chapter I have discussed experiences in the implementation of action research in a variety of different contexts. The types of contexts and the related challenges can essentially be divided into two broad fields – action research in settings that were primarily research oriented, and settings that were primarily development oriented. Although there is an overlap, the types of challenges arising in each of these broad settings are somewhat distinct and I will deal with them separately below. It is, however, to some extent an artificial distinction, as many of the research oriented projects took place in a development context and were substantially focused on change.

In the research/academic setting one challenge arises from the generally limited time frame within which action research takes place. This is especially true of research for Ph.D. or Masters degrees, where the time available for field work is limited, meaning that opportunities for multiple action research cycles are limited. This limits the possibilities of contributing to large-scale change. There is also a problem in deciding when to consider a project 'finished' from the point of view of the thesis project. It also limits the opportunities to institutionalize longer term learning practices within groups of collaborators after the research is ended.

There is no simple strategy for dealing with this problem when it occurs. There seem to me to be two main approaches. One is to make sure that the projects are not too ambitious, although that can be difficult as action research projects, by definition, evolve as they are implemented. The second approach is to encourage research students, as far as possible, to select projects that relate to their professional work outside their Ph.D. studies. In the case of Hawkesbury many of our Ph.D. students were mid-career professionals who intended to go back to the agencies within which they worked on completion of their Ph.D.s. This presented opportunities to continue at least some level of activity related to their project after their studies were completed.

Another challenge to action research in an academic context or in the context of a research organization is the fact that action research is sometimes seen as lacking credibility as a rigorous research methodology. Although there was often concern among Hawkesbury academics that Ph.D. examiners would object to theses on this basis, in fact I never experienced a case where an examiner raised this as a criticism. There were, however, occasions where examiners from specific disciplines had trouble with the multi-disciplinary nature of theses, usually because they felt that a particular discipline had not been adequately covered.

The practice of action research and adaptive collaborative management inside research organizations such as CIFOR also raises problems of academic or scientific credibility. At least in the early days of CIFOR's ACM project the ACM researchers felt very much under pressure to convince sceptics in CIFOR of the scientific utility of their research. In many ways I think the problem was the credibility of critical social research rather than ACM as such. For scientists and even social scientists applying quantitative methodology, any form of qualitative research presents a challenge. In my view, there is no easy answer to this, but I think that arguing on a philosophical basis about differing epistemology and paradigm differences tends to achieve nothing. The best approach is to write up research in such a way that connections between actions and change are told in the form of 'robust narratives' that emphasize 'plausible causal connections'.

It is important to recognize that some supposedly 'traditional' scientists find the idea of action research liberating. In a training workshop for ICIMOD with rangelands scientists from a number of countries in the Hindu-Kush Himalayan region it was clear that many of the participants found action research a useful way to contextualize their research projects.

The challenges confronting action research in development contexts are, I think, more substantial. The first important challenge arises from usually short project time frames. This not only limits the time to carry out iterative and exploratory research, but it also tends to lead to a loss of momentum when a project ends. The long time frame of the Nepal–Australia Forestry Project was one of the key reasons why action research was meaningful in that project, but such long time frames are unusual in development projects.

The alternative is to attempt to develop a learning approach in a larger programme rather than within individual projects. The difficulty here is that government programmes naturally tend towards standardized administrative practices and generally work against flexible action. In my experience action research (and learning approaches generally) tend to be most suited to project level activity. In this type of situation projects may act as 'policy experiments'. Frankly I cannot recall ever seeing action research operating at 'programme' level.

But action research within projects raises the usual problem with projects: how do you 'scale up' from project level? A related issue is that action research projects operating within a larger programme are subject to decisions made beyond the project scope. For example the Goilala project, while supported by various levels of the district, provincial and national bureaucracy, ultimately fell victim to the loss of funding due to a previously poor reputation for performance.

There are also problems in applying action research arising from donor policies and procedures. Although it is almost *de rigeur* for development agencies to talk about 'learning' and 'adaptive' approaches, in practice projects are designed around tight time frames and scheduled inputs and outputs associated with logframes. The rigidity of logframes is totally inconsistent with genuine flexible and adaptive learning. This is the major challenge facing action research in development.

There are two strategies that can go some way to creating space for action research, at least at a project level. These suggestions are hardly novel, but they are, I think, useful:

- Identify champions within government and donor bureaucracies who will allow action research oriented activities to take place at least on some sort of pilot basis. This will involve systematic attempts to keep the champions informed and engaged. They need to have ownership of the approach and of the project. Involvement of officials in training that introduces the approach and incorporating them in action plans arising from action research can help in encouraging support.
- Make sure there is a critical mass of people within a project and among its collaborators who understand action research and who are committed to practising it. This requires strong leadership within the project.

I have suggested that Mohammad Emadi's project with Iranian nomads and the Nepal–Australia Forestry Project made contributions to policy or, in the case of Iran, at least to the policy debate. These impacts illustrate the value of the two strategies. Along with a number of other bilateral projects and government forest officers, NAFP[11] contributed to the policy dialogue around community forestry in Nepal and provided case material that showed that community forestry could work. This influenced the eventual shape for the community forestry programme. In Iran, the action research project exposed actors in key government departments to alternative ways of understanding nomadic perspectives and training activities went some way towards wider promotion of this understanding and of participatory methodologies. This was a conscious attempt to 'scale up'. The long-term result was a change in the policy dialogue about nomadism.

Both of these cases illustrate the importance of leadership and of cultivating champions. NAFP closely engaged senior forest officials and obtained de facto licence to experiment. Emadi had close relations with the Deputy

Minister of the Forest and Range Organization who provided easy access to resources and staff and endorsed the 'new approach'. It is worth adding that Emadi was himself a fairly senior official before he commenced Ph.D. studies and this gave him access to many officials. So, in both cases we have champions and leadership as crucial factors.

All this is fairly obvious, but I think it is crucial to meeting the challenges facing action research in a development context.

Conclusion

In this chapter I have reflected on a variety of experiences in action research and ACM. One of my strongest impressions from work in a variety of contexts is that many people, researchers and development practitioners, find action learning empowering. As Kurt Lewin is credited[12] with saying: 'Nothing is so practical as a good theory.'

Acknowledgements

I am grateful to all those I had the privilege of working with in the action research and ACM activities described in this essay for the ideas and excitement shared. I especially wish to thank Don Gilmour, Mohammad Emadi, Solieng Mak and Hemant Ojha for commenting on aspects of the essay.

Notes

1. The Nepal–Australia Forestry Project (under different names) was a bilateral project of His Majesty's Government of Nepal and the Australian Development Assistance Bureau (subsequently AIDAB and then AusAID). Australian involvement in forestry began with occasional input from Australian foresters in the early 1960s and became formalized as a bilateral project as the Nepal–Australia Forestry Project in the mid-1970s. At the time of my involvement it was managed by ANUTECH Pty Ltd (the Australian National University's business arm) and was implemented technically by the Department of Forestry at ANU.
2. See Bawden (1991) on action research at UWS-H. The extent of the intellectual revolution that had taken place in the faculty, under the leadership of Richard Bawden and others, was such as that undergraduate students in agriculture were exposed to philosophical concepts such as epistemology and ontology.
3. The project was funded by the Australian International Development Assistance Bureau as it was then called, under a Development Research Grants Scheme.
4. In the background is the frequent concern of governments that nomads are a threat to state authority.
5. This was facilitated by Emadi, Robert Woog, a colleague from Hawkesbury and myself.
6. There was also some excellent Masters and Ph.D. work done in community development contexts, but these are outside the scope of this chapter.
7. Australian Ph.D. and Masters theses are examined by external examiners, not members of a supervisory panel.

8 Dick (1993) provides a useful guide to some of the issues involved in using action research as the basis of a research thesis.
9 A paper written for a workshop on protected areas (Fisher and Jackson 1999) presents some of the basic ideas of action research in a simple form applicable to protected area management.
10 I attended this meeting as a contributing observer, but not a steering committee member.
11 I want to stress here that there were many actors who contributed to the change. I focus on NAFP only because I had direct experience with the practice of action research within the project.
12 Lewin is credited with this remark but it does not appear in his writings (Greenwood and Levin 1998).

References

Bawden, R.J. 1991. Toward Action Researching Systems. In *Action Research for Change and Development*, Ortrun Zuber-Skerrit (ed.). Sydney: John Wiley.

Brown, L.D. 1993. Social Change through Collective Reflection with Asian Non-governmental Development Organizations. *Human Relations*, Vol 46 (2): 249–273.

Chambers, Robert 1986. Normal Professionalism, New Paradigms and Development. Institute of Development Studies, University of Sussex. Discussion Paper 227.

Dick, Bob 1993. You Want to Do an Action Research Thesis? Brisbane: Department of Psychology, University of Queensland. Downloadable from: http://www.scu.edu.au/schools/gcm/ar/art/arthesis.html (accessed 4 May 2011).

Emadi, Mohammad H. 1995. Pastoralists, Participation and Policy: An Action Oriented, Systemic and Participatory Approach to Improve the Relationships Between Pastoralist Nomads, Government and Natural Resources in Iran. Ph.D. Thesis, University of Western Sydney, Hawkesbury. http://ibcperu.org/doc/isis/5674.pdf (accessed 23 February 2012).

Emadi, Mohammad Hossein 2005. Research Through Action with Nomadic Pastoralists in Iran. In *Participatory Research and Development for Sustainable Agriculture and Natural Resource Management: A Sourcebook*, Volume 1. Doing Participatory Research and Development, Julian Gonsalves, Thomas Becker, Ann Braun, Dindo Campilan, Hidelisa de Chavez, Elizabeth Fajber, Monica Kapiriri, Joy Rivaca-Caminade, and Ronnie Vernooy (eds). International Development Research Centre. Downloadable from www.cifor.cgiar.org/livesinforests/publications/pdf_files/nomadic_pastoralists_inIran.pdf (accessed 25 May 2011).

Emadi, Mohammad H., R.J. Fisher and Robert Woog 1992. Action Research as an Approach to Research and Development Among Iranian Nomads. Paper presented at International Conference Nomadism and Development: Survival Strategies and Development Policies, Shar-e-kord, Iran. 1–6 September 1992.

Fisher, R.J. 1991. Research at Hawkesbury: Reflections of a Newcomer. Paper prepared for Inward Bound Workshop. Faculty of Agriculture and Rural Development, University of Western Sydney, Hawkesbury. 19 July 1991. [Published in *Inward Bounding*, an occasional paper published by the Faculty of Agriculture and Rural Development, University of Western Sydney, Hawkesbury. December 1991.]

Fisher, R.J. 2002. Review of *Introduction to Action Research: Social Research for Social Change* (Davydd J. Greenwood and Morten Levin). *Practicing Anthropology*, Vol 24 (1): 41.

Fisher, R.J. 2003. When is Action Research Participatory? *ACM News*, Vol 4 (2), April 2003 (CIFOR ACM Project Newsletter).

Fisher, R.J. 2006. Collaborative Learning for NREM Policy Development. In *Proceedings of a Regional Seminar on Natural Resources and Environmental Management in the Greater Mekong Sub-region: The Role of Higher Education in Policy Design and Implementation*, Nguyen Luong Bach, J. Keith Syers and Okitsugu Fujiwara (eds). Chiang Rai: Mae Fah Luang University.

Fisher, R.J. and W.J. Jackson 1999. Action Research for Collaborative Management of Protected Areas. In *Collaborative Management of Protected Areas in the Asian Region: Proceedings of a Workshop held at Royal Chitwan National Park, Sauraha, Nepal, May 25–28, 1998*, Krishna Prasad Oli (ed.), pp. 235–243. Kathmandu: IUCN Nepal.

Fisher, Robert, Ravi Prabhu and Cynthia McDougall (eds) 2007. *Adaptive Collaborative Management of Community Forests in Asia: Experiences from Nepal, Indonesia and the Philippines.* Bogor, Indonesia: Center for International Forestry Research (CIFOR).

Friere, Paulo 1982. Creating Alternative Research Methods: Learning To Do It By Doing It. In *Creating Knowledge: A Monopoly?* B. Hall, A. Gillette and R. Tandon (eds), pp. 29–37. New Delhi: Society for Participatory Research in Asia. [Reprinted in Kemmis and McTaggart 1988b.]

Gilmour, D.A. and R.J. Fisher 1991. *Villagers, Forests and Foresters: The Philosophy, Process and Practice of Community Forestry in Nepal.* Kathmandu: Sahayogi Press.

Gilmour, D.A., G.C. King and R.J. Fisher 1988. Action Research into Socio-economic Aspects of Forest Management. In *Role of Forest Research in Solving Socio-economic Problems in the Himalayan Region.* Proceedings of IUFRO Symposium held in Peshawar, Pakistan. Pakistan Forestry Institute. 17–27 October 1987.

Greenwood, Davydd and Morten Levin 1998. *Introduction to Action Research: Social Research for Social Change.* London: Thousand Oaks and New Delhi: Sage Publications.

Griffin, D.M.1988. *Innocents Abroad in the Forests of Nepal: An Account of Australian Aid to Nepalese Forestry.* Canberra: Anutech Pty Ltd.

Jackson, W.J. 1999. The Dynamics of *Lekh* Forest Use in the Middle Hills of Nepal. Ph.D. Thesis, University of Western Sydney, Hawkesbury.

Kemmis, Stephen and Robin McTaggart (eds) 1988a. *The Action Research Planner.* Deakin University, Australia: Deakin University Press (3rd edition).

Kemmis, Stephen and Robin McTaggart (eds) 1988b. *The Action Research Reader.* Deakin University, Australia: Deakin University press (3rd edition).

Kolb, D. 1984. *Experiential Learning: Experience as the Source of Learning and Development.* Englewood Cliffs, New Jersey: Prentice-Hall.

Lewin, Kurt 1946. Action Research and Minority Problems. *Journal of Social Issues*, Vol 2 (4): 34–46.

Mak, Solieng 1997. Rainfed Lowland Rice Farming in Cambodia: A Study of Agricultural Change. Ph.D. Thesis, University of Western Sydney, Hawkesbury.

Rhoades, Robert E. 1984. *Breaking New Ground: Agricultural Anthropology.* Lima, Peru: International Potato Center.

Ritchie, Helen. 1998. Beyond the Fences: Co-ordinating Individual Action in Rural Resource Management through Landcare. A Case Study of Managing Non-point Source Discharges to Water in Waikato, New Zealand. Ph.D. Thesis. University of Western Sydney, Hawkesbury.

Sriskandarajah, N. and R.J. Fisher 1992. A Participatory Approach to Improving Rural Livelihoods of People in the Goilala District of Papua New Guinea. Report on a research project funded by the Australian International Development Assistance Bureau under the Development Research Grants Scheme.

Toafa, Tevita 1994. Action Research to Improve the Pumpkin Industry in Tonga. M.Sc. (Hons) Thesis, University of Western Sydney, Hawkesbury.
Weekley, Paul 1999. Improving Sahelian Food Security Through Facilitating Action Learning Networks: A Case Study Among the Fulbe Jelgobe of Northern Burkina Faso. Ph.D. Thesis, University of Western Sydney, Hawkesbury.
Whyte, William Foote (ed.) 1991. *Participatory Action Research.* London: Sage Publications.

9 Confronting challenges in applying adaptive collaborative approaches

Lessons from practice

Hemant R. Ojha and Andy Hall

Overview

The aim of this chapter is not to describe the challenges of natural resource governance in general, but the challenges of applying adaptive collaborative approaches (ACA) to improve governance and innovation. Moreover, these are narrated from the perspective of actors who are practically concerned with issues of natural resources management and human well-being, and more importantly who value the action orientation in research and a move towards a more adaptive and collaborative approach. The chapter also pools diverse ways through which the ACA actors confronted these challenges in practice as part of their attempts to improve governance and facilitate innovation. So the analyses here seek to capture the perspective of ACA actors – who have themselves become hybrid agencies, with multiple origins – from research, training, policy and implementation backgrounds, but all struggling with the process of change in the 'context of application', beyond 'classical academic extractivism' or 'scientism'. For this reason, the structure and flow of the text includes quotes, narratives and interpretations made in the previous chapters, with the aim to distil actionable insights from the experience of having put the ideas into practice.

We have identified eight key domains of learning and collaboration in which ACA actors have had to muddle through. These are as follows:

1 *Tackling institutional constraints within which ACA practice is organized:* Cultivating ACA within an international institution with biophysical bias (Chapters 3 and 4), undertaking ACA from an NGO platform (Chapters 5 and 7), undertaking ACA through a local unit of international organization (Chapters 4, 5 and 6), and handling the frequent change of trained ACA facilitators within the local partner institutions (Chapter 7).
2 *Tackling reductionist scientific attitude and practice:* In the international domain (Chapters 3 and 8), and in local domains (Chapters 6 and 7).

3 *Working with local communities* (Chapters 4, 5, 6 and 7).
4 *Balancing research and action processes:* Getting academic researchers to do action research (Chapter 8), collecting data and facilitating interventions in the field (Chapters 3, 6 and 7).
5 *Working within and through the national policy regime and authority structure:* Acting politically towards change in the context of highly inequitable socio-political systems (Chapters 6 and 7).
6 *Facilitating cross-scale linkages:* From local through meso and national levels (Chapters 3, 5, 6 and 7) and creating horizontal linkages among local communities (Chapters 5 and 6).
7 *Managing sponsorship and funding:* Raising funds for a multi-country programme (Chapter 3), for specific ACA methodology (Chapter 4), for continuing small-scale ACA work in national contexts (Chapters 6 and 7), for covering the transaction cost of local communities to act collectively to influence policy (Chapter 5), and dealing with donor bias toward positivist evaluation of ACA initiatives (Chapters 3, 4, 6 and 7).
8 *Sustaining results and ensuring innovation pathways:* Community networking (Chapters 5 and 7), knowledge production for influencing wider discourses (Chapter 3), political mobilization of local people and communities (Chapters 4, 5 and 7), and helping local communities institutionalize learning oriented processes (Chapters 5 and 6).

We discuss each of these points below drawing on the insights from the previous chapters.

Tackling the institutional constraints

The evidence reported in this volume demonstrates that the institutional context of developing and applying adaptive collaborative approaches has profound influence on the development and success of ACAs. For instance, the experience of developing the ACM programme within CIFOR indicates that institutions focusing on biophysical research pose added difficulties in ACA oriented researches, through their preference for traditional, extractive research (see Chapters 3, 6 and 7). Resonating with this, Sherwood *et al.* (Chapter 4) report that managers of agricultural research institutions are often driven by the desire to produce large-scale hard scientific evidence or large-scale quantifiable results (Chapter 4), which is not consistent with ACA's emphasis on learning in particular, and a concern for helping local actors to bring about needed change. As described in Chapter 3, CIFOR was established partly to cater to the demand for research and technology in forest and natural resource management, and the organization is oriented to maximize its achievements in terms of peer-reviewed scientific publications, leaving the role of facilitating actual change on the ground to its research beneficiaries and collaborators. As CIFOR was institutionally dedicated to conducting hypothesis-testing, experimental, biophysical research in the forest sector, it was not easy for the ACM team to create space for engaged local research that aims to connect forest management and livelihoods of local people. CIFOR's forestry orientation was also shaped by the widely held techno-scientific assumptions and values shared and upheld by the global forestry community. Although CIFOR has significantly restructured itself as inter-disciplinary research organization, Colfer's account of the struggle of the ACM team provides interesting insights into what it takes to try this at a time when there is strong institutional resistance (Chapter 3).

Colfer (Chapter 3) recounts her experience of working with a CIFOR Director General:

> [H]e would reiterate his skepticism about anthropology, descriptive methods, participatory approaches, and holistic analyses. He reminded us of the Board's insistence that all research at CIFOR test hypotheses . . . Besides recurrent explanations and dialogues with our DG, I was grilled twice a year at Board meetings (unlike other programmes), throughout the first four years of our efforts.

The ACM team at CIFOR faced, as Colfer recounts, comments such as 'How will you scale up from single cases?' and 'But your results won't be generalizable.' ACA researchers have to confront these challenges within their home research institutions. Another way the ACM team's choice was restricted by CIFOR management was through encouragements and instructions to the ACM team to partner with national forest research organizations, which were even more technocratic, and less appreciative of adaptive collaborative approaches. While there are successful cases in which the organizational management was also supportive and actively engaged in learning based approach, as in the case of a forestry project in Nepal (Fisher, Chapter 8), we need a more nuanced understanding of what it takes to mainstream ACA in larger institutional regimes.

The institutional challenge continues all the way to local communities. As Sultana and Thompson (Chapter 4) report, the leaders/members of local community based organizations found it difficult to apply some of the new ideas they received through participating in the ACA process facilitated by researchers. The individuals who began learning with others developed different visions that were not easily appreciated in the local institutional regime. This demonstrates that while there is a possibility of some actors going 'against the stream' regarding learning and change, and also undertaking some experimental actions, there is still little insight as to how such small-scale innovations can be absorbed by the wider institutional systems.

Another institutional dimension is how members of an ACA team organize collective actions. Like the challenges faced by the ACA team in relation to management, working internally within an ACA team has also been found to be challenging. Within the team different roles have to be organized: conceptualization, abstract thinking, acting, engaging, documenting, interpreting, marketing, fundraising, networking. Questions to be dealt with include authorship, representation, forging team spirit, enhancing skills, ensuring inclusion and gaining credibility of the work. At times, the ACA researchers themselves find it difficult to internalize and communicate the idea of ACA clearly to the local communities and other collaborators, as it requires innovative techniques of communication and critical and fruitful engagement across disciplinary beliefs and assumptions (Chapters 3, 6 and 7).

Moreover, how ideas and knowledge are created within the team matters; and the more democracy of knowledge the better (of course subject to constraints such as urgency and transaction cost). To share ownership of the ACA ideas and strategies among the team members, Colfer drafted 17 versions of the project concept note and circulated it to the whole team (Chapter 3).

The ACA team usually involves partners of different institutional history and disciplinary backgrounds, and this can bring both opportunities and challenges in the process of learning and action. If the partners bring divergent management strategies and financial policies, that can create confusion when working with beneficiaries or other collaborators, as in the case of multiple institutions working in Nepal (Chapter 7). Banjade (Chapter 7) provides examples of how he faced difficulty in balancing field-work with the expectations of international research collaborators. In order to create an organic team for ACA, Fisher reports that joint review and reflection in action research projects should become part of organizational culture, and this has been possible in his experience (Chapter 8).

Unlike traditional project management, ACA actors conceive results as work in progress in the pathways to innovation, moving beyond the input–output model of change. Some see as cycles of learning and innovation, not just the personally focused approach of Kolb (Kolb 1984), but also more socially engaged processes, linked to multiple scales. And they also see outcomes at multiple stages of the process of change – much beyond the linearly conceived output–outcome–impact framework. Fisher reports his experience of having used action-oriented research as the modus operandi of various projects facilitating action learning (Chapter 8). All this suggests that ACA is likely to take root more easily in organizations that are open to new, flexible and innovative forms of management outcomes.

Yet the diversity of experience reported in the book suggests that ACA oriented research and action can emerge in any institutional home – international research organizations, national public research institutes, NGOs, and extension organizations – provided there are actors interested in such an approach. It can also be a combination of these institutions working together to promote ACA. It is not usual to have the management or the executives of the organization start the process through some decisions – they are more likely to act, under the best scenarios, to accept and endorse ACA oriented practices. Interests and initiatives by some staff member of an organization are sufficient, and the capacity and motivation of such initiating actors do matter. More importantly, the success of such innovative practices depends on how they confront the institutional environment and engage the organizational management to a point when the latter actually begins to appreciate the approach itself.

The cases reported here indicate the diversity in terms of institutional openness to ACA oriented actions, and the structure of incentives also varies greatly, as the cases show. In the two cases reported here (Chapters 5 and 8), university based researchers appear to enjoy a significant degree of freedom and flexibility when they are willing and committed to move beyond hypothesis-testing scientistic research – as in the works of Fisher (at the University of Western Sydney) and Sultana and Thompson (University of Middlesex). Likewise, working with a locally engaged research NGO, Banjade enjoyed a learning-oriented and flexible learning environment at ForestAction, and yet faced some notable constraints while he was trying to move adaptively at the local situations (such as the one case in which his decision to hire local actors was revamped by higher management in the project). Sherwood *et al.* present the case of multiple institutions and multiple groups of ACA actors working within institutions, demonstrating varied and complex challenges faced by the ACA oriented individuals – from the coercive supervisory power to large scale and mechanistic 'technology' diffusion interests of the organizational management. CIFOR's ACM team faced several questions from board members, such as 'What were the experimental designs?', 'What were the controls?', etc. Still in another unfavourable action of the organizational management, CIFOR's ACM team had to face an evaluation that was 'surreptitiously and informally arranged' for an additional, independent review of their work, urging the reviewer 'to be as critical as possible' at a time when the programme leader was on an approved sick leave (Colfer, Chapter 3). Inflexible budgets and output-based control also sometimes hampered ACM performance at CIFOR.

Unlike the CIFOR ACM case in which an attempt was being made to introduce a new adaptive and action-oriented research, the case of Farmer Field Schools (FFS) in Ecuador reported by Sherwood *et al.* (Chapter 4) involved stories of confrontation between FFS facilitators and a different organizational environment. One of the FFS professionals working on behalf of a national research organization of Ecuador (INIAP) was concerned with the methodological erosion of FFS while the higher organizational management was fast moving towards engaging the private pesticide industry to achieve efficiency in the extension process. Despite the staff's enthusiasm about a learning based approach to FFS, he was forced to implement a technology-centred version of FFS. In another instance, an FFS facilitator became technology-centric, moving away from a learning based approach, despite higher management's encouragement to remain learning-focused. Yet another FFS worker, coming from a community background, was less constrained by the organization, and expanded the practice of FFS towards the realms of people, linking with farmer movements in the area. An adaptive learning and innovation strategy like FFS thus became diversely packaged and sold in different institutional environments. What is most noteworthy is that when large organizations adopted FFS, 'they undermined the people-centred intentions of FFS' (Chapter 4).

Within the domain of organizational micro-politics, Fisher (Chapter 8) also encountered a situation in which he was prevented from calling himself an anthropologist, and had to accept the title of 'community worker'. While the title of community worker may have helped in certain ways (to work with communities and with local stakeholders), this was clearly a bias against anthropological knowledge in forestry contexts. It is not uncommon for development projects to discourage the use of particular terminology and the involvement of particular disciplines, favouring the language and conventional expertise of the development industry.

Given all these diverse confrontations of ACA actors, what strategies have worked in advancing ACA in different institutional settings? At least 10 specific points are identifiable:

1 Up to a certain point, a critical mass of like-minded people and teams have worked together to drive the process through institutional inertia (Chapter 3).
2 Constant contestations and dialogues between ACA researchers and the institutional managers have helped to transform the perceptions of intuitional managers (Chapters 3 and 7).
3 Setting up independent advisory committees have contributed to institutional uptake through creating a sphere of dialogue and reflections 3 and 7).
4 Creating alliances among ACA oriented actors working with different organizations (Chapters 6 and 7).
5 Training, nurturing and capacitating internal champions of change (Chapters 3, 6, 7 and 8).
6 Identifying and capturing relevant aspects of institutional goals and strategies and leveraging them for advancing ACA activities. For instance, some institutional elements such as CIFOR's commitment to multiple scales and interdisciplinary work were used as favourable windows of opportunity by the ACM team (Chapter 3).
7 Wherever feasible, mobilization of support by seniors following action research and innovation perspective (Chapters 7 and 8).
8 Identifying and articulating relevant wider political economic drivers in justifying the need for more adaptive and collaborative approaches. For instance, mobilization of the public response to harmful effects of pesticides in favour of adopting FFA. Likewise, the wider context of agricultural modernization and involvement of private and non-governmental organizations in agricultural extension was also a conducive factor (Chapter 4).

9 Thinking beyond the organizations and looking at how networks of institutions and actors shape the process of learning and innovation – such as the socio-technical regime (Chapters 4 and 5).
10 Using action research as the modus operandi of the whole organization (Chapters 7 and 8).

Tackling reductionist scientific attitude and practice

Given the long history of traditional *reductionist and extractive scientific approaches* in the field of natural resource management, deeply rooted in the institutions of the state itself, the move towards adaptiveness and collaboration has not been easy. In the wider field of science and policy, ACA oriented researchers continue to face challenges to demonstrate their value added in contrast to the existing dominant model of 'research and development'. The case of CIFOR as a site of struggle between ACM scientists and traditional biophysical scientists (Chapter 3), obviously following two paradigms, demonstrates that for an institution with a biophysical orientation, it is not easy to allow and nurture action-oriented science (see Chapter 2 for a conceptual orientation of action-oriented science). The leadership at CIFOR came from a biophysical scientific paradigm, and they only rarely appreciated the ACM approach, even though some of the members of the governing board clearly recognized its value (Chapter 3). The contestations between the two approaches lasted through the end of the ACM programme within CIFOR. As of today, there is still limited conceptual and practical development to reconcile the two approaches, in ways acceptable to the advocates of both approaches, and to the satisfaction of its collaborators.

Another threat from reductionism emerges when there is an attempt to scale up a learning and collaboration oriented practice beyond the domain of innovation. Sherwood *et al.*'s (Chapter 3) analysis of Farmer Field Schools (FFS) shows that the participatory, people oriented character of FFS was taken over by the 'Mode 1 science' (dominated by expert-centric learning) in the course of scaling out in Ecuador, when large scale organizations wanted to scale up the process widely, and when the national policy system also encouraged the private sector to deliver agricultural extension services. This means that the farmer-centred small-scale innovation was taken to scale without changing the underlying institutional regime, and in the process the very innovative character of FFS was lost. This raises a fundamental challenge about how we can go to scale from small-scale intensive innovation, implying a need for looking at adaptive and collaborative approaches at larger scales of governance.

A key form of struggle by the ACM team within CIFOR was over the meaning of research itself, and was oriented to justify moving beyond traditional, extractive research, and creating space for action within research programmes. The ACM team had to 'straddle' the established regime of binary projections that tended to disfavour action orientation in their research – 'research and advocacy, research and development, and quantitative and qualitative'. Clearly, their struggle was geared towards opposing such binary interpretations of reality and processes, and to bring in an action-oriented and holistic approach. As the CIFOR ACM team had to compromise with the stringent management conditions, ACM researchers in the field had to encounter the remnants of reductionist bias coming to them as the requirements for ACM project outputs. As Banjade, who was one of the leaders implementing ACM in Nepal, recounts: 'in a project of less than three years, improving wood production was an unrealistic goal'. Besides, they had to spend more time in data gathering than in building relationships despite the adaptive learning emphasis of the project (Chapter 7).

ACA researchers have sought to develop an innovative approach to the study of science, and hence to reveal the limits of reductionism and create space for an action-oriented approach. Sherwood *et al.* (Chapter 4) offer an institutional approach to the study of how a reductionist science faces challenges, followed by a stage in which spaces are open for a Mode 2 version of scientific practice, and then a regressive phase emerges, moving the innovation back to Mode 1 science.

Even when there is continuation of an ACA oriented initiative, the participants face diverse constraints and situations that can lead them to drop the learning and collaboration related core values in the process. FFS methodology explicitly emerged in response to the adverse consequences of modern, industrial era rice farming in Asia, and it was proposed that FFS could provide an entry point for profound institutional change: a shift from expert-led, technology-centred agricultural science and development practice towards more people-centred designs, emphasizing local knowledge and community-based innovation. But as Sherwood *et al.* (Chapter 4) discovered from the close observation of FFS trainers and facilitators in Ecuador, this ideal of FFS was utilized for diverse purposes, involving consequent adaptations to the techno-scientific approach.

Several innovative strategies have been deployed to reframe research as a part of an adaptive collaborative approach to innovation. First, it is important to clarify the links and overlaps between social and technical research. In the context of natural resources management, there is still a tendency to dichotomize between social and technical research. ACA related contributions in the volume recognize that there is a need to go beyond this dichotomy. As Fisher asserts,

> While it is clear that action research is a *social process*, and while it is often about organizational and social change, . . . action research could also incorporate cycles of 'technical' research (such as research into appropriate silviculture arising from action research into project implementation).

Second, ACA attempts have focused on the study of the politics of science, including juxtaposing strengths and weaknesses as well as orientations and interests of Mode 1 and Mode 2 science (Sherwood *et al.*, Chapter 4). Mode 1 emphasizes abstract, specialized and externally based knowledge production, whereas Mode 2 science aims at knowledge production that is an applied activity. Colfer has done the same based on her own experience of working through ACM and encountering the other group of scientists at CIFOR (Table 3.1). Colfer has characterized how ACM practitioners differed with their traditional counterparts on various aspects – preference, beliefs, methodological emphasis and inquiry goals. In both these formulations, we see that ACA oriented researchers are in serious disagreement with the dominant reductionist scientific community, and they are struggling to develop conceptual models that justify their claim in the wider scientific community.

Third, a major headway to move beyond biophysical reductionism resides in the reflexivity potential of the biophysical scientists themselves. Colfer recounts that some of the early attempts to bring social science (with qualitative and constructionist orientations) was actually by young foresters; unfortunately this was not strong enough to challenge the counter attack from senior biophysical scientists. As Colfer recalls:

> In 1995, some of the 'social science' research was planned by an interested young forester. I witnessed one early seminar where his sensible suggestions were vituperatively attacked by senior biophysical scientists whose critiques actually showed more ignorance than the scientific rigour they purported to demand. Without the strong social science background, the young forester could not confidently defend his good ideas to such attackers. Prestige within the institution at that time also favored the established biophysical scientists, though that gradually changed.

This is especially significant, as the forestry sector is slow to accept holistic and collaborative action research approaches compared to agricultural circles.

Banjade (Chapter 7), who himself comes from a technical forestry background, makes the point in an even more illuminating way. As he reflects upon his own case, he acknowledges the difficulty of recognizing and internalizing social learning processes in forest management. The experience of ForestAction as a locally engaged research organization, as Banjade indicates, was about extending cooperation with institutions having similar approaches, and not necessarily the same language. As a result ForestAction had worked with a number of themes – action learning, self-monitoring, social learning, adaptive management, and the like, but there was some degree of continuity in the basic process of learning and action.

A similar reflection is found in the case of Zimbabwe. Forest scientists were found to realize the limit of their reductionist approach when they were unable to impose regulations on the forest without the cooperation of local people (Chapter 6). This opened up spaces for a more adaptive and collaborative approach to forest governance.

Fourth, ACA leaders looked for ways and means to articulate their ideas, capitalizing on the existing practices and initiatives already engaging reductionist scientists, rather than transplanting anew from the top. At CIFOR, ACM team leaders built on the existing projects with traditional scientific methods and added elements of action, reflection, interdisciplinarity, contextual grounding and collaborative learning (Chapter 3). Banjade also provides a rich account of how researchers at ForestAction always tried to reuse ACA related experience in successive project planning.

Fifth, ACA researchers also looked for larger patterns of how the previous uses of traditional science create problems for society. As Sherwood *et al.* assert, one way in which the traditional scientific paradigm is questioned is when the impact of a reductionist approach becomes visible in the wider society, and people come forward to challenge it (Chapter 4). As with the larger environmental movement, during the 1990s, the harmful consequences of agricultural modernization in Ecuador – in particular, severe health problems associated with pesticide exposure, degrading soils, and declines in productivity – made a growing number of people concerned over industrial era technology, leading to rising waves of protest and questioning of public policy. This paved the way for a more adaptive and farmer-centred approach at least for some years after the incidents.

Sixth, as Fisher notes (Chapter 8), developing a culture of review and reflection within the organizations comprising both biophysical and social scientists can enable two-way communications, and especially bring the unreflected core of beliefs and prejudices in the discursive domain to pave way for more ACA oriented research and action.

Finally, as opposed to the assumptions of the traditional scientific community, conflicts among stakeholders can at times provide an opportunistic moment for an ACA oriented practitioner to intervene. As the ACA work in Zimbabwean forestry reveals (Chapter 6), intense conflict over resource access between the local communities and the Forestry Commission actually provided space for ACM project facilitators to come to the scene, perhaps more easily than a situation fully under the grip of the Forestry Commission (Chapter 6).

To summarize, in order to respond to the reductionist bias involved in the process of learning and innovation in natural resource governance, ACA researchers and practitioners have reported a number of specific initiatives:

1 Discarding the methodology of experimental and control communities as they were considered unethical (Chapter 3).
2 Using elements of traditional research (such as in the form of 'context studies') partly to persuade the biophysical scientists' expectations and partly as a hybridization of methods (Chapter 3).

3 Allowing autonomy to use a variety of conceptual and analytical elements in different locations as the team prefers and can articulate (Chapters 3, 4 and 8).
4 Allying with community based movement and spontaneous networks (such as Canastas Solidarias, Society for Water Resources Management and FECOFUN) (Chapters 4, 5 and 7).
5 Employing participant observation with extended periods of time in the field talking to people (Chapter 6, 7 and 8).
6 Exploring opportunities for potential for action research to address both social and 'technical' change in a rather eclectic manner (Chapter 8).
7 Locating and allying with professionals working in a field context, whether as development workers or researchers, who are open to the new approach (Chapters 6, 7 and 8).
8 Structuring an experiential process to frame analysis that makes sense to the biophysical scientists and also add rigor (Chapters 3 and 8).
9 Building the capacity of actors to change their perceptions on their own role and to appreciate the value of collaboration and joint learning (Chapters 6 and 7).

Working with local communities

A key goal of ACA oriented initiatives has been to improve the human well-being of the peoples living in and around natural resources by engaging and empowering the local communities in the learning processes. In view of this, while working at multiple scales has remained a key aspiration, direct collaboration with local communities has usually remained a priority. Unlike the experiences of service delivery projects (which offer some tangible services immediately) or the traditional research projects (which do not engage people apart from via collecting information), a number of unique challenges have emerged pertaining to the ACA projects. First, there is a scale perception gap between the ACA facilitators and local communities. The ACA facilitators tend to see the problem of resource management and human well-being as a cross-scale phenomenon, and hence the need to analyze issues and generate evidence for meso level and national policy deliberations. On the contrary, members of local communities tend to prioritize immediate fulfilment of needs and solution of the problems they are facing. Complex trajectories of change involving learning and collaboration are also not an immediate priority of local communities who have pressing livelihood needs. For these reasons, all contributors report that developing relationships of trust with local actors was not always easy. The local stakeholders did not clearly see what benefits could emerge from it (Chapters 6 and 7).

Communities follow largely tacit ways of learning, and there is always some level of resistance to a research process that exposes unreflected assumptions

(Chapters 7). A particularly critical challenge reported by Banjade is one that emerged while working with the leaders of local communities at different scales. The leaders of the federation at different levels were in competition with each other to access any resources available. The competition hampered the process for a long time, and the ACA professionals have had to spend significant time in mediating the relations, and arriving at amicable partnership relations. Banjade further recalls his experience with some local communities in Nepal:

> [L]ocal people perceived the researchers as NGO employees, among thousands of others, whom they considered as having lucrative jobs and being interested only in 'dollar *pachaune*' (literally, 'digesting dollars coming from foreign sources'), rather than being committed to enhancing community well-being. As such, the researchers became victims of the dominant image of NGOs in Nepal, which are commonly blamed for (mis)using foreign support in the name of people.

This was largely animated by the people's frustrations over the failure of the previous 50 years of the development industry in Nepal, which had only strengthened social inequality (Metz 1995). Besides, efforts to generate research and address issues that are beyond the immediate concerns of local communities are particularly likely to meet with local resistance.

Second, while ACA facilitators support communities in better identifying issues and developing visions, communities also appear to take the plans forward to implementation. In several situations, forming task groups and engaging locally based facilitators has become a key strategy. Working with a select group of local 'change agents' or one or two group representatives from within the communities was found to be useful, but ended up with limited institutionalization of learning processes (Chapters 5 and 7). Nonetheless, as the planning and learning processes intensify through ACA-based interventions, it is not straightforward to define, negotiate and agree on who is going to participate in the actual action process – even when the process is being done at the local level (Chapter 8). As Fisher reflects upon his experience, applying the idealistic notion of participation as having all participants in all aspects of the research is 'quite unclear and problematic'. It is particularly perplexing to organize meaningful participation in a project operating at multiple locations.

Third, on the beneficiary side, when leaders/members of CBOs work with ACA facilitators, they gain new insights but find it difficult to persuade their organizational leaders. Community level ACA actors are then confronted with the existing organizational decision system while proceeding with applying learning. In the floodplain CBOs of Bangladesh, the new lessons brought by the leaders were discussed and reviewed in broader meetings. As Sultana and Thompson report (Chapter 4), 'After obtaining general body responses, the CBOs revised their management plans finding room to change their decisions on the basis of not only their own experiences but also experiences of the other CBOs as shared in the workshops.'

ACA processes have had to face additional challenges when local communities changed their leadership. In the Nepal forestry case, tensions and conflicts erupted between local change agents trained by the ACA researchers and the community leadership. The local resource persons did not get enough of a handle on the institutional dynamics of CFUGs, particularly because they were considered junior to reigning community leaders (and thus facing apparent difficulty in communicating new knowledge to the senior leaders in the prevailing culture) (Chapter 7).

Given these challenges of working with local communities in advancing an adaptive collaborative approach, several ways in which ACA facilitators sought improvements can be summarized as follows:

1 Supporting local groups to develop their own action plans, for which ACA practitioners provide necessary facilitation (Chapter 5).
2 Facilitating networking among community groups so that they can themselves augment and articulate voices and concerns (see the federation of forest groups in Chapter 7, the adaptive learning network of floodplains groups in Chapter 5, and the Canastas movement in Chapter 4).
3 Facilitating farmer-centred learning processes in agricultural development planning and management (Chapter 4).
4 Allowing autonomy to develop responsive strategies within the ACA management team (Chapter 3).
5 Sensitizing government officials to visit and respond to community concerns raised through local level learning and reflective processes (Chapters 6 and 7).
6 In order to minimize transaction costs associated with a more adaptive and collaborative approach, carefully plan interactions and events in terms of time and space (Chapters 5).
7 Forming a task team or implementation sub-committee for effective management of local level innovations. ACA facilitators can go for on-site visits and follow-up support (Chapters 6 and 7).
8 Enhancing the ability of ACA facilitators to catalyze learning processes remained a key management concern (Chapter 3).

A major shift is suggested in the relationship between ACA actors and the local stakeholders – away from a traditional service provider–recipient relationship to partners in 'cooperative inquiry' (Dewey 1916/1966). Issues of transparency, trust and compatibility have been reported, at least at the beginning of the ACA process, in particular locations (Chapters 6 and 7). Evidence has confirmed that when ACA actors operate fully in the 'context of application' (Nowotny 2003), and not in the context of extraction, chances of collaboration among NRM actors will be higher than without such an orientation.

Integrating research and action

A key element underpinning ACA is learning *in and from* action (see Chapters 2 and 8). But it has not been easy to achieve this duality. Organizing time and efforts to balance research and action is reported to be particularly challenging, as the two functions often require diverse and competing sets of competencies, skills and efforts, as well as institutional mechanisms. Also, ACA actors do differ in their own orientations and inclinations towards research or action, depending on their earlier experience and professional backgrounds. The effect of the institution – discussed earlier in the section – further compounds the ACA strategy to integrate research and action, as the existing institutions are either research- or action-oriented, and much less ACA oriented.

The research part of ACA is sometimes seen as extractive and not relevant for the local stakeholders and even policy makers. Likewise, the methodological frameworks of research still tend to carry the legacies of the positivist paradigm, with little thinking on how experiential insights can be tapped and interpreted. Additionally, another critical challenge experienced is when ACA actors are seen as advocates of particular standpoints in specific governance debates, and not as independent researchers doing 'neutral research'. There is indeed a basic dilemma here – how ACA researchers can make a legitimate choice for their role in the spectrum between research and advocacy.

Contributors have found several specific strategies to integrate research and action in varying extents.

1. Find novel ways to present research contributions to practice and policy. To do this effectively in the context of the dominance of the positivist research paradigm, as Fisher argues, 'the best approach is to write up research in such a way that connections between actions and change are told in the form of "robust narratives" that emphasize "plausible causal connections"' (Chapter 8).
2. Communicating the stance and perspectives continuously in everyday discourses and in the deliberations within wider scientific communities of practice, that ACA actors want to clearly go beyond the traditional approach to research and development (Chapter 3). For instance, CIFOR's ACM team argued that they wanted to 'see changes during the course of our project, rather than waiting until others could (or would) implement our findings'.
3. Treating rural people as partners and learning from 'their energy, motivation, goals, knowledge – that had been largely ignored' (Colfer, Chapter 3).
4. Anchoring ACA related processes with a strong action-orientation at some local institutions with commitment to facilitating local level change through research (Chapter 7).

5 Focus on specific practical issues relating to communities and natural resources, and then define learning questions to guide action, monitoring and innovation (Chapters 6 and 7).

Working within and through national policy regimes

Adaptive collaborative approaches are equally concerned with national policy regimes as much as they are with local level issues of resource governance. This policy orientation makes them politically engaged and invites several challenges. But the actual challenges faced vary depending on the type of ACA actor and the country contexts. The ACA initiatives reported here are mostly designed by international actors who supported national and local actors to catalyze learning and collaborative processes in respective countries and sites. The sites reported in the book also represent very diverse political situations of stability, conflict and democratic governance.

For ACA actors, working with government institutions – at both the policy making and the field level – has remained an arduous job. This is especially so as the public sector natural resources institutions are quite fragmented and discipline bound (for instance, see Chapter 4 on divergent institutions on Bangladesh floodplains), and also structured techno-bureaucratically. While there are plenty of discretionary spaces to experiment with new innovations, to a certain extent, getting individuals to engage in the process with a positive attitude is critical so that they can later ventilate ACA insights widely within the government and policy making system.

Apparently, the Zimbabwean context at the time of the ACM study was politically fragile and the team realized that it was too sensitive to engage stakeholders politically, despite appreciating the need for the same (Chapter 3). As Colfer asserts, 'We, like most actors . . . were, at some level, dependent on the good will of the government in the countries where we worked to maintain our effort.' An additional constraint she reports 'on our political activity was the potential implications for the people living in the communities where we worked. Overt political activity can be dangerous – in our experience, particularly with the Zanu-PF regime in Zimbabwe and the Maoist and governmental conflicts in Nepal'. Local ACM teams had to face tough times facilitating interactions on issues of rights and governance in such contexts.

Obviously the question is how far the ACA facilitators can go in coping with such adverse situations. This also depends to a significant degree on the agency and identity of the ACA actors themselves, as there could be different advantages and disadvantages of ACA practitioners coming from the same or different countries and regions. Interestingly, observations made by Mutimukuru-Maravanyika (that CIFOR ACM work in Zimbabwe was not sufficiently political) (Mutimukuru-Maravanyika 2010) have triggered some critical reflections among CIFOR team members – which is itself a healthy process of reflective engagement, beyond the project period.

Even in the normal political situation, with no stringent restrictions on freedom of speech and movement, it is not easy and feasible for outside ACA actors to influence local policy, without building local demands for the same. As Sultana and Thompson report from the experience of Bangladesh (Chapter 5):

> [W]e focused on a horizontal learning process between CBOs . . ., first helping the poor to organize and develop trust and a common platform and voice before they are thrown into the frame of formally interacting with traditionally more powerful actors (government and local elites).

The horizontal link was also a crucial element of ACM in CIFOR, but this was only one part of the three important elements – vertical links, horizontal links and iterative process, all of which were simultaneously pursued for bringing about learning and innovation in resource management (Chapter 3).

Colfer, as the first team leader of the ACM programme at CIFOR, recalls that ACM sought to recognize 'empowerment' as the central goal, of the men and women in forest communities, and in interaction with others. Despite this, they had to tone down their rhetoric, as they feared that 'emphasizing "empowerment", in our strategic planning documents, was probably too politically sensitive, that it might frighten the governments with whom we knew we would have to work: we used different words'.

As Colfer recalls (Chapter 3): 'It does seem probable that our "home" within CIFOR reduced our capacity to act in the overtly political manner' called for by those who identified the need for more active political engagement on the part of ACM members (as argued by the authors of Chapter 7). Colfer's reflections demonstrate that moving ahead with ACA is likely to meet with resistance from organizational management, especially when the organization has a research mandate. Her team had to repeatedly face problems from the management. And changes in management had direct bearing on the ACM team.

The question of how to go about dealing with power still remains pertinent, and requires further intervention through social learning and innovation processes. Locally engaged ACA actors with significant legal-political rights (such as citizenship) to act can make some difference. But pre-existing identity comes into effect and complicates the process. The experience reported by Banjade is evidence of what happens further when ACA actors attempt more politically engaged facilitation. He reports that the process of attempting to challenge power relations in local forest management contexts triggered significant conflicts and resistance. Banjade reflects upon his own personal career and identity of being a technical forester as well as a political activist. He found his political disposition, developed through activism, was more amenable to ACA practice, than to the way he was brought up as a technical forester.

The process of learning and innovation becomes political, right from the moment of entry of an ACA actor and hence changes the configuration of power. Such changes in power relations may be conducive for learning and deliberative processes, as the engagement of local actors alone may not lead to the desired level of openness and reflexivity, and constructive dialogues. This is as simple a case as when an intermediary has more leverage to act in a conflict situation, than do the conflicting parties themselves.

As natural resource management systems are organized at multiple scales – from local communities to bio-regional scales – it is equally important that local communities have clear and strong tenure rights over natural resources. While some see that adaptive and collaborative approaches to governance at higher scales can only build on strong local community institutions (Chapters 5 and 7), others see that it is indeed through such processes that strong tenure rights can be established and institutionalized (Fisher *et al.* 2007).

In the context of the farmer-centric version of FFS having degenerated into an expert-led research method in Ecuador, Sherwood *et al.* report that the evolution of the Canastas movement is a more politically engaged incarnation of an ACA, moving away from the technocratic set back in FFS, to directly engage people in the politics of learning and innovation (Chapter 4).

To address how ACA actors can become more political, yet remaining within the broad ACA framework, contributors have suggested the following more specific insights:

1 Before starting the process, analyze the local power structure and attend to the realities and priorities of both marginalized members (women, lower castes, despised ethnic groups) and elites (Chapters 4, 6 and 7).
2 Remain sensitive to micro-politics within and among local groups and others with whom ACA work is done, as there can be significant dangers (even death) to local people if care is not taken (Chapters 6).
3 Take sides with marginalized political actors, as a strategy of what Fisher calls 'emancipatory action research' (Chapter 8).
4 Engage local change agents in the process of learning so that new ideas and knowledge immediately find their way into practice (Chapters 5 and 7).
5 Engage external consultants to do snap-shot assessments of the situation and process while the ACA oriented researchers are into the PAR process (Chapters 6 and 7).

Facilitating cross-scale linkages

While cross-scale linking or vertical coordination is widely recognized as a key normative element of ACA, the contributors report that this aspect of ACA is particularly challenging. Two conceptions of cross-scale links are dominant in the resource governance literature – as an institution and as an adaptive system – both of which are problematic, as they downplay the chaotic processes of interactions, as well as how agency interacts with systems and structures in the process of change or reproduction.

We consider that the linking process has at least four axes and is not as structured an institution as some find it. We conceive cross-scale linking processes as 'transactional flows' in terms of a) information flow, b) legitimating processes, c) value chain linkages, and d) exercise of power and counter-power. Attempts reported here are related to addressing multiple combinations of these diverse transactions, but practitioners report that success was much more limited than what they were able to achieve internally at local community levels.

All these processes are linked to whether or not the management of natural resources takes a system-wide view and provides an enabling environment for local innovations. A key gap in ACA oriented initiatives is that they still fail to question and counteract the mentality of 'invent and diffuse across scales', rather than continuous innovations, which requires taking a learning based approach at all stages, and at all scales (from community to national and beyond), affecting the practice of natural resource management. The 'meso sphere', between local and macro, has been found to be an important layer in the linking process; this is more like 'forums', rather than well-formed institutions (Chapter 7). The first phase of CIFOR's ACM work in Nepal started at the community level, and then ended up with a greater realization of the need to work at the meso level, and then link to the national level. CIFOR's multi-country programme allowed different national teams to start working on any level considered suitable and then to proceed for vertically linking (Chapter 2).

For ACA researchers and practitioners, handling the issue of scale has remained a key. As Colfer recounts (Chapter 3): 'Scale presented another dilemma. Early on, most of us assumed we would be focusing primarily on a very local scale for the PAR work – though all recognized and agreed that our "vertical" interests were also key.'

The enthusiasm to forge cross-scale linkage is marked by a fear regarding the domination of collaborative processes by the powerful actors. The experience of Bangladesh floodplains is a case of horizontal scaling out and linking of community organizations into a network, to secure community access to resources (Chapter 4). The strategy involves first empowering the local communities and then linking them with state agencies. This has helped overcome power differences while moving up the scale.

In the context of natural resource management in the developing world, the aid priorities are still governed by the colonial legacies of state-centric management. This approach also strictly separates various institutions and innovation functions into boxes and chambers hampering collaboration and shared learning (Chapter 5).

While methodologies and tools need constant adaptation, the pressure to scale out across a wider region creates disincentives for learning and innovation (Chapter 4).

Scale has two aspects – horizontal and vertical. Horizontal relates to expanding the useful idea across wider space and time. Vertical scaling means creating enabling conditions for successful operation and expansion of a useful practice and idea. Contributors offer a wide range of insights across both the dimensions. The question either way is – once we have some innovation or changes at a particular local domain, how do we proceed to scale, as ACA actors?

Most processes of resource management start at a local level (such as a community or a household), and then there is an attempt to link vertically with other layers. But the problems and challenges faced while scaling up are qualitatively different and more complex than those faced at the time of developing an innovation at a relatively small scale under more intensive care. Networking of actors – and hence strengthening these transactions – can help develop a system-wide view of management, away from the focus on particular resource areas and sectors (Chapter 5).

By the time the ACM project was conceptualized in CIFOR, there was already some appreciation and experience of working at multiple scales. This made it easier conceptually and institutionally for the ACM team to have an element of strengthening the *vertical links* between communities and wider scale actors (government, industry, conservation projects). The links involved not only flow of information but also the more substantive aspects of management cooperation and accountability – including, as Colfer asserts, 'the ideas of local level empowerment, the potential for local people to act as monitors of political actors at broader scales, and for better local access to outside resources (expertise, funds, links to others, support for local initiatives)' (Chapter 3).

Farmer Field Schools, as discussed by Sherwood *et al.*, provide a case of what can happen when a niche level methodological innovation is taken up by organizations for wider scaling up. Sherwood *et al.* observe (Chapter 4), 'While FFS scaled in name, its fundamental principles often were lost in translation. Instead of enabling desired institutional change, scaling exposed vulnerabilities that led to the transformation of FFS.'

The experience of Bangladesh floodplains is also a case of horizontal scaling, and with limited considerations for vertical scaling – to secure community access to resources (Chapter 5). The initiative attempted to scale up the learning units, by linking up CBOs that were unique and acting in isolation from each other so that a more integrated approach to flood plain management could be initiated. Their approach, to focus on horizontal scaling and learning, was based on the tenure-insecure environment of Bangladesh, in which they believed that any vertical linking process could be successful only when sufficient horizontal networking CBOs had been achieved. Moreover, this could also minimize the potential risk of powerful actors in the higher scale getting information and resources to further entrench existing power relations. Contributors make persuasive claims that this approach marks a break away from research-centred adaptive co-management discourse in developed countries to a more people-centred and bottom-up approach (Chapter 5).

In the context of the well-functioning groups of community forestry in Nepal, as Banjade asserts, ACA approaches went beyond an emphasis on local control, and sought to forge interaction of multiple sources of knowledge, empower the disadvantaged actors, and facilitate reflexive learning among actors at different scales of governance (Chapter 7). Perhaps it was a bit ambitious, and this is the reason why 'at the meso level, however, the processes and outcomes were more erratic', with far more limited success than initially anticipated (Chapter 7). The diverse goals of different actors and their mandates would often limit the effectiveness of ACA processes at this level. The actors even failed to arrive at any workable definition of the meso sphere in the context of forest governance, apart from outlining issue-based forums. They also had limited success in formulating conditions that would work effectively. Nevertheless, the research identified various issues and challenges in working within this sphere in different contexts.

Banjade further recounts his experience of working with ForestAction Nepal to scale up and scale out ACM related innovations from a few sites to over 60 sites horizontally, and vertically through policy engagement. Despite initial enthusiasm of the lead actors, the project team had to struggle hard to reach minimum levels of success, not to mention innovation. This was because the work continued to be seen as political and threatening by the government forest officials; there was poor networking and advocacy capacity of the community forest federation, and a lack of accountability among the federation leaders.

Mutimukuru-Maravanyika and her colleagues also had to face pressure for linking local innovations directly into the policy process (Chapter 6). They organized policy roundtables on emerging issues to inform the policy makers. How far this went in changing policy is not clear.

Managing sponsorship and funding

Almost all cases of ACA practice reported by the contributors were or are operating within internationally funded development projects. Situated within the field of development means that such initiatives have to face the legacies of post-World War II development – of western control over power, knowledge and financing (Escobar 1995). This posed particular challenges for ACA actors, who have to confront not only the existing donor assumptions of linear planning and management but also the heightened material expectations of local stakeholders. For instance, developing relationships of trust with local actors was not always easy for an ACA oriented project, mainly because the local stakeholders did not clearly see what benefits could emerge from it (Chapters 6 and 7). Besides, much of ACA work has been framed by *development* sponsors (including all the works reported by authors of this volume, and even this chapter and the origins of the book!). Heavy sponsorship means there is still little endogenous thinking and innovation around different ways and possibilities of adaptiveness and collaboration. But a bit of good news is that sponsors often do not have full control of the innovation pathways, and if situations are favourable, more locally grounded processes of innovation and learning can occur (Chapters 3 to 8).

Contributors to this volume report a common challenge that adaptive collaborative initiatives were short of public funding support. Sultana and Thompson (Chapter 5) report that in Bangladesh, it was necessary to provide some modest funding and active facilitation support to enable the community based resource user groups to come together in adaptive learning networks. Sherwood *et al.* highlight the danger associated with large-scale buy-in of the idea of FFS by government, donors and even private service providers (Chapter 4). All wanted to replicate but without the costly process aspects of FFS. There is usually a mismatch between funding cycles and temporal pathways of innovation. The contributors to this volume report that the types of results, and the way results themselves are conceived, are starkly different from the mainstream technocratic paradigms of development.

The other challenge related to the development industry is about the politics of evaluation. Donors continue to look for tangible results – usual evidence of welfare changes. ACA actors, however, do not seem willing to compromise the quality of the process and sustainability of the results in order to satisfy the short-term needs of donors for success stories to placate their own domestic political constituencies. Contributors report several instances of struggles with the donor and manager-commissioned evaluations (see especially Chapters 3, 4 and 6).

The effort to address the challenges of natural resource management cannot be left to the 'invisible hands' of the market. Those benefiting from such innovations are the poorest people and also future generations in the case of sustainability. Creating innovations in resource management, and also facilitating *innovative pathways* to innovation require societal investment. Yet, developing countries are less likely to earmark funding for such 'soft development', given their emphasis on creating tangible development outcomes. In such situations, foreign aid continues to be a key source of financing for innovation in natural resource governance, at least in the foreseeable future.

While aid has provided laudable support in emergency relief and basic services such as health and education, it has remained largely insensitive to understanding and facilitating how societies can learn their own way to development and environmental sustainability. In the context of natural resource management, the aid priorities are still governed by the colonial legacies of state-centric management, still continued primarily through national public institutions for research, management and resource control (or supporting another extreme of marketization). What is missing is a broad view of innovations in which research can be one function closely embedded within the process of innovation itself. Following the environmental turn of the development debate and more recently the climate change crisis penetrating resource management approaches, the level of public funding (including international aid) has increased. Yet the priorities focus on either delivery of services or technology transfer and are still guided by a positivist scientific ideology.

Cases reported in this volume include donors supporting the initial process, but then putting intense pressure to show 'return on investment' in direct, tangible and immediate returns (all empirical chapters from 3 to 8). ACA leaders have made sustained efforts to convince donors about the high cost of the processes needed to progress in the face of complexity and democratic engagement. They have however failed to persuade the technocratically oriented donors and public investors of this all too apparent reality. Part of the reason is that the policy makers and fund allocation managers are also under pressure to demonstrate the development outcomes of government spending and taxpayers' money.

The problem is not that ACA-like initiatives do not yield return on investment, but that their gestation period is longer, and we need to conceive a different evaluation methodology to ascertain the aid-induced change. In several instances, funders have frustrated the ACA practitioners through their conventional quantitative cost-benefit approach to evaluation, focusing on what quantitative tools can measure, but ignoring more fundamental socio, cultural and institutional outcomes of the initiatives. This difference at times has resulted in some desperate cases – involving mismatch of expectations and even early termination of projects (Chapters 3 and 7). Contestations around the meaning, scope and epistemology of evaluation therefore represent a key bottleneck in securing sponsorship for adaptive and collaborative approaches to natural resource management.

As Fisher summarizes (Chapter 8), aid programmes have become even more resistant to learning and collaborative approaches despite their (sometimes lip service) acceptance of participatory approaches. He claims that 'While project designs sometimes talk about learning approaches, the same documents generally include targeted inputs and outputs that create a momentum that neutralizes any real adaptive or action learning'.

Given these entrenched challenges on sponsoring ACA initiatives, very few successful interventions have been reported in this volume. ACA oriented researchers and activists will have to work further to find ways to persuade the development donors to recognize the social, cultural and institutional outcomes as part of development. They also have an uphill struggle in influencing public spending policies in the developing countries to supporting adaptive learning and innovation processes to address contested issues of natural resource governance.

Sustaining results and innovation pathways

A short-term ACA based action can result in improvement in processes, but may not quickly lead to the desirable outcomes – such as human well-being and environmental change (Chapter 3). Case studies demonstrate that a series of specific human well-being related outcomes have been realized at the local level – such as increased sale of products, improved access to benefits and the like (Chapter 6 and 7). But such results were vulnerable to quite delicate institutional support mechanisms. For instance, many of the processes initiated at the community level were not sustained after the death of an involved Forestry Commission staff member (Chapter 6).

All the six case study contributors report a trajectory of governance innovation, initiated with high enthusiasm, but followed by one or other aspects of 'erosion' of the ACA process. When the innovation process has remained in the hands of its founding leaders, it has survived. When scaled up, it has eroded. The project mode of innovation development – typically within the context of aid – could not catalyze innovation processes mature enough to survive beyond the project. When the innovative action began to penetrate the core areas of institutions and governance, it was defeated or dissipated.

Among the pathways of innovation that have been sustained and that have also been absorbed to some extent by the policy system is Nepal's community forestry. This was in part possible because of the increasingly supportive and democratic political system that paralleled the evolution of community forestry in the country, and the proliferation of community networking and federation building activities. Specific innovations within community forestry – such as application of self-monitoring techniques at the community level – succeed when they occur in areas where local groups with strong commitments to such approaches hold sway. The CBO network of Bangladesh has come a long way in expanding the voice, knowledge and power of community groups, and now faces challenge to effectively interact with the state and markets, if it has to demonstrate livelihood benefits to the member communities. Farmer Field Schools in Ecuador, once thrown into the social wild, did not thrive, at least not with the same potency of learning that was there earlier. It was later subjected to commercial and technical manipulations.

ACM interventions in the difficult times in Zimbabwe created some local innovations in specific aspects of resource management (such as promotion of broom grass by local communities), but these were not effectively taken up in national policy. As a result they did not survive long. As the initiatives were facilitated by teams funded by international organizations, the innovation process did not find a locally grounded institutional home. The process was also found to be inadequately political to link the local process with the national. Fisher's prolonged involvement in multiple action learning projects confirms that institutions that seek to support action learning processes with communities should themselves make action research at least one of their modus operandi.

Contributors report an aspect that requires caution. Sometimes an ACA based process may enhance the agency of local communities and stakeholders to imagine a positive future – through visioning exercises and interactive engagements. Leaving the local stakeholders alone to look for ways to realize the newly conceived goals will create frustrations as local people may not always be capable of mobilizing resources at the expected level. This indicates a need for strengthening collaborative aspects of the process so that local communities can get access to needed services and inputs. The other option is to think about the issue of achievability and capacity of the beneficiaries to plan and organize the type and degree of change, clarifying the inevitable lags between vision and actuality.

After all, the criteria of success and failure should be part of the discussion and negotiation of the adaptive collaborative way in natural resource management (Chapters 3, 6 and 7).

Key results valued and monitored by ACA actors for sustaining innovation in natural resource governance are as follows:

1. Ensuring ownership of all involved and affected in the learning process (Chapter 3).
2. Challenging the perceptions of farmers and local communities about the use of technology as a superior form of practice without looking more holistically at the pros and cons (Chapter 4).
3. Maintaining communicative engagement with multiple actors (Chapters 6 and 7).
4. Addressing issues related to inclusion in decision-making and equity in benefit sharing (Chapter 7).
5. Explicitly identifying and addressing issues of conflict management (Chapter 6).
6. Making the culture of review and reflection routine (Chapter 8).
7. Strengthening the capacity of beneficiary organizations to seek services from public agencies (Chapter 5).
8. Balancing local action outcomes with globally relevant knowledge generation (Chapters 3 and 8).
9. Fostering mutually supportive teamwork (Chapters 3 and 7).
10. Ensuring strong leadership in the process and harnessing commitment of different agencies (Chapters 3 and 7).
11. Going beyond dissemination of a successful innovation to explore and transform the entire socio-technical regime governing the innovation (Chapter 4).
12. Undertaking critical research to examine multiple expressions of innovations in different contexts and unravel any erosion in the quality and rigour of the learning package (Chapter 4).
13. Facilitating networking and federation for more political expressions of local communities (Chapters 4, 5 and 7).

Concluding remarks

The above-mentioned eight domains that the ACA researchers and practitioners muddled through represent a complex set of challenges as well as emerging innovations. What has become clear is that notions of participation, learning and innovation are easier said than done. Yet the encouraging part is that there is now a wealth of experience and insights gained as part of engaged and adaptive learning processes in diverse settings around the world. In the next and final chapter, these insights from the world of practice will be used as a basis to explore future directions towards transforming natural resource management policies and practices.

References

Dewey, J. (1916/1966) *Democracy and Education: An Introduction to the Philosophy of Education.* New York, The Free Press.

Escobar, A. (1995) *Encountering Development: the Making and Unmaking of the Third World.* Princeton, NJ, Princeton University Press.

Fisher, R., R. Prabhu and C. McDougall, Eds (2007) *Adaptive Collaborative Management of Community Forests in Asia: Experiences from Nepal, Indonesia and the Philippines.* Bogor, Indonesia, Center for International Forestry Research (CIFOR).

Kolb, D. A. (1984) *Experiential Learning: Experience as the Source of Learning and Development.* Prentice-Hall, Englewood Cliffs, NJ.

Metz, J. J. (1995) *Development in Nepal: Investment in the Status Quo.* GeoJournal 35(2): 175–184.

Mutimukuru-Maravanyika, T. (2010) *Can We Learn our Way to Sustainable Management? Adaptive Collaborative Management in Mafungautsi State Forest, Zimbabwe.* Ph.D. Thesis, Wageningen University.

Nowotny, H. (2003) *Democratising expertise and socially robust knowledge.* Science and Public Policy 30(3): 151–156.

10 Moving forward with adaptive collaborative approaches

Hemant R. Ojha, Andy Hall and Rasheed Sulaiman V

In the previous chapter, we synthesized insights presented by contributors in confronting the challenges faced in the process of enhancing adaptiveness and collaboration in natural resource governance. This chapter draws broader meanings from these insights and weaves them into a future direction, including rethinking the current, standard way of understanding and facilitating change. To start with, we recapitulate the key insights of the contributors in Box 10.1.

Box 10.1 Moving forward through adaptive collaborative approaches

- Influence established regime of mainstream institutions and actors (Chapter 4), such as through creating national coalitions with advocacy groups and media (Chapter 7).
- Empower local people to monitor political actors (Chapter 3) through empowerment training (Chapter 6), training and visits (Chapter 5) and ongoing assistance in reflections and deliberation (Chapter 7).
- Recognize gender differences in learning and participation in the adaptive learning processes (Chapter 5).
- Create a network of local communities and facilitate horizontal learning among them and then enable the network to effectively interact with the policy regime (Chapter 5).
- Adopt an iterative strategy (Chapter 3).
- Create supportive senior management for field level processes (Chapter 5), and ignore anti-ACA instructions of management (Chapter 3).
- Adopt a two-pronged strategy of research and action (Chapters 3 and 8), including opportunistic use of research methods (Chapter 3) and snap-shot studies before and after PAR (Chapters 3, 6 and 7).
- Balance multiple and conflicting expectations of actors (Chapter 3).

- Establish and communicate plausible connections between ACA processes and outcomes (Chapter 8).
- Strengthen capacity to integrate research into practice and communicate across the innovation pathways (Chapters 3 through 8).
- Strengthen processes of self-evaluation (Chapter 3).
- Be more political (Chapters 6 and 7).
- Commit to longer-term action, learning and change (Chapters 3 through 8).
- Provide some financial subsidies for adaptive learning and experimentation (Chapter 5).
- Discuss and clarify any suspicions and past legacies of development projects to prepare the ground for a genuinely learning based approach (Chapter 6 and 7).
- Identify specific management issues and constitute specific task groups for conducting participatory action research to understand and resolve the problems and realize new possibilities (Chapters 6 and 7).
- Once lessons and evidence are generated at the local level, bring those to higher levels of policy making (Chapters 3, 5, 6, 7 and 8).
- Link learning processes (usually involving cycles of visioning, planning, implementing, monitoring, review and reflections) with specific economic opportunities (Chapter 6) or commonly perceived issues of governance such as exclusion (Chapter 7).

As is clear from Box 10.1, the experience reported in the case study chapters reconfirms evolving wisdom about taking complexity and uncertainty seriously in the learning and innovation process (Lee 1993). Yet it also throws out more and subtler challenges that require rethinking how we can act, learn, innovate and change. There are plenty of experiments and enough evidence to prove that we have to learn the way out, manage conflicts among ourselves, and create space for collaboration for the common good. But because we failed too frequently, there is a need to fundamentally rethink how we learn, collaborate, link across scales and empower the people deserving fair access to information and resources. This question becomes central if we are serious about finding ways to improve the lives of poor people dependent on natural resources or staying close to the natural systems.

Rethinking, revisiting, reframing

Paradigmatic rethinking is needed to move beyond biophysical reductionism, a linear conception of change, a thin view of participation, technical views of innovation and learning framed within existing assumptions and underlying institutional boundaries and regimes. We suggest that we need to explore how ACA can be better grounded as the modus operandi of policy and institutional development as well as management. Table 10.1 identifies possible aspects of rethinking needed for a shift towards a more adaptive and collaborative approach.

These insights imply the need for revisiting a number of assumptions that guide current resource management policy and practices. First, we need to consider reframing research more seriously, not just in relation to the needs and concerns of disadvantaged groups or the ultimate clients of a development project, but more importantly as a catalytic and reflective learning activity among a wide range of actors participating in the regimes of resource governance and pathways of innovations. While we recognize the value of more basic research and philosophical reflections at one end of social learning, we need to think of how research can be part of innovation and social change processes, in the more pragmatic conception of science popularized by Dewey (Bohman 2002) – that is to say, science for and by society. We do not deny the role of carefully planned technical and more sophisticated research as an integral process of innovation (such as a new variety or a new silvicultural technique). The business-as-usual of developing technical solutions and then using extension systems to put the research into use does not work for everything. Public funding should treat practice-based innovations as new forms of research and provide institutional support to groups of innovators on the ground.

Second, participatory discourse has concealed a technocratic regime in practice. There is still an unanswered question – how can disadvantaged people really claim their stakes in decisions, innovations, policies, management and practices? There is certainly a 'learning paradox' (Armitage *et al.* 2008) in relation to who learns and how, but the experience reported here clearly demonstrates that there is a visible 'power–learning nexus' – as

Table 10.1 Key shifts needed to promote ACA

Dimensions of change	*Rethinking through adaptive collaborative approach*
Cultural dimension	Develop a culture of review, reflection, sharing among groups, within organizations and wider communities of practice
Management dimension	Reframe management from controller to facilitator of learning systems and collaborative processes
Technological dimension	Link research and technology development with larger social learning and deliberative systems
Political dimension	Recognize political inequality, and explicitly align the ACA process towards changing the political status quo
Economic dimension	Value and reward the intrinsic processes of learning, social interactions, and the more sustainable and enduring forms of change

powerful groups and established institutional regimes fail to open up to learning processes. Such learning would essentially involve challenging the spaces for established order (see Chapters 3 and 4), which means that ACA researchers and facilitators should more directly engage with the power–learning nexus rather than just cherish the ideals of learning and collaboration (see Chapters 6 and 7).

The need to become more political in the process of catalyzing ACA brings even more challenges. In many political regimes of the developing world, it is too sensitive to link research with the empowerment of disadvantaged groups. Yet, it is clear from the experience reported here that there is no politically neutral research, facilitation or even service delivery; and neutrality itself is a political position. For change, we need to start taking sides in research and innovation processes. The good news is that it is possible to go beyond the ideological impasse (such as between modernist development versus indigenous and local management), through action-oriented and innovation-focused research and development.

Third, we need to see cross-scale process not as an institution or adaptive system but as a forum, involving the transaction and interchange of information, economic resources and also the processes of symbolic legitimation. Though the issue of cross-scale linkages has resurfaced again (Berkes and Turner 2006), we need to better ground a learning-based approach to understand and facilitate the linkages among all three key dimensions – information, resource flows and symbolic legitimation. Creating cross-scale links can also address the concern that direct participatory processes may not be able to acknowledge the deeper and distant functions of the socio-ecological system.

Fourth, learning should not be confined to operational or local systems; all decision makers can have an additional plan to test assumptions through some form of monitoring system. Transferring rights and empowering the disadvantaged are the keys, but these are not enough, unless there are deliberative links across scales, and transactional flows across value chains, all contributing to innovation in material, institutional/political and symbolic domains of natural resource governance. Indeed, our findings confirm the view that public policy systems could be redesigned as learning systems (Hall 1993; Schon 2010), along with the scope for traditional scientific analysis and public debate.

Fifth, the paradigm of 'invention and scaling up' should be rejected, and a conception of dynamic learning, innovation, sharing and reflection should be promoted. The problems and challenges faced while scaling up an innovation are qualitatively different and more complex than those faced at the time of developing the new innovation, which is done at a relatively small scale under more intensive care. We need to conceptualize a universe of action and learning through which people can connect, speak to, and engage with each other in the process of learning and collaboration, not just in the simulated social chambers of development projects, but in the 'social wild' (see Chapter 4).

Finally, we need to go beyond traditional categorization of actors as researchers, trainers, activists and policy makers, at least when we are engaged in the context of application. Indeed, all have to be learners, collaborators, innovators and political actors. How much of these one has to enact in a particular practice depends on the personal trajectories, existing partnerships and institutional setting. There is a need to think of hybrid actors taking different combinations of various functional roles.

Transitioning to adaptive collaborative approaches

The next question is how we can conceive of an effective transition to adaptive collaborative approach in practice. We suggest both deconstructive and reconstructive strategies. Deconstructive processes include interrogating the language, meanings, science, truths and claims made by multiple actors in the domain of natural resource management and innovation dynamics. And this process should itself be democratized so that people and actors can engage in every domain of natural resource management practice. The whole idea is to take all actors to a new level of reflection on their own mindset, practice, institutions and history so that they can better appreciate the need to go beyond existing disciplinary frames, institutional boundaries, and entrepreneurial self-interests.

Reconstructive processes should include developing a number of signposts and helping hands for the innovators facing the challenges on the ground. This can also be done at different levels, but primarily via the public policy system and facilitative regime. Processes of reconstruction can occur through: incentives, learning environments, creating space for experimentation while managing, creating spaces for managing to learn while acting in problem-solving mode, defining parameters and structures of accountability to hold research and action useful to innovation that benefits society fairly, learning platforms, and linking different ways of innovation and social change. The experience reported here suggests a new set of assumptions and worldviews to enable ACA to take root in natural resource governance (Box 10.2).

Box 10.2 Key insights for transitioning to adaptive collaborative approaches to natural resource governance

- *Embracing hybrid agency:* Actors in real world situations are different from actors on paper or the logical discourse. A researcher is also an activist, an activist is at the same time a researcher and entrepreneur, etc.
- *Harnessing the potential of reflexivity:* Actors can change if they learn to be more reflexive of their own practice, dispositions, relations with others, etc.

- *Becoming a more engaged researcher:* We need researchers who are more passionately involved in the real world contexts. More thought is needed on how this approach can be promoted.
- *Translating conflicts and difference into learning resources:* We should not be frustrated by diversity, conflict, and differences. If we can properly internalize an adaptive and collaborative approach, we can use these as learning resources to make a difference.
- *Reframing the facilitative regime:* Learning and innovation can be seen as the property of a system to self-organize and evolve, but this can be catalyzed much faster and with much better results, in terms of fairness and equity (unlike taking a social Darwinian approach), through appropriate facilitative regime.
- *Thinking of a new third way – action beyond the ideological divide:* A social learning and innovation paradigm, involving action-oriented and critical science, and a pragmatic orientation to involve agencies in the process of change, can demonstrate a way out of ideological stalemates – between left and right, state-centric and market oriented, etc. For every problem context, there is a possibility of acting together if we think through a learning and innovation paradigm and follow adaptive and collaborative approaches.
- *Reconceptualizing cross-scale linking – beyond institutional and systems approaches:* We need to think of cross-scale linkages as a more chaotic situation, involving all different kinds of transactions. We should aim to harness them as forums for imaginations, negotiations, conflict management, etc., rather than well-defined forms of governance.
- *Adopting multiple planning horizons:* In order to facilitate change simultaneously at different temporal and spatial scales, we need to conceive, facilitate and support multiple and overlapping planning processes.
- *Taking a value adding approach:* ACAs should not be seen as a total replacement for an old system, but as a value adding approach – combining elements that are working under a more holistic framework of learning, collective action and creating change.
- *Moving beyond methodology:* While methodology development and application is an important part of the adaptive collaborative approach, ACA actors should not fall into the trap of methodology, thus being detached from the wider regimes and political economy in which such methodologies are framed and created (see Chapter 4).
- *Cultivating local champions of change:* Ending change can rarely be injected from outside; they must nurture local partners and champions of change locally and inside the institutions that need to be changed.

Having discussed the usefulness of ACA in natural resource governance, we should also be mindful of its limits. Three major limitations include: a) the inherent limits of local action and the supplementary role of wider movements and structural change; b) not every aspect of society is consciously learned and discovered, there are socio-ecological cycles of change and transformation operating and beyond the conscious domain of individuals and groups; and c) power relations and social asymmetry always create unequal learning fields and we can rarely rely only on the learning disposition of human agents to change political relations. Moreover, there are also some preconditions in which ACA can flourish better (Armitage *et al.* 2007), but our experience suggests that some of the conditions can be taken as areas of intervention (Chapter 3).

Messages for key actors

At the risk of being too prescriptive, the above conclusions do allow us to draw up some broad 'ways forward' that are relevant to all key actor groups involved in natural resource management in the developing world (Table 10.2).

Four cross-cutting areas of improvement require urgent attention to facilitate effective transition to adaptive collaborative approach to natural resource governance. First, local level resource management plans and practices should more explicitly have an element of learning, monitoring and review, and based on that, local resource managers should be able to generate evidence and insights not only to improve their own management practice but also to provide feedback to higher scale institutions (vertical linkages) and to other similar actors (horizontal linkages). We need to nurture and support new types of researchers who bridge, broker and facilitate change around specific poverty–NRM issues and at the same time connect the processes across the wider policy and institutional systems.

Second, the policy system should itself be seen as a learning system, and its participants should develop ways to organize action-reflection-review in relation to their own domains of planning and decision-making, and also proactively seek out ways to enable cycles of lower-scale learning processes, across diverse threads of innovation. Learning emphasis should be reflected in treating emerging policies as experiments – and hence having a strong element of monitoring, review and reflections. To realize this shift, there is a need to revisit current practices of policy evaluation and accountability. The policy evaluation system should not just take a retrospective analysis of outcomes, but also form a prospective view of exploring possibilities and identifying emerging and promising threads of innovations. There is also a need to reverse the current of evaluation being solely sponsored and undertaken from the perspective of policy makers and donors. Public policy and wider institutions should be oriented to enable and support local innovations,

Table 10.2 NRM actors and suggested ACA oriented changes

Actors	*ACA oriented changes*
National policy actors	• Remain sensitive to emerging innovations and provide enabling environments for them to flourish. • Periodically review the institutional structure of policy making and implementation in the light of beneficiary expectations. • Provide space and incentives for learning oriented initiatives and experiments and absorb any costs associated with failures. • Along with the strategies of decentralization and community based resource management, develop collaborative and learning oriented strategies for the management of large-scale natural resource reserves such as protected areas, large blocks of forests, river basins and fisheries.
Research groups	• Focus research to understand and facilitate innovation pathways. • Politically align research agendas with the rights of disadvantaged people. • Focus on understanding cross-scale linkages and the dynamics of change through ACA.
Extension and capacity development service providers	• Collaborate with other service providers, researchers, and stakeholders to facilitate system-wide processes. • Facilitate networking among community groups in NRM for knowledge and empowerment. • Recognize the value of learning with communities and include these in the curricula of education and training among extension service providers.
Transnational knowledge networks	• Work with local partners and collaborators and build their capacity. • Provide added political and knowledge back up at national and international policy processes.
Local community groups and their associations	• Recognize experimental activities that go beyond their established system (such as innovating new community institutions for watershed management). • Address issues of internal accountability, representation and governance.
Donors	• Appreciate and recognize the learning oriented processes for change at local and national levels and align funding to support those processes, rather than new institutions and processes. • Use the principle of shared accountability to both recipient societies and donor country taxpayers, rather than unilateral accountability to taxpayers. • Reframe planning, monitoring and evaluation approaches to go beyond measuring and quantifying variables to develop shared understanding of the innovation processes and develop feedback to channel back to the actors involved (not just the donor programme managers).

and citizens acting locally should be empowered to review policies and exercise their agency to influence policy for the kind of change they need.

Likewise, another aspect of reframing policy as a learning system is to develop agreed frameworks of accountability for public agencies in enabling local innovations in practice. Accountability for the policy system to the larger citizenry should include an audit of how the system is improving over time, including the processes of learning. This is essential to ensure that public investment is put into effective use and saved from the current, continuous series of failed attempts.

The third aspect needing urgent attention is the promotion of a multi-scale learning approach – with simultaneous learning processes employed at different scales. There should be efforts to facilitate linkage among learning processes at all levels of NRM, as well as linking them vertically and horizontally, through interactive institutional platforms, and two-way communication mechanisms. Innovation and positive changes in NRM and livelihoods should be seen as outcomes of dynamic learning systems, and not as something resulting from the application or replication of technological packages.

Finally, the modus operandi of development aid has to change to support adaptive collaborative innovations in natural resource governance and management. International aid strategies should become more responsive to local innovation processes and pathways rather than to predefined frameworks and outputs. They should not confine themselves to the delivery and scaling out of standardized service packages or generating consultant recommendations to public policy change. They should instead support how each project, region and country learns to develop innovative solutions to its own problems. A key aspect of reframing aid strategy in natural resource governance is about evaluation of international development funding. There is still a focus on evaluation from the perspective of the donor, which has resulted in learning and policy processes oriented to conforming to the logic and demands of donor organizations. Aid project evaluation should be organized in the light of 'shared accountability' – meaning the aid in a particular social context should be accountable to both donors and local society. In other words, the effectiveness of development funding should be judged/assessed jointly by donors and the stakeholders of the recipient country. Moreover, the performance of development projects should be seen not in terms of immediate, measurable values of tangible outcomes in the short run, but on the basis of promising plausible connections found between action, learning and possible outcomes, in the long run. That is to say that the prime purpose of international aid should be to help recipient countries learn their way into new capacities for innovation and change, rather than to deliver short term welfare results. Collaborative learning and reflection should be key elements of programme evaluation.

Final remarks: towards an era of democratic innovation

This book's account of the emergence and progress of learning-based adaptive management approaches has resonance beyond natural resource management. It reveals what we believe is one of the greatest paradoxes that society faces in the twenty-first century. This is the paradox of public innovation programmes disabling innovation process and pathways emerging from practice. Innovation, in its broadest sense of cultural, institutional and technical learning and change, has been and will always be the means by which societies generate social and economic well-being and sustainability. However, the main instruments used to promote innovation that have served us well in the past are actually the main bottlenecks for the sorts of innovation processes needed to take us into the future.

These old instruments include modes of policy making, research and development assistance that are expert driven; that value generic solutions over context relevance; that create static or slow-changing policy and institutional frameworks; and that hive off learning into specialist elite intellectual enclaves. These instruments worked – or at least appeared to work – in the more stable social, economic and environmental conditions of the nineteenth and twentieth centuries. But these conditions no longer exist – or at least we can no longer pretend that they do. The world we live in today is a fast moving, interconnected and chaotic world. In this situation the best we can do is muddle through, continuously looking for new ways to enhance adaptive capacity and achieve human well-being, relying on any and every source of information that might help us learn our way through the situation that we find ourselves in. ACA is more than a collection of analytical tools and processes but a whole way of life through which champions of change muddle through complexity and uncertainty, engaging people and institutions, acting politically, in pragmatic yet critical and transformative ways.

Yet the key conclusion from this book is that conventional views of policy, technology, and research and development assistance reinforce each other to form an almost impenetrable regime that resists the very processes of learning and change that it seeks to enable. The gatekeepers of this regime are often senior figures who mistakenly believe that maintaining existing professional standards and roles is a proxy for using public resources more effectively for society's good. Much of this book is about the political encounters between these gatekeepers and the champions of ACA.

The ways forward suggested by the book – new spaces for policy learning, decentralized and user driven monitoring and evaluation, the blending of research and practice – are actually about democratizing the innovation process. That is to say, giving all those with a stake in the process an opportunity to contribute in a meaningful way to a process of learning and change. And this is not just about the patronizing acknowledgement of indigenous or local knowledge and innovation – important though these are. This is about everyone concerned having a say in the policy learning process that shapes the environment of innovation and development, and also ensures societal adaptation to large-scale environmental risks such as climate change. After all, innovation is about defining and propelling workable development pathways to achieve what society desires and this is no longer something that can be designed by experts alone. The fast-moving pace of events, and the changing climatic regime, mean that policy formulation increasingly needs to rely on 'just good enough' practice-derived information if it is to support the sorts of innovation processes that society needs and with the required degree of urgency.

Is this just another well-meaning pipe dream of disinterested intellectuals? No, it's probably the last hope that the existing national regimes of public policy have of avoiding the catastrophic social costs of civil unrest. We are, after all, in an era where public dissent can be marshalled and organized as a powerful force for change in ways that have never before been seen. And there is no doubt that existing policy and institutional regimes shape development pathways, affecting peoples' lives in fundamental and often unacceptable and unsustainable ways. In the case of development assistance it could easily slide into a rapid decline, becoming an irrelevant historical curiosity left over from the colonial era.

The essence of this new process of democratic innovation is set out in this book. It involves radically changing the current institutional arrangements that structure learning. It is not just calling for better ways of learning and change in the immediate domain of rural communities. Rather, it is recognizing learning and change as a multi-level activity and that unless learning takes place at all levels from the policy regime down to the level of natural resource management practices, development progress can never be made in a significant and widespread way. ACA provides a glimpse of what this new paradigm might look like – as an approach to learning based and inclusive innovation practice. It hinges on blending research and practice and building bridges to the policy making process to drive reflexive learning and entails professional transformation of many involved in the process.

This will certainly require new structures and capacities to allow communities, researchers and policy makers to 'tool up' for democratic innovation. International donors could play a crucial role in helping build and strengthen these new capacities. Yet as many of the experiences in the book have shown, for donors, charity will truly need to begin at home, as they must first appreciate that the role of development assistance is first and foremost about building capacity for innovation and change and not merely about transferring ideas and resources to poor people and counting how many remain poor.

All of this will be challenging, time-consuming and expensive. But it is increasingly apparent that the cost to society of doing nothing and continuing with business as usual could well be immeasurable.

References

Armitage, D. R., F. Berkes and N. Doubleday, Eds (2007) *Adaptive co-management: collaboration, learning, and multi-level governance*. Toronto: University of British Columbia Press.

Armitage, D., M. Marschke and R. Plummer (2008) *Adaptive co-management and the paradox of learning*. Global Environmental Change 18(1): 86–98.

Berkes, F. and N. J. Turner (2006) *Knowledge, learning and the evolution of conservation practice for social-ecological system resilience*. Human Ecology 34(4): 479–494.

Bohman, J. (2002) *How to make a social science practical: pragmatism, critical social science and multiperspectival theory*. Millennium: Journal of International Studies 31(3): 499–524 (426).

Hall, P. A. (1993) *Policy paradigms, social learning, and the state: the case of economic policy making in Britain*. Comparative Politics 25(3): 275–296.

Lee, K. N. (1993) *Compass and gyroscope: integrating science and politics for the environment*. Washington DC: Island Press.

Schon, D. (2010) *Government as learning system*. Social Learning Systems and Communities of Practice. C. Blackmore, The Open University Press: 5–16.

Index

Note: Page numbers in **bold** are for tables, those in *italics* are for figures.